AF335050

COMPUTATIONAL SYSTEMS BIOLOGY OF SYNAPTIC PLASTICITY

Modelling of Biochemical Pathways Related to Memory Formation and Impairment

Series on Advances in Bioinformatics and Computational Biology – Volume 10

COMPUTATIONAL SYSTEMS BIOLOGY OF SYNAPTIC PLASTICITY

Modelling of Biochemical Pathways Related to Memory Formation and Impairment

Don Kulasiri

Yao He

Lincoln University, New Zealand

World Scientific

NEW JERSEY · LONDON · SINGAPORE · BEIJING · SHANGHAI · HONG KONG · TAIPEI · CHENNAI · TOKYO

Published by

World Scientific Publishing Europe Ltd.

57 Shelton Street, Covent Garden, London WC2H 9HE

Head office: 5 Toh Tuck Link, Singapore 596224

USA office: 27 Warren Street, Suite 401-402, Hackensack, NJ 07601

Library of Congress Cataloging-in-Publication Data

Names: Kulasiri, Don, author. | He, Yao, PhD., author.

Title: Computational systems biology of synaptic plasticity : modelling of biochemical
    pathways related to memory formation and impairment / by Don Kulasiri
    (Lincoln University, New Zealand), Yao He (Lincoln University, New Zealand).

Description: New Jersey : World Scientific, [2017] | Series: Series on advances in bioinformatics
    and computational biology ; volume 10 | Includes bibliographical references and index.

Identifiers: LCCN 2017007701 | ISBN 9781786343376 (hc : alk. paper)

Subjects: LCSH: Biological systems--Computer simulation. |
    Biological systems--Mathematical models. | Computational biology--Methods. |
    Neuroplasticity. | Computational neuroscience.

Classification: LCC QP363.3 .K85 2017 | DDC 570.1/13--dc23

LC record available at https://lccn.loc.gov/2017007701

British Library Cataloguing-in-Publication Data

A catalogue record for this book is available from the British Library.

Desk Editors: V. Vishnu Mohan/Mary Simpson

Typeset by Stallion Press
Email: enquiries@stallionpress.com

Printed in Singapore

Dedication

Don Kulasiri dedicates this book to his late mother, Jayawathie Sinhabahu, and to his late father, David Gamalath, for their tireless efforts to educate him.

Yao He dedicates this book to his grandfather, Nianyou He, who is now suffering from dementia, his parents, Qiang He and Lorna Xiaole Liu and his wife, Yanzhen Rao.

Preface

During the last two to three decades, we have seen an unprecedented increase in neurobiological research and we know more about how brain functions at neuronal level than ever before. It is important to realize that the fact that we can remember the past and recall the past events, facts and images even in a vague sort of way creates human societies and cultures. Without the capacity to have a memory, we simply cannot function. Various forms of dementia show us clearly the important role memory plays in our lives. Therefore, it is not surprising that we have been trying to understand the formation of memory and its impairment through many disciplines.

We feel that it is timely to understand the memory from a systems biology perspective. We have a wealth of latest experimental research on the pathways at a molecular level related to the dynamics within neurons and synapses that "connect" them. This book explains the latest research on modelling of biochemical pathways related to intra-cellular networks within the synapses. We introduce the biology of synapses in an introductory manner and provide a succinct background of the mathematical methods related to systems biology in general.

There are diverse ways of modelling the processes associated with synaptic plasticity and a number of models with different approaches that deal with the core synaptic plasticity. The models based on a single synaptic component, called single component models, highlight its potential functional and structural properties associated with the dynamic behaviour of synaptic plasticity. Simplified models concentrate on the interactions between certain synaptic proteins which form simplified synaptic circuits. These chosen proteins are believed to be critical for synaptic plasticity. Simplified models are theoretical frameworks; the encapsulated relationships between synaptic proteins may not have direct biochemical associations with experimentally discovered interactions. However, the simulation results are useful to understand how essential a certain synaptic protein is and to investigate its potential role in the interacting networks of synaptic plasticity. Complete pathway models similarly focus on emergent properties of synaptic networks, but differ from simplified models in two ways: (a) complete pathway models contain the full spectrum of synaptic proteins of a section of, or complete, functional synapse; mathematical frameworks of the associations between these proteins are developed based on their experimentally discovered biochemical reactions or plausible biochemical reactions; and (b) the results of complete models often explain the experimental data better, and thus the predictions made by them are highly credible. Caveat associated with these models is the difficulty of estimating a large number of parameters. In this book, we attempt to choose the level of details for models to be at an "essential optimum" based on the biochemistry of the pathways. We also use the techniques such as Markov Chain Monte Carlo (MCMC), Global Sensitivity Analysis (GSA) and partial ranking correlation coefficient (PRCC) method in arriving at suitable sets of parameter values.

This book will be valuable to researchers in systems biology in general, postgraduate students who are looking for directions for their research and broadly as a reference book. Our hope is that this book will spur the reader on to new research in computational and mathematical biology related to memory so that the future experimentalists have a large number of hypotheses to test in their laboratories.

We would like to acknowledge Dr. Jingy Liang at our centre in helping us to write the chapter on dementia (Chapter 7) and thank her for her support in completing this book. We are also grateful to Professor Sandhya Samarasinghe of Lincoln University, New Zealand for her insights and comments on our work. We also acknowledge Lincoln University, New Zealand for supporting us many ways during the writing of this book. The first author would like to thank the colleagues, especially Prof. Philip Maini, in Wolfson Centre for Mathematical Biology, Mathematical Institute, University of Oxford, UK, for hosting him many occasions during the past few years.

Don Kulasiri and Yao He
Centre for Advanced Computational Solutions (C-fACS)
Lincoln University, Christchurch, New Zealand

About the Authors

Professor Don Kulasiri obtained his PhD related to Bioengineering from Virginia Tech, Blacksburg, USA. He had been a visiting academic to Stanford University, Princeton University, USA, and has been visiting Centre for Mathematical Biology, Mathematical Institute, University of Oxford, UK, regularly since 2008. He is the Professor of Computational Modelling and Systems Biology, a personal Chair, at Lincoln University, Christchurch, New Zealand since 1999. He founded and directs the Centre for Advanced Computational Solutions (C-fACS) at Lincoln since 1999. He is a fellow of the Modelling and Simulation Society of Australia and New Zealand (MSSANZ).

Dr. Yao He obtained his PhD in Computational Systems Biology at Lincoln University, New Zealand, and he also has a Bachelor of Science degree in Computer Science and Mathematics from University of Canterbury, New Zealand. He has been engaged in systems biology research since 2010, and works as a research scientist in the Centre for Advanced Computational

Solutions (C-fACS). His expertise is on modelling the mechanisms of synaptic plasticity, parameter sensitivity analysis and stochastic modelling. His current research focuses on understanding the linkage between presynaptic release and postsynaptic plasticity using stochastic modelling.

Contents

Contents

Chapter 6.   Uncertainty Quantification of Models Related to Synaptic Plasticity   255

List of Tables

List of Figures

Chapter 1

Introduction

1.1. Synaptic Plasticity and Related Dynamics

The neuron is known as the basic component of the complex neuronal network in the brain. Neurons are involved in many neurological processes, such as cognition, learning, and memory. The major roles of neurons are rapid transmission and processing of a large amount of signals induced by extracellular stimuli in order to support the neurological response. A synapse is the transmission tunnel connecting two neurons (Fig. 1.1). Consequently, the transmission taken place at a synapse is called synaptic transmission, which is a one-way, unidirectional movement that delivers information between neurons efficiently and flexibly. Therefore, the proper functioning of the synapse is essential for normal operation of the complex and widely distributed neural networks. The principles of synaptic function are explained in detail in many books. Purves *et al.* [1] and Bear *et al.* [2] give excellent explanations not only on the synapse, but also on the other discoveries in neuroscience.

Each neuron is an independent unit for signal transmission and processing, and a single neuron is usually involved in many signal circuits. The transmission speed of neurons is very fast (a few nanoseconds) in order to acquire the high-speed response of the neurological system. However, neurons cannot fire continuously; there is a refractory period between two transmissions. This means that neurons are capable of rapid switching states to satisfy the different conditions for transmission and refractory, respectively.

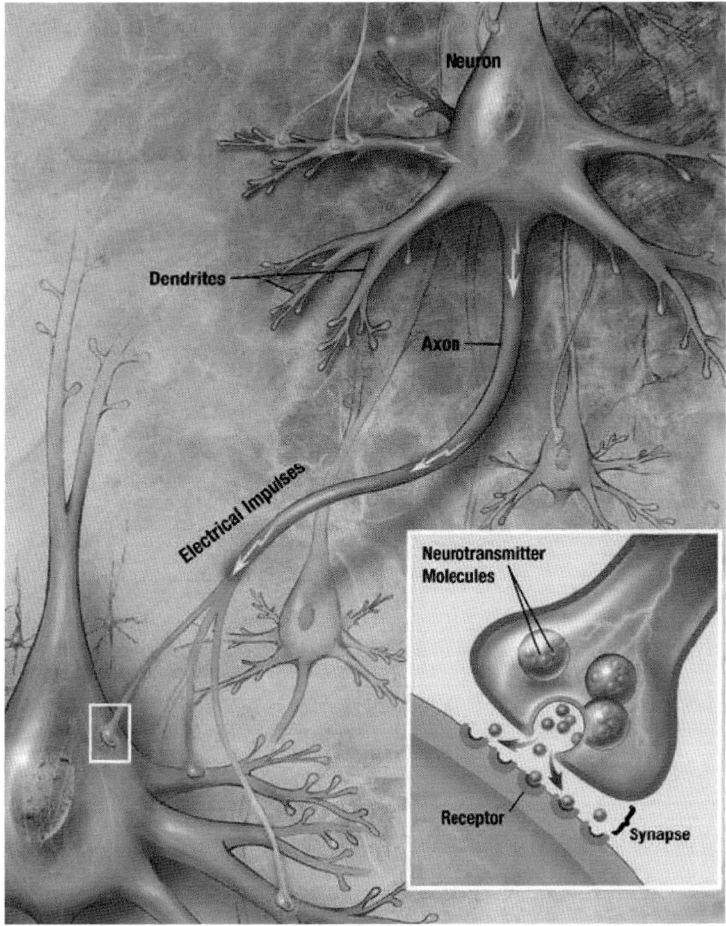

Fig. 1.1.   Neuron and synapse (public domain in US).

This gives rise to the question of how neurons can achieve the rapid switching of states and can maintain its robustness at the same time.

We need to understand the dynamics of synapses based on current experimental evidence. A synapse is a gate between two neurons and it consists of three parts: an axonal terminal of the presynaptic cell, a presynaptic cleft and a postsynaptic dendrite of the postsynaptic cell. Synapses deliver information between neurons through synaptic transmission. Synaptic transmission is a process with repeated sequential phases. Synapses have developed mechanisms for each phase of the synaptic transmission to ensure the transmission is correct and efficient. Moreover, these mechanisms help the neurons to maintain their states at different phases of synaptic transmission.

A large amount of research focused on the dynamics of the synapse and a large body of evidence indicate that synapses play an important part in maintaining the robustness of brain functions. The alteration on synapses may be highly correlated to memory formation. More importantly, the alteration on synapses can be controlled by a set of activities that are originally built into the neurological network [3]. Moreover, deficiency in functioning of synaptic components leads to neuronal degenerative diseases [4]. Those discoveries indicate that neurons act more than simple transits in the neural network; they are also involved in many high-level functions of the brain.

Synaptic plasticity is a process integrating the dynamics as well as the interactions of the three parts of the synapse that leads to alteration of the core components of synaptic transmission [5]. The mechanisms of synaptic plasticity involve neurotransmitter dynamics in the presynaptic cell; neurotransmitter–receptor interactions in the synaptic cleft; receptor trafficking, membrane dynamics and intracellular protein–protein interactions in the postsynaptic cell. These complex mechanisms of synaptic plasticity give flexibility to synaptic transmission in order to provide a dynamic transmission rate with respect to different circumstances and support memory formation.

Two sides of a synapse are named pre- and postsynaptic terminals indicating the normally pre to posttransmission direction (Fig. 1.2). There is a gap called synaptic cleft located between the pre- and postsynaptic terminals. The synaptic cleft is filled with a matrix of fibrous extracellular protein. One function of this matrix is to adhere the pre- and postsynaptic membranes tightly to each other.

The presynaptic terminal is usually the axon terminal of the neuron that initialises synaptic transmission. Initialisation of

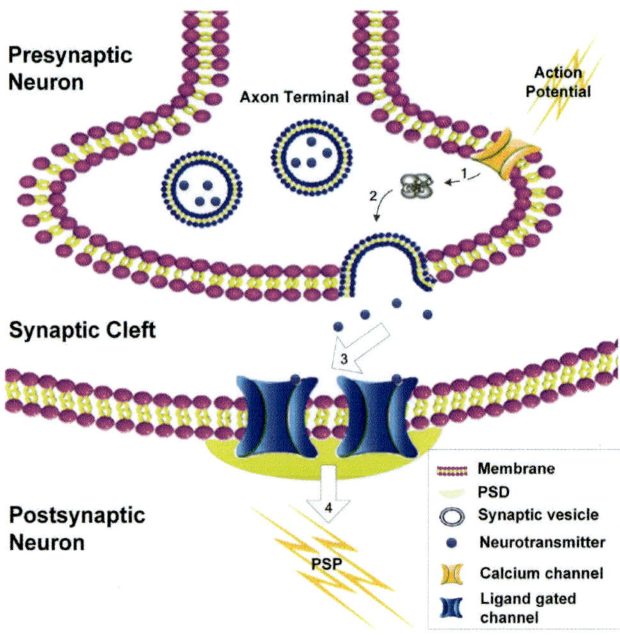

Fig. 1.2. Synaptic transmission. (1) Action potential depolarises the presynaptic membrane to open the calcium channel, which allows an influx of Ca^{2+} ions. (2) Ca^{2+} ions trigger the release of neurotransmitters into the synaptic cleft by exocytosis at the active zone. (3) The released neurotransmitters diffuse across the synaptic cleft and bind to ligand-gated receptors at PSD. The binding opens the embedded ion channels of the receptors. (4) The opening of the ligand-gated ion channels allows an influx of ions to induce a postsynaptic potential (PSP).

transmission appears to be the release of synaptic vesicles, which contain neurotransmitters, into the synaptic cleft. Synaptic vesicles are membrane-enclosed spherical objects and are accumulated close to the active zone, where neurotransmitters are released. Neurotransmitters are pumped into synaptic vesicles by vesicular transporters embedded in the membrane of the synaptic vesicle.

The two regions, active zone and PSD, located at the opposite ends of the synaptic cleft, are found to be protein-rich, reflecting their heavy duties in neurotransmitter releasing and neurotransmitter receptor anchoring, respectively. Understanding of the roles of these proteins in these two regions sheds light on synaptic transmission, which is a neurotransmitter-dependent process that converts signals between electrical and chemical forms. Consequently, the whole journey begins at the synthesis of neurotransmitters. Among the different types of neurotransmitters, amino acids and amines are the most prevalent molecules with respect to their appearance in the fast synaptic transmission. Amino acids and amines are small molecules that are stored in and released from synaptic vesicles. They are synthesised differently from each other although they are released the same way. Amino acids are elementary blocks of proteins that are abundant in all cells, while amines are only synthesised by the neuron releasing them. Both amino acids and amines are synthesised in the axon terminal by their synthesising enzymes, and are taken up by synaptic vesicles shortly after their synthesis [6]. From this stage, neurotransmitters are ready to be released.

An incoming action potential triggers neurotransmitter release through depolarising the presynaptic membrane. The action potential can be generated in different ways: by sensory nerve activation in response to stimuli, or by interneurons in response to the signals from their presynaptic neurons. Once generated, the action potential travels towards the axon terminal and depolarises the presynaptic membrane at the active zone. There are many voltage-gated Ca^{2+} channels embedded in the presynaptic membrane at the active zone. These voltage-gated Ca^{2+} channels are sensitive to the membrane potential; they are opened as soon as the membrane is depolarised. A strong inward driving force is induced across the membrane in reaction to the opening of Ca^{2+} channels due to the great Ca^{2+} gradient across the membrane. All those actions result in a strong increase in the intracellular Ca^{2+} concentration that serves as the signal of neurotransmitter release [7].

Neurotransmitter release occurs very rapidly through a process called exocytosis. Synaptic vesicles fuse with the presynaptic membrane at the active zone upon receiving the releasing signal and release their contents, neurotransmitters, into the synaptic cleft. In order to respond fast to the influx of Ca^{2+}, synaptic vesicles are firstly transported to and docked at the active zone, then fastened by a protein belonging to the soluble NSF attachment protein family [6].

This protein family allows the fusion between the two membranes. How Ca^{2+} triggers neurotransmitter release is still unknown, but it is believed that a vesicle protein called synaptotagmin 1 senses Ca^{2+} and rapidly triggers fusion of the vesicle membrane [6]. After the release is completed, synaptic vesicles re-enter the axonal terminal by endocytosis.

Released neurotransmitters induce postsynaptic potentials through binding to their target transmitter-gated ion channel receptors in the postsynaptic membrane. Transmitter-gated ion receptors consist of four or five subunits that join side by side into a structure in which a closed pore is formed in the middle. Binding of a neurotransmitter leads to a conformation change in the receptor so that the subunits are twisted in order to make the pore open [8]. Although the pore only opens for a very short period, the influx of ions into the terminal is large enough to induce a transient postsynaptic potential due to the large ion concentration difference across the postsynaptic membrane. The type of postsynaptic potential induced is dependent on the ion permeability of the ion channel. The membrane of the postsynaptic cell is depolarised towards the threshold of generating an action potential when the channels are permeable to cations. This transient depolarisation of the postsynaptic membrane induced by the release of presynaptic neurotransmitters is called excitatory postsynaptic potential (EPSP). Depolarisation of the postsynaptic membrane beyond the threshold would cause a current that initializes the action potential in the spike-initiation zone, which usually locates in the axon hillock of the postsynaptic cell. On the other hand, the membrane is hyperpolarised away from the threshold of generating an action potential when the open channels are permeable to anions. This transient hyperpolarisation of the postsynaptic membrane induced by the release of presynaptic neurotransmitters is called inhibitory postsynaptic potential (IPSP).

Many factors can influence the induction of the postsynaptic potential. These include the binding probability between neurotransmitters and transmitter-gated ion channel receptors; the conductance of the transmitter-gated ion channels and the activation of the non-transmitter-gated ion channels. These factors formulate the base of dynamic control of the postsynaptic potential by a small protein, called G-protein. A large number of G-protein-coupled receptors are embedded in the PSD. The major role of G-protein-coupled receptors is to activate postsynaptic G-protein in the presence of neurotransmitters. The G-protein-coupled receptors can influence the dynamics of the postsynaptic cell by three ways. Firstly, G-proteins directly influence the synaptic potential by the opening of G-protein-gated ion channels [9]. Like the transmitter-gated ion channels, the net effect (whether to depolarise or hyperpolarise the membrane) of G-protein-gated channels is largely dependent on its ion permeability. Secondly, G-proteins activate effector proteins, which serve as the secondary messengers either to active downstream enzymes or to directly influence the postsynaptic potential. For example, a downstream enzyme like protein kinase C (PKC) phosphorylates transmitter-gated receptors to increase the conductance of the embedded ion channels and results in more calcium influx and further depolarisation of the membrane toward the threshold. Thirdly, a single G-protein-coupled receptor can activate multiple G-proteins each time, which amplifies the net effects onto the postsynaptic potential.

Neurotransmitters are cleared from the synaptic cleft at the end of the transmission in order to give space for the next transmission. There are two ways to remove neurotransmitters out of the synaptic cleft. One way is presynaptic reuptake. Presynaptic reuptake is a process where the released transmitters are transported across the membrane back into the presynaptic terminal by the membrane transporter proteins located in the presynaptic membrane. Once in the terminal, neurotransmitters are either degraded rapidly or reloaded into synaptic vesicles. The other way is the enzymatic destruction of neurotransmitters in the synaptic cleft. After the enzymatic destruction, neurotransmitters usually become inactive and decomposed.

Once the neurotransmitters are removed out of the synaptic cleft, one cycle of synaptic transmission is completed. In this cycle, the message carried by the action potential is first converted into a chemical signal, the neurotransmitter. This chemical signal is released into the synaptic cleft and captured by the postsynaptic receptors. Then, the chemical signal is converted into an electrical signal, the postsynaptic potential. At this point, the message is successfully delivered across the synapse to the postsynaptic cell. In the last step, the postsynaptic cell filters this message so that an action potential is induced to carry the corresponding message to the next destination if certain requirements are satisfied.

One possibility of message encoding during transmission is proposed to be capsulated by the firing frequency of the incoming action potentials. Experiments show that the continuous depolarisation current induces a burst of action potentials and the fire frequency of the burst is decoded by the magnitude of the depolarisation current [2, 3]. However, the fire frequency cannot exceed 1000 Hz (1000 impulses per second), which is based on the existing absolute refractory period between two action potentials.

1.2. Importance of Synaptic Plasticity

Why is synaptic plasticity important? Direct evidence shows that antagonists or gene mutations related to synaptic plasticity impair memory formation [10–16]. The exact relationship between synaptic plasticity and memory formation is far from completely understood. But, current understanding suggests that memory formation is experience dependent and has a strong correlation to synaptic activities. The memory may store in the synaptic strength of the complex nervous network of the brain, which consists of billions of neurons connected by trillions of synapses [17, 18]. One hypothesis based on Hebbian theory [19] states that the brain "remembers" the experiences of an environmental stimulus as memory by selectively adapting synaptic strength of the synapses among the activated neurons in response to the stimulus [20]. As a result, the neurons are wired up as a cell assembly [2]. The cell assembly recognises the environmental stimulus at the next occurrence. A good example is the hippocampal place cells, which fire at restricted locations, called place fields, to orient the location of the rat [2, 21]. Although the hypothesis may be too simple to account for different types of memories, it indicates, together with the experimental observations, neuronal functions rely on the connections (synapses) among the neurons within the cell assembly. Hence, synaptic plasticity, which participates in the synapse formation [22, 23], is critically important to neuronal functions.

The processes involved in synaptic plasticity have drawn wide interests from neuroscientists over the last two decades. Synaptic plasticity has multiple forms depending on the brain region and neuron type, while the mechanisms, expressing sites and expressing targets are different among them [9].

The most studied form is hippocampal N-methyl-D-aspartate receptor (NMDAR)-dependent synaptic plasticity due to the extensive experimental investigations in this critical area of the memory system. For this particular form, synaptic plasticity induces long-lasting bidirectional modulations of the postsynaptic response when the area is stimulated by electrical stimulations [24–26]. The postsynaptic response is defined as the magnitude of the postsynaptic receptor-mediated current (EPSC) or excitatory postsynaptic potential (EPSP), in relation to their basal levels *in vivo* [24, 26], and the major postsynaptic receptors in hippocampal neurons are glutamatergic receptors, including NMDAR and a-amino-3-hydroxy-5-methyl-4-isoxazolepropionic acid receptor (AMPAR).

NMDAR-dependent synaptic plasticity requires the activation of NMDAR to induce a sequence of postsynaptic activities to alter the property of AMPAR (see Chapter 3 for the explanation), which controls the magnitude of the postsynaptic response induced by the synaptic transmission in hippocampal neurons [27–29]. The activated NMDAR triggers a rapid Ca^{2+} influx leading to an elevation in the intracellular Ca^{2+} level. The degree of the Ca^{2+} elevation determines the direction of the alteration on AMPAR and hence causes the corresponding increase or decrease of the magnitude of the postsynaptic response [20, 30, 31]. Specific terms are used to describe the bidirectional modulations: potentiation denotes an increase of the postsynaptic response, and depression denotes a decrease of the postsynaptic response.

Two subforms of NMDAR-dependent synaptic plasticity contain the later stages that last very long from hours to days that are named long term potentiation (LTP) [24] and long term depression (LTD) [26]. The characteristic of the later stage is the outlasting activities after the vanishing of the Ca^{2+} signal. During the later stage, it is known that an alteration on gene expressions is required to maintain the changes on the synaptic system by the earlier stage and the alteration on gene expressions may remodel the structure of the synapse [22, 32–43], possibly through modulating actin dynamics. Actin is the primary cytoskeletal components of the synapse. Actin exists in a dynamic equilibrium between the monomeric globular form (G-actin) and the filamentous form (F-actin): G-actin is polymerised into F-actin and F-actin is depolymerised into G-actin. Importantly, F-actin is involved in morphogenesis of dendritic spines to stabilise spine structure and in recruiting various postsynaptic proteins [44–47]. F-actin is also shown to control the number of synaptic contacts [48]. Furthermore, evidence shows the modulation of actin dynamics by synaptic plasticity [34, 38], where LTP signal shifts the equilibrium towards F-actin and LTD signal shifts the equilibrium towards G-actin. Usually, the distinct periods, early phase and late phase, are defined to distinguish the time frames with or without alterations on gene expressions during synaptic plasticity [25, 42, 43]. Evidence suggests that the early phase LTP (E-LTP) is initiated by a single train of high-frequency stimulation and lasts for less than 3 hours, while late phase LTP (L-LTP) requires multiple trains of high-frequency stimulations separated by more than 5 minutes and lasts for at least 10 hours [42, 49]. Several studies strongly suggest that the late phase of synaptic plasticity is accompanied by morphogenesis changes of dendritic spine and has a strong relationship to memory consolidation [22, 32].

Recent investigation has shown a strong correlation between LTP/LTD and memory formation: (1) both memory formation and LTP trigger the morphogenesis change of synapses [22, 23, 33, 39, 50]; (2) several studies show that rats with deficient in LTP, either by application of NMDAR antagonist, deletion of the NMDAR gene or altered expression of the key synaptic proteins involved in synaptic plasticity, diminishes in spatial learning. Rats in these experiments reveal normal synaptic transmission, normal retention of spatial information, but much slower in learning the hidden-platform of the water maze experiments [10, 12, 14, 16, 43, 51, 52]; and (3) memory loss related brain diseases are associated with the enhanced LTD, for example, a potential pathogen in Alzheimer's disease, Amyloid-β, facilitates LTD induction [53, 54]. Since the exact principles of the memory system are not yet completely understood, the emerging understanding of the protein networks associated with synaptic plasticity may provide insights into the biochemical basis of memory formation.

1.3. Modelling Single Cell Dynamics Associated with Synaptic Plasticity

A cell dynamically regulates its intracellular signalling, transcription, plasticity and fate [55]. The dynamics start with the detection of extracellular signals by the receptors at the cell surface. The consequent second messenger transmits through protein signalling pathways (usually involves feedback loops) and delivers the message to the nucleus, where the gene expression occurs. Resulting effects involve activations of specific proteins and alterations on multiple gene expressions [55]. Signalling pathways may have complex behaviours facilitated by the emergent properties of the interacting networks [56]. However, understanding the complex behaviours is very challenging because the measurements of the interacting networks in real time are difficult with current technology. Even if the measurements are feasible, the complexity of the protein interacting networks associated with the signalling pathway involves too many biological details that may be difficult for an unbiased human mind to comprehend. With the support from mathematical modelling and computational methods [55, 57], we can develop mathematical models based on hypotheses for the cell dynamics and validate the hypotheses through the computational experiments of the established model against available experimental data. Thoughtful interpretation of the computational results can help us to better understand the complex behaviour and develop further hypotheses to be tested experimentally.

The dynamic participation of the synaptic proteins in the sophisticated interacting networks gives rise to synaptic plasticity. In one facet of systems biology, we aim to understand the system-level functions of synaptic proteins and the emergent properties of the protein interacting networks of synaptic plasticity through mathematical modelling based on biophysics. Thus, mathematical modelling based on biologically meaningful assumptions may provide insights into the dynamic behaviour of synaptic plasticity, which may be difficult to obtain through experiments. The models focus on three key aspects linking the synaptic components to the dynamic behaviour of synaptic plasticity [58]: (1) the structural and functional properties of individual synaptic protein that show high correlation to the expression patterns of synaptic plasticity; (2) the emergent properties of the protein interacting networks that support the functionality of synaptic plasticity; and (3) the robustness of synaptic plasticity with respect to the low copy numbers of proteins. The models of these categories help us to identify the essential modulators of synaptic plasticity as well as to understand distinct and dynamic roles they play in the complex set of networks.

There are many questions of synaptic functioning that may be critical for understanding synaptic plasticity, which could be answered through mathematical modelling. The detail level of the synaptic system considered in each model maybe different depending on the specific question asked.

What are the essential protein interacting networks underlying synaptic plasticity? Considering the distinct functions of LTP and LTD, which of the protein interacting networks induce the emergence of LTP and which of them induce the emergence of LTD? The answer should take into account the essential protein interacting networks in the emergence of synaptic plasticity. A minimal model integrating the essential interacting networks of synaptic plasticity is necessary to analyse for the essential emergent properties underlying the dynamic behaviour of synaptic plasticity. The model development should be started with a careful selection of the indispensable synaptic proteins to be used as the variables of the model. As a result, the model development should provide insights into the essential modulators of synaptic plasticity and the simulation results of the model should gain insights into the dynamic roles that the modulators play in synaptic plasticity.

How would the co-localisation of synaptic proteins impact synaptic plasticity? This problem deals with the co-localisation of synaptic proteins and spatial movement of synaptic proteins among synaptic compartments. Hence, the models of this category need to be developed with the consideration of the spatial information and co-localisation of synaptic proteins. The simulating results of the models should provide a more precise prediction on the effective timescales of the synaptic proteins and hence allow us to link synaptic proteins to the phases of synaptic plasticity.

What is the implication of interactions between modulators, specifically the binding between NMDAR and Ca^{2+}/CaM-dependent protein kinase II (CaMKII), in the dynamic behaviour of synaptic plasticity? The answer should take into consideration of the biochemical details of the modulators, possibly expanding previous models of the modulator to include the latest understanding of the interaction. Then, the new model can be used to test the functionality of the particular modulator as well as its interaction to other modulators.

With the abundant experimental data of synaptic components, modelling of synaptic plasticity enables the integration of the fragmented information of the synaptic components into a system level view to provide insights into the interactions and the roles of the main synaptic modulators contributing to synaptic plasticity [57]. The existing models related to synaptic plasticity help us to understand the structural and functional properties of synaptic proteins in the emergence of synaptic plasticity [59].

The development of mathematical models and computational methods, as mentioned before, helps us to understand the protein interacting networks at different detail levels in the emergence of synaptic plasticity. In this book, we discuss, for example, a simplified mathematical model of the NMDAR-mediated pathway to explore the bidirectional behaviour of synaptic plasticity. Specifically, we focus on understanding the essential modulators as well as the essential interactions among them, participated in the selective activation of the antithetic synaptic behaviours, LTP and LTD, by a common signalling upstream, the Ca^{2+} level. Analysis of the model reveals the main factors of the selectivity, which may be valuable to understand the important processes in different phases of synaptic plasticity and possibly uncover the biochemical mechanisms underlying the memory system.

At a more single molecular level, we discuss a theoretical model of the state transitions of CaMKII, which have long lasting activity after the vanishing of the Ca^{2+} signal [60] and play an important role in LTP as well as in memory formation [61]. CaMKII has several states and the state transitions among them during LTP can be simulated using this model. Understanding the relationship between the state transitions and LTP may not only give the knowledge for the roles of these states in LTP but also provide insights into memory formation. For instance, we use this model to understand the role of a particular state of CaMKII, the autophosphorylation, in the emergence of early phase LTP (E-LTP), which has diverse conclusions in the literature.

Our intention is to show that mathematical models based on biophysics and experimental biochemistry extend our understanding of the dynamic nature of synaptic plasticity. The modelling will help experimentalists to test competing hypotheses and develop new experiments based on the computational experiments. For mathematicians and biophysicists, this area provides a challenge to understand memory by developing models which provide insights into the phenomena that cannot be measured using current experimental technologies.

References

[1] Purves D. *et al.* (2008). *Neuroscience*, 4th edn. (Sinauer Associates, Inc, Sunderland, MA).

[2] Bear M.F., Connor B.W. and Paradiso M.A. (2007). *Neuroscience: Exploring the Brain*, 3rd edn. (Lippincott Williams & Wilkins, Baltimore).

[3] Kandel E.R. (2009). The biology of memory: a forty-year perspective. *J Neurosci*, 29, pp. 12748–12756.

[4] Selkoe D.J., Triller A. and Christen Y. (2008). *Synaptic Plasticity and the Mechanism of Alzheimer's Disease* (Springer-Verlag, Berlin).

[5] Stevens C.F. (2004). Presynaptic function. *Curr Opin Neurobiol*, 14, pp. 341–345.

[6] Südhof T.C. (2004). The synaptic vesicle cycle. *Annu Rev Neurosci*, 27, pp. 509–547.

[7] de Jong A.P.H. and Verhage M. (2009). Presynaptic signal transduction pathways that modulate synaptic transmission. *Curr Opin Neurobiol*, 19, pp. 245–253.

[8] Wollmuth L.P. and Sobolevsky A.I. (2004). Structure and gating of the glutamate receptor ion channel. *Trends Neurosci*, 27, pp. 321–328.

[9] Citri A. and Malenka R.C. (2007). Synaptic plasticity: multiple forms, functions, and mechanisms. *Neuropsychopharmacology*, 33, pp. 18–41.

[10] Bourtchuladze R. *et al.* (1994). Deficient long-term memory in mice with a targeted mutation of the cAMP-responsive element-binding protein. *Cell*, 79, pp. 59–68.

[11] Chang H.P. *et al.* (1999). Impaired memory retention and decreased long-term potentiation in integrin-associated protein-deficient mice. *Learn Mem*, 6, pp. 448–457.

[12] Davis S., Butcher S.P. and Morris R.G. (1992). The NMDA receptor antagonist D-2-amino-5-phosphonopentanoate (D-AP5) impairs spatial learning and LTP *in vivo* at intracerebral concentrations comparable to those that block LTP *in vitro*. *J Neurosci*, 12, pp. 21–34.

[13] Giese K.P., Fedorov N.B., Filipkowski R.K. and Silva A.J. (1998). Autophosphorylation at Thr286 of the α calcium-calmodulin kinase II in LTP and learning. *Science*, 279, pp. 870–873.

[14] Morris R.G. (1989). Synaptic plasticity and learning: selective impairment of learning rats and blockade of long-term potentiation *in vivo* by the *N*-methyl-*D*-aspartate receptor antagonist AP5. *J Neurosci*, 9, pp. 3040–3057.

[15] Silva A.J., Stevens C.F., Tonegawa S. and Wang Y. (1992). Deficient hippocampal long-term potentiation in alpha-calcium-calmodulin kinase II mutant mice. *Science*, 257, pp. 201–206.

[16] Silva A.J., Paylor R., Wehner J.M. and Tonegawa S. (1992). Impaired spatial learning in alpha-calcium-calmodulin kinase II mutant mice. *Science*, 257, pp. 206–211.

[17] Anderson J.R. (2000). *Learning and Memory*, 2nd edn. (Wiley, New York).

[18] Isaac J.T.R., Nicoll R.A. and Malenka R.C. (1995). Evidence for silent synapses: implications for the expression of LTP. *Neuron*, 15, pp. 427–434.

[19] Hebb D.O. (2002). *The Organization of Behavior: A Neuropsychological Theory* (Psychology Press).

[20] Lisman J.E. (1989). A mechanism for the Hebb and the anti-Hebb processes underlying learning and memory. *Proc Natl Acad Sci USA*, 86, pp. 9574–9578.

[21] Shapiro M. (2001). Plasticity, Hippocampal place cells, and cognitive maps. *Arch Neurol*, 58, p. 874.

[22] Engert F. and Bonhoeffer T. (1999). Dendritic spine changes associated with hippocampal long-term synaptic plasticity. *Nature*, 399, pp. 66–70.

[23] Morgado-Bernal I. (2011). Learning and memory consolidation: linking molecular and behavioral data. *Neuroscience*, 176, pp. 12–19.

[24] Bliss T.V.P. and Lømo T. (1973). Long-lasting potentiation of synaptic transmission in the dentate area of the anaesthetized rabbit following stimulation of the perforant path. *J Physiol*, 232, pp. 331–356.

[25] Bliss T.V.P. and Collingridge G.L. (1993). A synaptic model of memory: long-term potentiation in the hippocampus. *Nature*, 361, pp. 31–39.

[26] Dudek S.M. and Bear M.F. (1992). Homosynaptic long-term depression in area CA1 of hippocampus and effects of N-methyl-D-aspartate receptor blockade. *Proc Natl Acad Sci USA*, 89, pp. 4363–4367.

[27] Lee H.-K. *et al.* (2000). Regulation of distinct AMPA receptor phosphorylation sites during bidirectional synaptic plasticity. *Nature*, 405, pp. 955–959.

[28] Lüscher C. and Malenka R.C. (2012). NMDA receptor-dependent long-term potentiation and long-term depression (LTP/LTD). *Cold Spring Harb Perspect Biol*, 4, p. a005710.

[29] Ahmad M. *et al.* (2012). Postsynaptic complexin controls AMPA receptor exocytosis during LTP. *Neuron*, 73, pp. 260–267.

[30] Bear M.F., Cooper L.N. and Ebner F.F. (1987). A physiological basis for a theory of synapse modification. *Science*, 237, pp. 42–48.

[31] Yang S., Tang Y. and Zucker R.S. (1999). Selective induction of LTP and LTD by postsynaptic $[Ca^{2+}]i$ elevation. *J Neurophysiol*, 81, pp. 781–787.

[32] Bailey C.H. and Kandel E.R. (1993). Structural changes accompanying memory storage. *Annu Rev Physiol*, 55, pp. 397–426.

[33] Bozdagi O. *et al.* (2000). Increasing numbers of synaptic puncta during late-phase LTP. *Neuron*, 28, pp. 245–259.

[34] Fukazawa Y. *et al.* (2003). Hippocampal LTP is accompanied by enhanced F-actin content within the dendritic spine that is essential for late LTP maintenance *in vivo*. *Neuron*, 38, pp. 447–460.

[35] Maletic-Savatic M. (1999). Rapid dendritic morphogenesis in CA1 hippocampal dendrites induced by synaptic activity. *Science*, 283, pp. 1923–1927.

[36] Matsuzaki M., Honkura N., Ellis-Davies G.C.R. and Kasai H. (2004). Structural basis of long-term potentiation in single dendritic spines. *Nature*, 429, pp. 761–766.

[37] Nägerl U.V., Eberhorn N., Cambridge S.B. and Bonhoeffer T. (2004). Bidirectional activity-dependent morphological plasticity in hippocampal neurons. *Neuron*, 44, pp. 759–767.

[38] Okamoto K.-I., Nagai T., Miyawaki A. and Hayashi Y. (2004). Rapid and persistent modulation of actin dynamics regulates postsynaptic reorganization underlying bidirectional plasticity. *Nat Neurosci*, 7, pp. 1104–1112.

[39] Toni N. *et al.* (1999). LTP promotes formation of multiple spine synapses between a single axon terminal and a dendrite. *Nature*, 402, pp. 421–425.

[40] Trachtenberg J.T. *et al.* (2002). Long-term *in vivo* imaging of experience-dependent synaptic plasticity in adult cortex. *Nature*, 420, pp. 788–794.

[41] Zhou Q., Homma K.J. and Poo M.M. (2004). Shrinkage of dendritic spines associated with long-term depression of hippocampal synapses. *Neuron*, 44, pp. 749–757.

[42] Frey U., Huang Y.Y. and Kandel E.R. (1993). Effects of cAMP simulate a late stage of LTP in hippocampal CA1 neurons. *Science*, 260, pp. 1661–1664.

[43] Abel T. *et al.* (1997). Genetic demonstration of a role for PKA in the late phase of LTP and in hippocampus-based long-term memory. *Cell*, 88, pp. 615–626.

[44] Okamoto K., Bosch M. and Hayashi Y. (2009). The roles of CaMKII and F-actin in the structural plasticity of dendritic spines: a potential molecular identity of a synaptic tag? *Physiology (Bethesda)*, 24, pp. 357–366.

[45] Cingolani L.A. and Goda Y. (2008). Actin in action: the interplay between the actin cytoskeleton and synaptic efficacy. *Nat Rev Neurosci*, 9, pp. 344–356.

[46] Dillon C. and Goda Y. (2005). The actin cytoskeleton: integrating form and function at the synapse. *Annu Rev Neurosci*, 28, pp. 25–55.

[47] Sekino Y., Kojima N. and Shirao T. (2007). Role of actin cytoskeleton in dendritic spine morphogenesis. *Neurochem Int*, 51, pp. 92–104.

[48] Shoji-Kasai Y. *et al.* (2007). Activin increases the number of synaptic contacts and the length of dendritic spine necks by modulating spinal actin dynamics. *J Cell Sci*, 120, pp. 3830–3837.

[49] Bhalla U.S. (2013). *20 Years of Computational Neuroscience.* ed. Bower J.M. Chapter 9 "Still looking for the memories: molecules and synaptic plasticity" (Springer Science+Business Media, New York), pp. 187–205.

[50] Mayford M., Siegelbaum S.A. and Kandel E.R. (2012). Synapses and memory storage. *Cold Spring Harb Perspect Biol*, 4, p. a005751.

[51] Tsien J.Z. *et al.* (1996). Subregion- and cell type-restricted gene knockout in mouse brain. *Cell*, 87, pp. 1317–1326.

[52] Grant S.G.N. and Silva A.J. (1994). Targeting learning. *Trends Neurosci*, 17, pp. 71–75.

[53] Li S. *et al.* (2009). Soluble oligomers of amyloid Beta protein facilitate hippocampal long-term depression by disrupting neuronal glutamate uptake. *Neuron*, 62, pp. 788–801.

[54] Shankar G.M. *et al.* (2008). Amyloid-beta protein dimers isolated directly from Alzheimer's brains impair synaptic plasticity and memory. *Nat Med*, 14, pp. 837–842.

[55] Spiller D.G., Wood C.D., Rand D.A. and White M.R.H. (2010). Measurement of single-cell dynamics. *Nature*, 465, pp. 736–745.

[56] Weng G., Bhalla U.S. and Iyengar R. (1999). Complexity in biological signaling systems. *Science*, 284, pp. 92–96.

[57] Kotaleski J.H. and Blackwell K.T. (2010). Modelling the molecular mechanisms of synaptic plasticity using systems biology approaches. *Nat Rev Neurosci*, 11, pp. 239–251.

[58] He Y., Kulasiri D. and Samarasinghe S. (2014). Systems biology of synaptic plasticity: a review on N-methyl-D-aspartate receptor mediated biochemical pathways and related mathematical models. *Biosystems*, 122, pp. 7–18.

[59] Manninen T. *et al.* (2010). Postsynaptic signal transduction models for long-term potentiation and depression. *Front Comput Neurosci*, 4, p. 152.

[60] Bayer K.U. *et al.* (2006). Transition from reversible to persistent binding of CaMKII to postsynaptic sites and NR2B. *J Neurosci*, 26, pp. 1164–1174.

[61] Lisman J.E., Yasuda R. and Raghavachari S. (2012). Mechanisms of CaMKII action in long-term potentiation. *Nat Rev Neurosci*, 13, pp. 169–182.

Chapter 2

Mathematical Methods in Modelling Biochemical Networks in Systems Biology

Since the first genome sequence of *Haemophilus influenzae* was published in 1995 [1], many genome sequences have been completed, of which the sequencing of the human genome is the most widely known [2]. Deciphering the genome sequences of many organisms is an important step towards understanding cells at the system level. However, knowing the information encoded in these sequences does not necessarily mean knowing how a living cell works. In order to arrive at biological properties and behaviours that arise from a list of components, we need to know not only the information about the genome, but also the information about mRNA expression, interactions of DNA with protein, interactions of protein with protein, and other molecular interactions.

Several high-throughput experimental technologies have been developed recently that allow us to assess genome-wide expression. These methods include cDNA microarrays, which can be used to obtain thousands of temporal gene expression patterns for different cell types in response to specific stimuli simultaneously [3]; proteome chips, which can be employed for global analysis of protein activities [4]; and two-hybrid screens, which enable the construction of protein interaction maps [5]. The development of these technologies gives us a golden opportunity to view a cell as a system, rather than focusing on its individual cellular components. This has opened up

19

a new field in biology that aims to understand molecular biology as systems, called systems biology [6].

In fact, the study of the system-level understanding of biology has a long history. It started as early as the 1940s with the introduction of cybernetics, which aimed at describing animals and machines using control and communication theory [7]. This was the first attempt to establish the idea of interactions between systems theory and biological sciences. Since then, several similar attempts have been made to describe and analyse biological systems at the physiological level. The unique attributes of systems biology which distinguishes itself from the previous attempts are that it connects system-level descriptions to molecular-level knowledge. Three major issues within systems biology are (1) to generate quantitative high-throughput data by experimentation, (2) to integrate various kinds of data by data processing, and (3) to build mathematical models based on the data. We focus on mathematical modelling in this chapter, and the modelling in systems biology raises a number of key questions as a result of inevitable integration of experimental and modelling approaches in the process: (1) what are basic structures and properties of biological networks?; (2) how do biological networks behave over time under various conditions?; (3) how does a biological system maintain its robustness and stability?; and (4) how can we modify or construct biological systems to achieve desired properties? These are very difficult questions to answer given the complexity of the systems involved, and therefore, we focus on a subset of systems biology namely modelling of biochemical and genetic regulatory pathways in this chapter. However, these subsystems are interconnected and at times these interconnections are more important than the subsystems we focus on, a fact we should bear in mind in the study of systems at the molecular level.

The purpose of this chapter is to render a sufficient background of some current research issues in systems biology, as succinctly as possible, particularly in relation to modelling biochemical and genetic regulatory pathways and related noise to understand the significance and depth of the issues.

2.1. Relevant Background

Gene expression is regulated at several stages in the synthesis of proteins [8]. Apart from the regulation of DNA transcription and RNA translation, the expression of a gene may be controlled during RNA processing and transport (in eukaryotes), and the posttranslational modification of proteins. In addition, the degradation of proteins and intermediate RNA products can also be regulated in the cell. Gene expression in eukaryotes is much more complex than that in prokaryotes. The major reason is that prokaryotic cells consist only of a single compartment in which transcription and translation can occur simultaneously whereas in eukaryotes, transcription occurs in the nucleus, but the RNA products must be transported to the cytoplasm in order to be translated. Therefore in prokaryotes, the control of the rate of transcriptional initiation is the predominant way of regulation of gene expression while in eukaryotes, the regulation of gene expression is controlled at many different points.

The proteins that perform regulatory functions that direct the expression of a gene are produced by other genes or even by itself. This gives rise to a regulatory network, which consists of a set of DNA, RNA, proteins, other small molecules, and is structured by mutual regulatory interactions between these components. The network of gene regulation can be very complex, where one regulatory protein controls genes that produce other regulators that in turn control still other genes. The level of complexity can be increased as a protein can either activate or repress its own synthesis.

Gene expression is a tightly regulated process which gives the cell control over its structure and function. Any step of gene expression may be modulated, from transcription, RNA degradation, translation, post-translational modification of protein and protein degradation. Here, more details about the control of transcription are given because this process is the predominant site for control of gene expression [9].

Every gene consists of a coding region and a regulatory region. The coding region is the part that is transcribed into an mRNA, and the regulatory region is the part that contributes to the control of the gene. The transcription process begins when a RNA polymerase (RNAP), a catalytic protein, binds to the DNA upstream of the coding region, which is a part of the regulatory region called the promoter region. The RNAP separates the double-stranded DNA and then moves along a single strand, step by step, and transcribes the coding region into a RNA transcript. As the RNA transcript is constructed, the RNAP peels away and the DNA strands are rejoined. The transcription process stops when the RNAP reaches a termination site at the end of the gene.

Typically, RNAPs do not bind to the regulatory region of DNA alone, but in complexes with transcription factors (TFs). In simple prokaryotes, the regulatory region is typically short (10–100 base pairs (bp)) and contains binding sites for a small number of TFs. In eukaryotes, the regulatory regions can be very long (up to 10,000 or 100,000 base pairs (bp)) and contain binding sites for multiple TFs. Sometimes TFs are called trans-regulatory elements, and regulatory sites where TFs bind are called cis-regulatory elements [10].

The function of TFs is to control the rate of transcription. When TFs associate with the regulatory regions of their target genes and affect the affinity of RNAP for the transcription initiation site of the gene, they can function to induce or repress synthesis of the corresponding mRNA, and are called activators or repressors, respectively. Activators enhance the interaction between the RNAP and the promoter, therefore, increasing the gene expression rate. Repressors impede the RNAP's progress along the DNA strand, therefore, hampering the expression rate of the gene. Accordingly, the binding site is called an enhancer if bound with an activator, and a silencer if bound with a repressor [11].

2.1.1. *Cooperativity*

In biochemistry or molecular biology, cooperativity is a phenomenon displayed by enzymes or receptors that have multiple binding sites. In a genetic network, there are always multiple binding sites in a regulatory region where several TFs are able to bind. Therefore, transcriptional regulation tends to involve combinatorial interactions between several TFs, which allow for a sophisticated response to multiple conditions in the environment. Non-cooperativity occurs when TFs are independently bound to a regulatory region. Cooperative binding occurs when the affinity of the TF to a binding site depends on the amount of TFs already bound. The cooperative binding can be either positive or negative; indicating that the affinity is either increased or decreased by the binding of other TFs. Competition is also possible when two different TFs bind to one site.

2.1.2. *Feedback loops*

Because TFs are themselves the products of expressed genes, they too are under regulatory control, increasing the complexity of regulatory networks. A regulatory network is a collection of DNA segments, proteins and other metabolites in a cell which interact with each other and form feedback loops in the cell. For a single-feedback system, there are two major kinds of feedback — positive and negative. For a multiple-feedback system, there are combinations of positive and negative feedbacks.

In a negative feedback, specifically, a TF inhibits the transcription of its own gene by blocking RNAP binding at the promoter region. Negative feedback loops generally seem to provide stability. They are required for stable oscillations and some examples are the circadian rhythms [12, 13], cell cycles [14] and calcium waves [15].

In a positive feedback, a TF promotes the transcription of its own gene by enhancing RNAP binding at the promoter region. Positive feedback loops in a resource-limited environment normally lead to a tendency to reinforce the growth of a species until it reaches a value that cannot be sustained. It is required for a permanent shift in behaviour, such as differentiation or evolution towards one of two states of a system [16].

With multiple-feedback networks, biological systems display more complicated behaviours. For example, chaotic systems, deterministic but essentially unpredictable systems, frequently result from some form of positive feedbacks, usually mixed with negative feedbacks [17]. The Elowitz and Leibler oscillator [18] is based solely on negative feedback loops but is unstable. This system could be made comparatively stable and robust by incorporating positive feedback loops [19, 20]. Negative feedback loops are responsible for producing the oscillatory behaviour; in addition, the additional positive and negative feedback loops both increase the robustness of the system.

2.2. Mathematical Concepts

The model of a biological system is often constructed based on biochemical reactions among the proteins of the biological system, which facilitate the functionality of the biological system. These biochemical reactions are modelled through mathematical representations based on their reaction rates (or reaction velocities), which determine the change of the concentrations of the proteins of the biological system. The reaction rate depends both on the chemical nature and on the quantity of the participating reactants in the reaction mixture [21]. These mathematical representations are rate laws describing the relationships between the reaction rate and the quantity of the participating reactants and are different among different types of reaction mechanisms. The rate laws express that the rate of the ith reaction, V_i, is expressed as a function, f_i, in terms of the concentrations of the participating reactants, $x_1 \sim x_n$, as follows:

$$V_i = f_i(x_1, x_2, \ldots, x_n). \tag{2.1}$$

As specific to synaptic plasticity, three rate laws are usually applied: (1) mass action rate law [21]; (2) Michaelis–Menten rate law [22]; and (3) Hill rate law [23].

2.2.1. *Mass action rate law*

Mass action rate law describes the reaction rate of an elementary reaction, which involves a single mechanistic step/state transition to form the product. Mass action rate law generally states that the rate of a biochemical reaction is proportional to the quantity of the participating reactants [21]. For a sample elementary reaction as shown in the following reaction scheme:

$$\alpha A + \beta B \xrightarrow{k} \sigma S + \tau T, \tag{2.2}$$

where A and B are the reactants, S and T are the products. α, β, σ and τ are the numbers of corresponding molecules involved in the reaction, and k is the reaction rate constant indicating the pace of the reaction. The corresponding reaction rate (V) is as follows:

$$V = k[A]^{\alpha}[B]^{\beta}, \tag{2.3}$$

where $[A]$ is the concentration of A and $[B]$ is the concentration of B ($[x]$ denotes the concentration of substance x). The order of a reaction defines the relationship between the concentrations of substances in the reaction and the rate of the reaction. For the above reaction, α and β are the orders of reaction with respect to reactants A and B, respectively. The overall reaction order is $\alpha + \beta$. A first-order reaction has an overall reaction order of 1 and a second-order reaction has an overall reaction order of 2. The dynamic change of the concentrations of A, B, S, and T is described by a set of ordinary differential equations (ODEs) in terms of V as follows:

$$\frac{d[A]}{dt} = -\alpha V, \tag{2.4}$$

$$\frac{d[B]}{dt} = -\beta V, \tag{2.5}$$

$$\frac{d[S]}{dt} = \sigma V, \quad \text{and} \tag{2.6}$$

$$\frac{d[T]}{dt} = \tau V. \tag{2.7}$$

As shown in Fig. 2.1, the solution of Eqs. (2.4)–(2.7) shows that $[A]$, as the reactant, decreases and $[S]$, as the product, increases over the first 5 s of the reaction. The result means A (respectively, B) is consumed to produce S (respectively, T). The slopes of the decrease

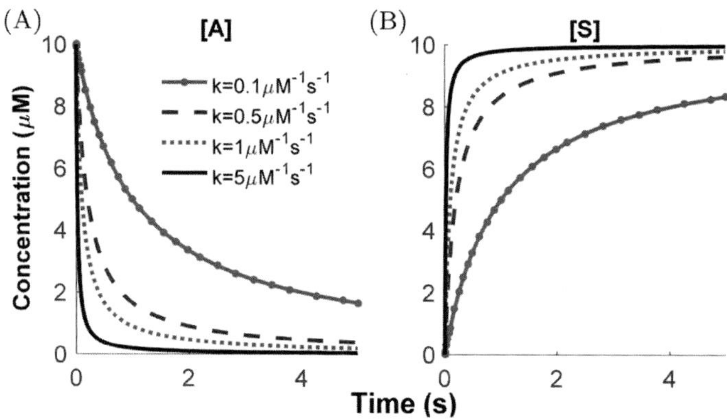

Fig. 2.1. The solution of reaction in Eq. (2.2). α, β, σ, and τ are set to 1 assuming a second-order reaction. The initial concentration for A, B, S, and T are 10, 10, 0, and $0\,\mu$M, respectively. Simulations of 5 seconds are shown in: (A) $[A]$, and (B) $[S]$.

and the increase depend on the rate constant, k. For faster reactions, the slopes are steeper.

2.2.1.1. *Factors for the final product of an elementary reaction cascade*

Consider a sequential elementary reaction cascade shown in the following reaction scheme:

$$A \xrightarrow{k_1} B \xrightarrow{k_2} P. \tag{2.8}$$

The reaction involves two first-order reactions: one converts substance A into substance B (reaction 1 with rate constant k_1) and the other converts substance B into the final product P (reaction 2 with rate constant k_2). Therefore, the reaction rates for reaction 1 (V_1) and for reaction 2 (V_2), are as follows:

$$V_1 = k_1[A], \quad \text{and} \tag{2.9}$$

$$V_2 = k_2[B]. \tag{2.10}$$

The ODEs for the dynamics of $[A]$, $[B]$, and $[P]$ are as follows:

$$\frac{d[A]}{dt} = -V_1, \tag{2.11}$$

$$\frac{d[B]}{dt} = V_1 - V_2, \quad \text{and} \tag{2.12}$$

$$\frac{d[P]}{dt} = V_2. \tag{2.13}$$

To understand the factors regulating the production of P, we solve the ODEs with respect to different rate constants. The results (with the initial concentrations of A, B, and P to be 10, 0, and 0 μM, respectively) are shown in Fig. 2.2. The red solid lines show the slope of the growth of $[P]$ with respect to time when the rates of the reactions are equal, and are used as the control for the further analysis. We fix k_1 and perturb k_2 for the first set of tests. As shown in Fig. 2.2(A), there is an insignificant impact on the slope of the growth of $[P]$ from the control when k_2 is greater than k_1. However, a significant impact on the slope of the growth of $[P]$ from the control is observed when k_2 is smaller than k_1. Both observations are independent of k_1. The similar observations are obtained when we fix k_2 and perturb k_1. As shown in Fig. 2.2(B), a greater k_1 than k_2 has an insignificant impact on the slope of the growth of $[P]$, while a significant impact on the slope of the growth of $[P]$ is caused when k_1 is smaller than k_2. Both effects are independent of k_2. The observations indicate that the production of P is not dependent on a particular reaction but limited by the slowest reaction along its upstream reaction cascade. Hence, the reaction in a slower timescale is dominating the cascade to produce the final product. Based on this property, a number of approximate descriptions of biochemical systems are formulated, i.e. Michaelis–Menten rate law and Hill rate law.

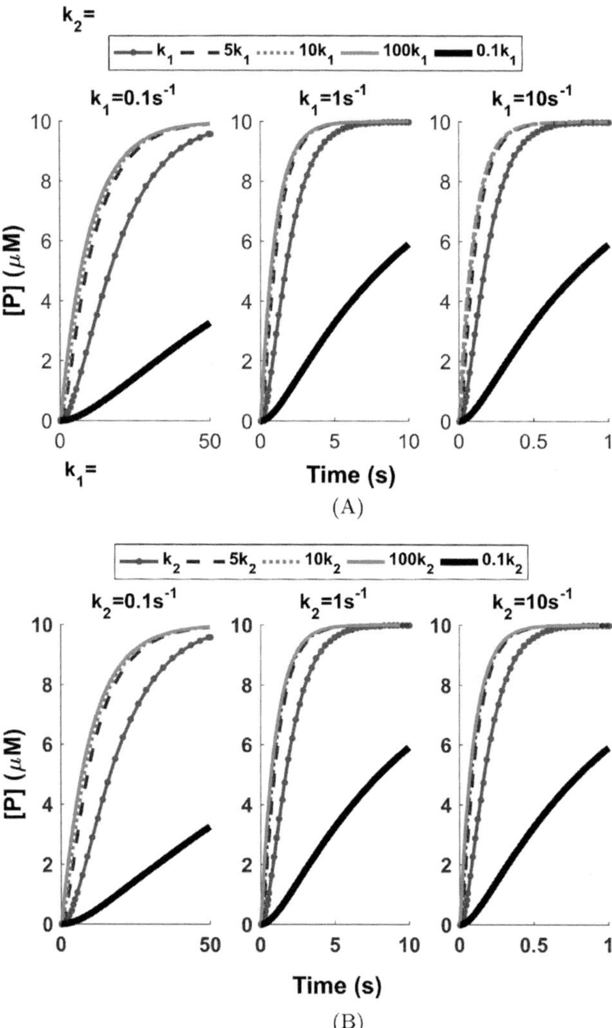

Fig. 2.2.   The growth of $[P]$ with respect to different ks. (A) k_1 holds constant in each of the three figures with values $0.1\,\mathrm{s}^{-1}$, $1\,\mathrm{s}^{-1}$, and $10\,\mathrm{s}^{-1}$ from left to right. In each figure, 5 values of k_2 are tested: same as k_1; 5 times greater than k_1; 10 times greater than k_1; 100 times greater than k_1; and a tenth of k_1. (B) Same procedures in (A) are followed with k_2 the constant given values $0.1\,\mathrm{s}^{-1}$, $1\,\mathrm{s}^{-1}$, and $10\,\mathrm{s}^{-1}$ from left to right and k_1 varies with 5 values: same as k_2; 5 times greater than k_2; 10 times greater than k_2; 100 times greater than k_2; and a tenth of k_2.

2.2.2. *Michaelis–Menten rate law*

Michaelis–Menten rate law provides an approximate description of enzyme-catalysed reactions [22] of the following reaction scheme:

$$S + E \underset{k_b}{\overset{k_f}{\rightleftharpoons}} ES \xrightarrow{K_{\text{cat}}} P + E, \tag{2.14}$$

where S is the substrate, E is the enzyme, ES is the substrate–enzyme complex or intermediate complex, and P is the final product. The reaction involves: (1) a reversible reaction by mixing S and E to form a complex, ES. The reaction rate constants are k_f for the forward reaction to form ES and k_b for the backward reaction to decompose ES into S and E; and (2) an irreversible reaction to convert ES into P and releasing E. The reaction rate constant is K_{cat} for the irreversible reaction. Hence, the ODEs of the dynamics of $[S]$, $[E]$, $[ES]$, and $[P]$ are given by Eqs. (2.15)–(2.18) based on mass action rate law:

$$\frac{d[S]}{dt} = -k_f[S][E] + k_b[ES], \tag{2.15}$$

$$\frac{d[E]}{dt} = -k_f[S][E] + (k_b + K_{\text{cat}})[ES], \tag{2.16}$$

$$\frac{d[ES]}{dt} = k_f[S][E] - (k_b + K_{\text{cat}})[ES], \quad \text{and} \tag{2.17}$$

$$\frac{d[P]}{dt} = K_{\text{cat}}[ES]. \tag{2.18}$$

2.2.2.1. *Equilibrium approximation*

The original assumption taken by Michaelis and Menten to approximate the reaction rate of the production of P is that S, E, and ES mix instantaneously to chemical equilibrium. i.e. $k_f[S][E] = k_b[ES]$. Therefore, $[ES]$ is given by

$$[ES] = \frac{k_f}{k_b}[S][E]. \tag{2.19}$$

According to the mass conservation law: $[E] + [ES] = [E_T]$, where $[E_T]$ is the total concentration of the enzyme, Eq. (2.19) can be modified to:

$$[ES] = \frac{1}{K_d}[S]([E_T] - [ES]), \tag{2.20}$$

where $K_d = k_b/k_f$ is the dissociation constant of the ES formation. Rearrange Eq. (2.20):

$$[ES] = \frac{[E_T][S]}{K_d + [S]}. \tag{2.21}$$

Substituting Eq. (2.21) into Eq. (2.18), the reaction rate of the production of P, V, is given by

$$V = \frac{d[P]}{dt} = \frac{V_{\max}[S]}{K_d + [S]}, \tag{2.22}$$

where $V_{\max} = K_{\text{cat}}[E_T]$ is the maximum reaction rate when all the enzyme binds to the substrate (ES). When $[S] = K_d$, V is half of $V_{\max}$.

The key of this approximation is that the reversible reaction is much faster than the irreversible reaction, i.e. $k_b \gg K_{\text{cat}}$, so that the irreversible reaction becomes the limiting factor of V.

2.2.2.2. *Quasi-steady-state approximation*

Briggs and Haldane [24] approximate the reaction rate, V, by taking an assumption that there is no change in the concentration of the intermediate complex at all times, i.e. $k_f[S][E] = (k_b + K_{\text{cat}})[ES]$. Hence, $[ES]$ is given by

$$[ES] = \frac{k_f}{k_b + K_{\text{cat}}}[S][E]. \tag{2.23}$$

According to the mass conservation law, Eq. (2.23) can be modified to:

$$[ES] = \frac{1}{K_m}[S]([E_T] - [ES]), \tag{2.24}$$

where $K_m = (k_b + K_{\text{cat}})/k_f$ is the Michaelis-Menten constant. Rearrange Eq. (2.24):

$$[ES] = \frac{[E_T][S]}{K_m + [S]}. \qquad (2.25)$$

Substituting Eq. (2.25) into Eq. (2.18), V is as follows:

$$V = \frac{d[P]}{dt} = \frac{V_{\text{max}}[S]}{K_m + [S]}, \qquad (2.26)$$

Similar to equilibrium approximation, $V_{\text{max}} = K_{\text{cat}}[E_T]$ is the maximum reaction rate when all the enzyme is presented as the complex with the substrate (ES). When $[S] = K_m$, V is half of V_{max}.

Equations (2.22) and (2.26) are similar, but based on different constant, K_d and K_m, respectively. However, since $k_b \gg K_{\text{cat}}$ (assumption of the equilibrium approximation), so that $K_d \approx K_m$. Hence, the general behaviours of the two approximation are similar. As shown in Fig. 2.3, there is a negligible difference between the two approximations.

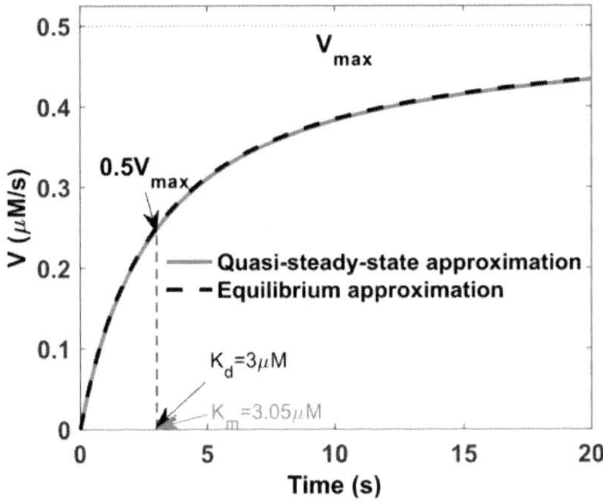

Fig. 2.3.   The approximations of the reaction rate, V, of an enzyme-catalysed reaction by equilibrium approximation and quasi-steady-state approximation. The rate constants of the enzyme-catalysed reaction are, $k_f = 2\,\mu\text{M}^{-1}\,\text{s}^{-1}$, $k_b = 6\,\text{s}^{-1}$, and $K_{\text{cat}} = 0.1\,\text{s}^{-1}$. $[E_T]$ is $5\,\mu\text{M}$. Hence, $K_d = 3\,\mu\text{M}$, $K_m = 3.05\,\mu\text{M}$, and $V_{\text{max}} = 0.5\,\mu\text{M}\,\text{s}^{-1}$.

2.2.3. *Hill rate law*

Hill rate law describes the cooperativity between two chemical species [23]. The law is often applied to mathematically formulate the binding between ligands and receptors by expressing the fraction of the bound receptor as a function in terms of the ligand concentration [25]. Moreover, Hill rate law has shown advantage in describing the dependence between chemical substances and is extensively applied in modelling the cooperative activation and inhibition in complex biochemical systems. Consider the situation where a receptor contains two ligand binding sites and the binding between the receptor and ligands is shown as the following reaction scheme:

$$L + R \underset{k_{1b}}{\overset{k_{1f}}{\rightleftharpoons}} RL, \tag{2.27}$$

$$RL + L \underset{k_{2b}}{\overset{k_{2f}}{\rightleftharpoons}} RL_2, \tag{2.28}$$

where L is the ligand, R is the receptor, RL_n is the ligand–receptor complex occupying n binding sites. The ODEs of the dynamics of $[R]$, $[L]$, $[RL]$, and $[RL_2]$ are given by

$$\frac{d[R]}{dt} = -k_{1f}[R][L] + k_{1b}[RL], \tag{2.29}$$

$$\frac{d[L]}{dt} = -k_{1f}[R][L] + k_{1b}[RL] - k_{2f}[RL][L] + k_{2b}[RL_2], \tag{2.30}$$

$$\frac{d[RL]}{dt} = k_{1f}[R][L] - k_{1b}[RL] - k_{2f}[RL][L] + k_{2b}[RL_2], \quad \text{and} \tag{2.31}$$

$$\frac{d[RL_2]}{dt} = k_{2f}[RL][L] - k_{2b}[RL_2]. \tag{2.32}$$

Now, if a final product, P, is produced at a rate proportional to $[RL_2]$ (i.e. ions pass through open ion channels), then the reaction rate, V, is given by

$$V = \frac{d[P]}{dt} = k_3[RL_2]. \tag{2.33}$$

To approximate V, we assume that the intermediate binding states (RL and RL_2) do not change on the timescale of V (quasi-steady-state approximation). Mathematically, this is given by

$$k_{1f}[R][L] - k_{1b}[RL] = 0, \quad \text{and} \tag{2.34}$$

$$k_{2f}[RL][L] - k_{2b}[RL_2] = 0. \tag{2.35}$$

Solving Eqs. (2.34) and (2.35), we have

$$[RL] = \frac{[L][R]}{K_1}, \quad \text{and} \tag{2.36}$$

$$[RL_2] = \frac{[L]^2[R]}{K_1 K_2}, \tag{2.37}$$

where $K_1 = k_{1b}/k_{1f}$ and $K_2 = k_{2b}/k_{2f}$. According to the mass conservation law, $[R] + [RL] + [RL_2] = [R_T]$, where $[R_T]$ is the total concentration of the receptor. Hence, the ratio of $[RL_2]$ to $[R_T]$ is given by

$$
\begin{aligned}
\frac{[RL_2]}{[R_T]} &= \frac{[RL_2]}{[R] + [RL] + [RL_2]} \\[2mm]
&= \frac{[L]^2[R]}{K_1 K_2 \left([R] + \frac{[L][R]}{K_1} + \frac{[L]^2[R]}{K_1 K_2}\right)} \\[2mm]
&= \frac{[L]^2}{K_1 K_2 + K_2[L] + [L]^2}.
\end{aligned}
\tag{2.38}
$$

Therefore, $[RL_2]$ is given as a function in terms of $[L]$ by

$$[RL_2] = \frac{[R_T][L]^2}{K_1 K_2 + K_2[L] + [L]^2}. \tag{2.39}$$

If we assume that there is cooperativity between the bindings; the first binding is very slow, but once it is bound, the second binding is very fast, i.e. $k_{1f} \to 0$ and $k_{1f} \to \infty$. Therefore, $K_1 \to \infty$, $K_1 \to 0$, and $K_1 K_2$ is a constant. Hence, Eq. (2.39) is modified to

$$[RL_2] = \frac{[R_T][L]^2}{K_{0.5}^2 + [L]^2}, \qquad (2.40)$$

where $K_{0.5}^2 = K_1 K_2$. Therefore, V is given by

$$V = \frac{d[P]}{dt} = \frac{V_{\max}[L]^2}{K_{0.5}^2 + [L]^2}, \qquad (2.41)$$

where $V_{\max} = k_3[RT]$ is the maximum reaction rate occurred when all receptors are double-bound by ligands. Equation (2.41) is the general expression of Hill rate law. The law states that V is dependent on the concentration of the ligand. If $[L]$ is large ($[L] \gg K_{0.5}$), V approaches the maximum, $V_{\max}$. If $[L] = K_{0.5}$, V is half of $V_{\max}$.

For a process involves n cooperative processes, Hill rate law becomes:

$$V = \frac{V_{\max}[L]^n}{K_{0.5}^n + [L]^n}, \qquad (2.42)$$

where n is called Hill coefficient indicating the degree of cooperativity. When $n = 1$, there is no cooperativity, and the equation is similar to Michaelis–Menten rate law.

An alternative form of Hill rate law describes the situation where V is inhibited by RL_n. That is, V decreases as $[RL_2]$ increases and is given by

$$V = V_{\max}\left(1 - \frac{[L]^n}{K_{0.5}^n + [L]^n}\right) = \frac{V_{\max}K_{0.5}^n}{K_{0.5}^n + [L]^n}. \qquad (2.43)$$

Here, V is maximum when $[L]$ is zero and deceases as $[L]$ increases. When $[L] = K_{0.5}$, V is half of $V_{\max}$. As shown in Fig. 2.4, the slope of V with respect to $[L]$ becomes steeper when the Hill coefficient increases. For $n > 1$, V exhibits a switch like behaviour that is a good framework to describe the activation/inhibition of signalling pathway in cellular dynamics. For instance, Eq. (2.42) is applied to describe the activation where a signalling pathway is turned on when the concentration of its controller protein passes the activating threshold, while Eq. (2.43) is applied to describe the inhibition where

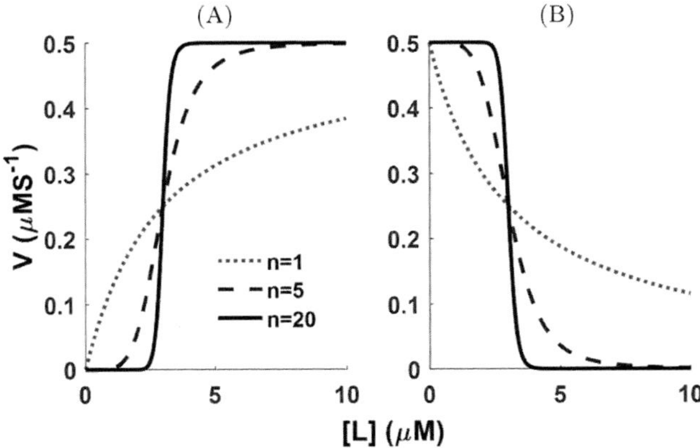

Fig. 2.4. The approximations of the reaction rate, V, by Hill rate law. The associated parameters are, $K_{0.5} = 3\,\mu\text{M}$ and $V_{\max} = 0.5\,\mu\text{M}\,\text{s}^{-1}$. Three values of Hill coefficients, 1, 5, and 20, are used. The left figure (A) shows the activation and the right figure (B) shows the inhibition.

a signalling pathway is turned off when the concentration of its controller protein passes the inhibiting threshold.

2.2.4. *Parameter estimation using Markov chain Monte Carlo (MCMC)*

Markov chain Monte Carlo (MCMC) is a sampling algorithm and is often applied to solve problems in large-dimensional space [26]. Parameter estimation is an optimisation problem over multiple dimensions, where each dimension corresponds to an unknown parameter. The estimation is very difficult to be sampled directly in large dimensions, but the goodness of a sample can be evaluated by calculating the error from the experimental data. MCMC generates samples for the parameters from a sample space, X, and the algorithm is constructed to draw samples from the most important regions (low error) which gives optimal parameters to describe the experimental data.

2.2.4.1. *Markov chain*

A chain here describes the evolution of a system/process among large but finite state spaces. The sampling process, x^i, (x^i denotes both the process and the ith sample) will draw samples from s discrete states $x^i \in X = \{x_1, x_2, \ldots, x_s\}$ according to a probability distribution, $P_i(X)$, which may be evolving over time. $P_i(X)$ evolution is a Markov chain only if the next state is dependent solely on the current state and a fixed probability governs the transitions among states as given by [26]:

$$p(x^{i+1} = x_n | x^i = x_m, \ldots, x^1) = p(x_n | x_m) \quad (m, n = 1, \ldots, s).$$

$$(2.44)$$

where $p(x_n|x_m)$ determines the probability that the next sample is x_n given the current sample is x_m. A transition matrix, T, contains the transitional probabilities, such that $p(x_n|x_m) = T_{mn}$ such that

$$\sum_{n=1}^{s} T_{mn} = 1.$$

Considering a simple Markov chain as illustrated in Fig. 2.5, which has a sample space of three discrete states ($X = \{x_1, x_2, x_3\}$). The transition matrix for the Markov chain is:

$$T = \begin{bmatrix} 0.2 & 0.8 & 0 \\ 0.4 & 0 & 0.6 \\ 0.9 & 0.1 & 0 \end{bmatrix}.$$

The chain is evolving from an initial distribution, $P_0(X) = (P1, P2, P3)$ such that the chance of getting a sample of x_1, x_2, or

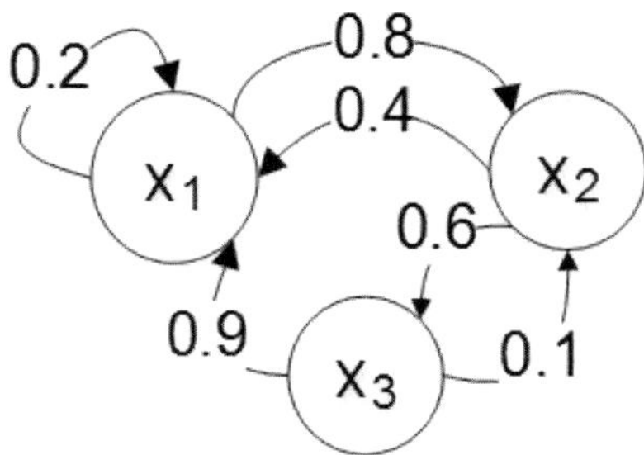

Fig. 2.5.   A Markov chain with fixed transitional probability among the three states.

x_3 is $P1$, $P2$, or $P3$, respectively. It follows that the $P_i(X)$ is evolved through the transformations by the transition matrix:

$$P_i(X) = P_{i-1}(X)T = P_{i-2}(X)T^2 = \cdots = P_0(X)T^i. \qquad (2.45)$$

If the size of the sampling is large enough, we have

$$\lim_{i \to \infty} T^i \to \begin{bmatrix} 0.4234 & 0.3604 & 0.2162 \\ 0.4234 & 0.3604 & 0.2162 \\ 0.4234 & 0.3604 & 0.2162 \end{bmatrix}.$$

As a result, the distribution of the samples drawn converges to an invariant distribution, $(0.4234, 0.3604, 0.2162)$, regardless of the initial distribution chosen.

The convergence is the central idea of MCMC and a Markov chain will converge to an invariant distribution, $P(X)$, if: (1) starting from any states, all other states are reachable (irreducibility), and (2) the chain is aperiodic.

The invariant distribution of the convergence must satisfy the detailed balance condition as given by [26]:

$$P(x_m)T_{mn} = P(x_n)T_{nm} \quad (m, n = 1, \ldots, s). \qquad (2.46)$$

Summing both sides of Eq. (2.46) over x_n, we have

$$P(x_m) = \sum_{x_n} P(x_n)T_{nm}. \qquad (2.47)$$

Testing the sample Markov chain in Fig. 2.5 with its invariant distribution, $P(X) = (0.4234, 0.3604, 0.2162)$. Let $m = 1$, we have

$$P(x_1) = 0.4232 \approx \sum_{n=1}^{3} P(x_n)T_{n1} = 0.42322.$$

2.2.4.2. *The Metropolis–Hastings algorithm*

If we can construct a Markov chain, which converges to the optimal distribution of parameters, then we can obtain the optimal distribution of parameters directly by drawing samples from the converged chain. The most popular MCMC algorithm for constructing the Markov chain is the Metropolis–Hastings algorithm [27, 28].

Supposing an aperiodic and irreducible Markov chain, C_1, with s discrete states of possible parameters space ($X = \{x_1, x_2, \ldots, x_s\}$) and a transition matrix, Q, given by

$$q(x^{i+1} = x_n | x^i = x_m) = Q_{mn} \quad (m, n = 1, \ldots, s), \qquad (2.48)$$

does not converge to the optimal distribution of optimal parameters, $P(X)$:

$$P(x_m)Q_{m,n} \neq P(x_n)Q_{n,m}. \qquad (2.49)$$

However, C_1 can be transformed by a function, A, in terms of x_m and x_n into C_2, so that C_2 satisfies the detailed balance condition:

$$P(x_m)Q_{m,n}A(x_m, x_n) = P(x_n)Q_{n,m}A(x_n, x_m), \qquad (2.50)$$

and importantly, C_2 converges to $P(X)$. The simplest form of A is:

$$A(x_m, x_n) = P(x_n)Q_{n,m}, \quad \text{and} \quad A(x_n, x_m) = P(x_m)Q_{m,n}. \quad (2.51)$$

Therefore, the transition matrix of C_2, Q', is

$$q'(x_n | x_m) = Q'_{mn} = Q_{m,n}Q_{n,m}P(x_n). \qquad (2.52)$$

A is called the acceptance probability, which declares the probability of accepting a state transition occurring in C_1 to occur in C_2 as well.

However, A may be so small that C_2 may take very long time (or samples) to explore the whole state space for converging to $P(X)$. Therefore, if we enlarge A by a factor, k, which satisfies the conditions given by

$$kA(x_m, x_n) \leq 1, \ \ kA(x_n, x_m) \leq 1 \quad \text{and} \quad k \geq 1, \qquad (2.53)$$

then we have a new Markov chain, C_3, which converges (as given by Eq. (2.54)) faster than C_2:

$$P(x_m)Q_{m,n}A(x_m, x_n) \times k = P(x_n)Q_{n,m}A(x_n, x_m) \times k. \qquad (2.54)$$

The speed of convergence becomes maximum when either $kA(x_m, x_n) = 1$ or $kA(x_n, x_m) = 1$, without breaking conditions given by Eq. (2.53). Therefore, the maximum of $kA(x_m, x_n)$ is

$$\min\left\{\frac{A(x_m, x_n)}{A(x_n, x_m)}, \ 1\right\}.$$

Now, the procedure of the Metropolis–Hastings algorithm is given as follows:

(1) Initialise $x^0 \in X = \{x_1, x_2, \ldots, x_s\}$
(2) For $i = 0$ to $N - 1$

    (a) Generate a random number, u, from uniform distribution, $U(0,1)$

    (b) Then, take a sample, $y \in X = \{x_1, x_2, \ldots, x_s\}$, according to $Q^i_{x,y}$.

$$\text{If } u < \min\left\{\frac{A(x^0, y)}{A(y, x^0)}, \ 1\right\}, \quad x^{i+1} = y, \text{ Else, } x^{i+1} = x^i$$

Although the Metropolis–Hastings algorithm may be incfficient for sampling over large dimensions, the basic ideas of MCMC are well encapsulated in this algorithm and our discussion on the algorithms of MCMC ends here. For the details of more efficient algorithms of MCMC, we refer to [26, 29].

2.2.5. *Parameter sensitivity analysis*

The robustness of a biological system reflects its ability to retain normal functioning with a modest change to the working environment [30] which may lead to highly variable parameters potentially resulting in undesirable levels of some species in the system. These stochastic effects are exacerbated due to the low copy numbers of proteins. Sensitivity analysis methods quantify the effects of variation of parameters on the system dynamics. Sensitivity analysis methods rely on sampling algorithms, such as Latin hypercube sampling (LHS) [31], which feeds the model with a large number of sets of parameters that are varied within a realistic margin. Then, sensitivity analysis methods map the variation of responses of the model to the parameters. In general, the highly weighted parameters in this map are very sensitive parameters.

2.2.5.1. *Parameter sensitivity analysis using PRCC*

Partial ranking correlation coefficient (PRCC) is a global sensitivity analysis method for quantifying the sensitivity of the model output with respect to the variation of parameters. To quantify the relation between a model output, y, and a parameter, $x_j (j = 1, \ldots, n)$, of n chosen parameters, correlation coefficient (CC), r_{yxj}, is usually used and given by

$$r_{yxj} = \frac{\sum_{i=1}^{N} (x_{ij} - \overline{x_j}) (y_i - \overline{y})}{\sqrt{\sum_{i=1}^{N} (x_{ij} - \overline{x_j})^2 \sum_{i=1}^{N} (y_i - \overline{y})^2}}, \tag{2.55}$$

where N is the total number of sets of parameters, x_{ij} the value of ith sample of x_j, which has y_i as the corresponding model output, $\overline{x_j}$ and $\overline{y}$ are the means of samples of x_j and outputs, respectively. Correlated cases x_j and y have r_{yxj} values close to -1 or 1, which means strong negative or positive correlation, while uncorrelated cases have r_{yxj} values approaching 0. A p-value is calculated according to a null hypothesis, which states that x_j and y are uncorrelated. A small p-value (smaller than the significance level, which is normally set at 0.05) leads to rejection of the null hypothesis, and the conclusion is made that x_j and y are correlated.

CC has two major problems: (1) sampling algorithms like LHS perturbs all parameters simultaneously, hence, CC may be affected by the variation of parameters other than x_j; and (2) since parameters are usually sampled according to a percentage around their reference values, the magnitude of the reference value may have an impact on CC.

A good solution for problem one is the partial correlation coefficient (PCC), which quantifies the relation between y and x_j after removing the linear effects on y from parameters other than x_j. A linear regression model predicts y and x_j in terms of the n chosen parameters in a form given by

$$x_j = \varepsilon_1 + \sum_{\substack{i=1 \\ i \neq j}}^{n} \alpha_i x_i, \quad \text{and} \quad y = \varepsilon_2 + \sum_{\substack{i=1 \\ i \neq j}}^{n} \beta_i x_i, \qquad (2.56)$$

where α is the regression coefficients for x_j, β is the regression coefficients for y, ε_1 and ε_2 are error terms. The linear effects from parameters other than x_j can be removed to obtain two residues $\hat{x}_j$ and $\hat{y}$ as follow:

$$\hat{x}_j = x_j - \sum_{\substack{i=1 \\ i \neq j}}^{n} \hat{\alpha}_i x_i, \quad \text{and} \quad \hat{y} = y - \sum_{\substack{i=1 \\ i \neq j}}^{n} \hat{\beta}_i x_i. \qquad (2.57)$$

PCC is the CC between $\hat{x}_j$ and $\hat{y}$.

The rank-transformed data (Fig. 2.6) is a good solution for problem two. The parameters and the output are first rank-transformed and then the PRCC is calculated by taking PCC between the ranked x_j and y. As a result, the parameters and the output are transferred from real values into ranks (integers from 1 to N) so that the magnitude of the reference value is removed.

2.3. Intracellular Fluctuations

2.3.1. *"Noise" in networks*

Although the number of studies, both theoretically and experimentally, on noise is still limited, positive initial efforts have revealed important features of noise (stochasticity). In this section, we summarise the noise related literature in the context of gene regulatory networks and attempt to provide a coherent picture of noise at the gene expression level. We use the terms "noise" and "stochasticity" interchangeably to allude to the fact that the molecular fluctuations inherently require probabilistic interpretations while acknowledging the use of the word "noise" implies the existence of "clean signals", which may not be the case in many situations.

It has long been recognised that genetically identical cells under the same environmental conditions can have significant variations in

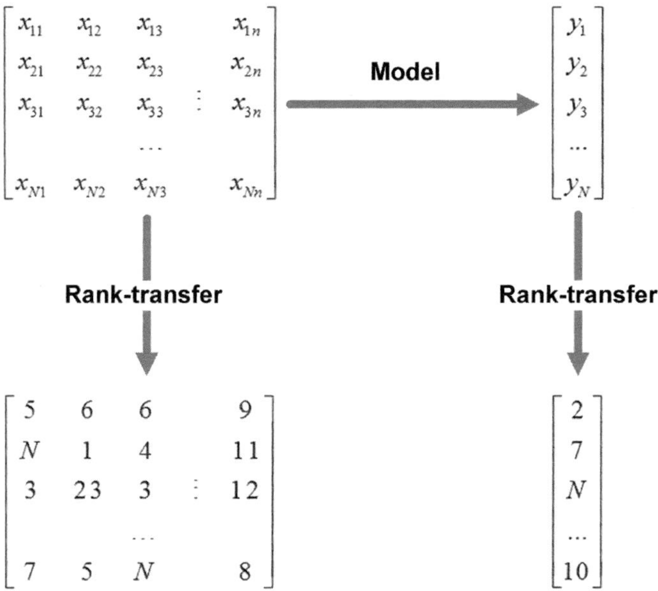

Fig. 2.6. Rank transfer of parameters and model output. A matrix x contains the LHS generated parameters. Each column of x comprises N samples of a parameter, and each row of x is a set of n parameters chosen. For each set of parameters, an output is simulated and stored into a vector, y, in the same row as that set of parameters. The rank-transformation ranks each column of matrix x (N samples of a parameter) or vector y, so that each column of matrix x (or y) comprises N integers from 1 to N and each of the integers appears once.

phenotypic characteristics [32]. Such variation has been observed in the cells of organisms ranging in complexity from bacteria to mammals, and is believed to be an important factor in the development and function of many living organisms, physiologically and evolutionarily. It is, therefore, of great interest to study the implication of stochasticity in gene expression for cellular regulation and non-genetic individuality [33–36]. Only in recent years, new experimental techniques in molecular biology, such as fluorescent reporters, allow stochastic gene expression to be quantified *in vivo* [37–43]. These elegant experiments, along with theoretical studies on stochasticity in gene expression [44–48], have greatly facilitated our understanding of the sources and consequences of such stochasticity in GRNs.

Stochasticity in the dynamics of molecular and cellular behaviour, in principle, are categorised as intrinsic noise and extrinsic noise. Although the definition of both is somewhat relative, in general, the intrinsic noise is confined in the system and the extrinsic noise is due to the changes in the surrounding environment [45, 46]. Research has been carried out by different groups trying to separate one type of noise from another theoretically [45, 46], and experimentally [37, 39]. In many cases, the evidence has shown that the extrinsic noise dominates the intrinsic noise and sets cell-to-cell variation in both prokaryotes and eukaryotes [38, 40].

2.3.2. *Intrinsic noise*

Elowitz and colleagues [38] initiated the efforts by conducting an experiment that enabled them to distinguish between intrinsic and extrinsic noise. This is done by using two green fluorescence protein (GFP) reporter genes under the control of promoters regulated by the *Lac* repressor. The genes encode the cyan and yellow forms of GFP, which are quantified by the fluorescence intensity of their respective emission peaks. Intrinsic noise is referred to as the difference in gene expression that arises between two identical copies of a gene expressed under precisely the same conditions. Therefore, intrinsic noise emerges from fluctuations inherent in the biochemical reactions involved in a gene's expression process, such as reaction events leading to transcription and translation of that gene. In a GRN, specifically, reacting molecules must first find each other through diffusion in a cell and their motion is driven by random collisions. Even if we ignore the diffusion process and assume that all molecules are well-mixed in the cell, reactions may occur with a finite probability per unit time, instead of continuously and deterministically. Such stochastic effects are especially important when mean numbers of the molecules are low, which is always the case in GRNs. Unlike metabolic processes, biochemical processes in GRNs are generally in a small volume and have low concentrations of molecular species [49]. For example, only ten molecules of the *Lac* repressor, on average, are present in *Escherichia coli* (*E. coli*) cells [8]. Therefore, the stochastic nature inherited from GRNs in cells often leads to intrinsic noise which cannot be negligible.

2.3.3. *Extrinsic noise*

On the other hand, extrinsic noise is defined as the cell-to-cell variation in expression levels of copies of identical genes, given that they are expressed in different cells [38]. As a result, extrinsic noise arises from fluctuations in other molecular components and factors involved in gene expression than those that give rise to intrinsic noise. Essentially, intrinsic sources of noise are dependent on the gene whereas extrinsic sources of noise arise independently of the gene but act on it.

The external factors leading to fluctuations in kinetic parameters in biochemical reactions, in turn, influence the expression of the gene of interest; and one of the most obvious external factors is the thermal fluctuations [28]. External factors could also come from the internal processes of the cell, including the variance of number of RNAPs, ribosomes and degradosomes, the timing of gene expression in different stages of cell cycles, the quantity of proteins, and energy demand [45].

2.3.4. *Notable experiments*

Over the years, many researchers have been interested in the implications of stochastic nature of gene expression [36, 50–53]. A classic and notable example is the lysis/lysogenic switch of lambda-phage infecting bacteria *E. coli* in which noise randomly dictates which phenotype the cell ultimately adopts [54, 55]. Noise is not only exhibited in cells with genetic differences, but also in an isogenic cell population, even when the cells have been exposed to the same environments and have the same history [38, 39, 56]. Recent single-cell experiments, in which noise can be quantified, provide even more convincing evidence of the presence of stochastic fluctuations at the gene expression level [37, 38, 57]. Typically, in these studies, small, simple and easily manipulated gene networks were artificially engineered in the cell, which were usually coupled with theoretical stochastic models formulated to explain the observations and furthermore, to give predictions. For example, Elowitz and co-workers [38] constructed strains of *E. coli* to detect noise using two different fluorescent proteins expressed from identical promoters. This study was the first quantitative demonstration and analysis of the existence of noise during gene expression in *E. coli*. In a complementary study, Ozbudak *et al.* [37] built a single-gene system in *Bacillus subtilis* in which they were able to vary transcriptional and translational rate independently and measure total noise. More recent single-cell experiments have provided further understandings into cell-to-cell variation by assessing noise dependence on the efficiency of gene expression processes such as transcription and translation [39], examining the sources and measuring relative contributions of different types of noise in bacteria as well as eukaryotic gene expression [39, 40, 56, 57], investigating how noise propagates through synthetic cascades of regulatory genes [39, 41], analysing the contribution of global noise and noise specific to pathway towards cell-to-cell variation in a cell-fate decision [58], and quantifying the noise frequency content and its relation with gene circuit structure [42].

Many experimental and theoretical studies mentioned above have focused on investigating the relative contribution of intrinsic and extrinsic noise to the overall noise in gene expression in prokaryotes as well as eukaryotes. Elowitz and co-workers [38] showed that, in bacteria, intrinsic noise decreases monotonically as transcription rate increases. This is not unexpected as higher transcription rate results in a higher number of corresponding molecules (e.g. mRNA transcripts), which tends to reduce fluctuation. However, under the same control condition, extrinsic noise initially increases, peaks at intermediate transcription level and then declines at higher transcription rate. Furthermore, it was revealed that extrinsic fluctuation is the dominant source of noise in bacterial cells *E. coli* and *Bacillus subtilis* [37, 38]. Rosenfeld *et al.* [56] examined the origin of variability of the gene regulation functions in bacteria *E. coli* and detected only a minor contribution from intrinsic factors, which also implied that extrinsic factors are the prominent noise sources. A similar conclusion was also arrived for the systems in eukaryotic organisms: for example, studies in *S. cerevisiae* indicated that gene expression is dominated by extrinsic noise [39, 40].

Several studies have gone further and looked into the origins of intrinsic and extrinsic stochasticity in gene expression [37, 39, 42, 59]. The question was which process, transcription or translation or other processes (e.g. chromatin remodelling), is responsible for causing noise, and to what extent? The effects of built-in efficiency of transcription and translation on noise were examined. It has been known that both transcription and translation occur in random size bursts [59]. The random activation and deactivation of a gene lead to the production of mRNA molecules in bursts of random sizes. In turn, translational bursting is due to the random lifetime of mRNA molecules during which several protein copies can be produced. Transcriptional and translational burstings are thought to contribute considerable fluctuations in the protein product.

Using experimental methods along with theoretical modelling, Ozbudak and colleagues [37] concluded that in prokaryotic gene expression, noise strength is more sensitive to variation in translational efficiency than transcriptional efficiency. This means for genes expressed at similar levels, low transcription rate coupled with the high translational efficiency result in more noise compared with a combination of high transcriptional rate and low translational efficiency. Because the average number of proteins and the cell volume are kept fixed, the increase in gene expression noise is attributed to the increased fluctuation in mRNA abundance, causing the increased fluctuation in the rate of protein synthesis [59]. This concept of increased noise due to low transcription rate coupled with high translation rate is usually referred to as the translational bursting mechanism and has been confirmed in both bacteria and eukaryotes. In a recent large-scale study, Bar-Even *et al.* [60] measured the variation in abundance of 43 fluorescently labelled proteins expressed across 11 different environmental conditions in *S. cerevisiae*. They interestingly found in most cases that there is an inverse relationship between protein noise and mean protein abundance, with a proportionality factor of ~1200. The authors further determined, after analysing the large experimental data sets, that this factor is consistent with the number of proteins produced per mRNA, which suggests that translational bursting is a major source of protein noise that could explain the observed scaling behaviour.

On the other hand, it was shown that slow promoter activation (transcriptional bursting) owing to chromatin remodelling also has an important role in generating stochasticity in eukaryotic gene expression [40]. Furthermore, the position of the genes along the chromosome can be more significant than the number of transcripts or proteins in terms of giving rise to protein fluctuations [61].

2.3.5. *Noise: nuisance or necessity?*

Stochasticity is generally perceived as being undesirable and unpredictable to cell functions. Healthy cells rely on the reliability and precision of cellular regulation, which ensures the production of involved molecules to be within a biologically acceptable range inside the cells. This is particularly crucial for essential molecular components such as key transcriptional factors or enzymes. In case the intracellular concentration of such molecules drops below or exceeds a threshold value, disease could be induced. The fluctuations in protein levels caused by noise, if large, can bring protein concentration to dangerously low level. It can also distort intracellular signals, negatively affecting cellular regulation or even bringing detrimental consequences to cells such as randomising developmental pathways, thereby disrupting cell cycle control. Furthermore, transmission of noise results in loss of coordination among cells.

However, fluctuations in living systems may be much more than just a nuisance. There are many studies, in which randomness in gene expression has been proposed as a useful mechanism for generating phenotypic and cell-type diversification [37, 49, 55, 62, 63]. This is particularly beneficial to microbial cells that need to adapt efficiently to sudden changes in environmental conditions. Because of the larger pool of physiological states due to phenotypic heterogeneity, fluctuations could help to increase the probability of survival against environmental cues, without the requirement of genetic mutation. This idea has been theoretically examined in a study [64] which used a stochastic population model to show that a dynamically heterogeneous bacterial population can sometimes achieve a higher net growth rate than a population with homogeneous cells under fluctuating environments. Moreover, intrinsic noise in gene expression has been hypothesised with increasing support to be subject to natural selection [65]. Stochasticity has also been suggested to have a positive role in the development and cellular differentiation [66, 67].

In short, noise in gene expression in particular or noise in biological systems in general has two-fold effects on cellular existence and functions. It is considered harmful in some cases but beneficial in other cases depending on a particular system of interest. Nevertheless, living systems are still functioning in the presence of noise. An important question to ask, therefore, is how they suppress or control noise when it is harmful and how they utilise or exploit noise when it is of benefit.

2.3.6. *Noise propagation in networks*

Individual genes in gene regulatory networks are subjected to several possible noise sources. Apart from intrinsic and extrinsic noise, a gene downstream of a cascade might also be influenced by noise that is transmitted from upstream genes or regulators. Noise can propagate through gene cascades within regulatory networks. Due to the complexity of gene regulatory networks with nonlinear interactions of molecular components, the study of noise propagation is considered an interesting but challenging topic. Insights into how noise transmits between molecular components will help to facilitate a deeper understanding of how noise arises in the first place. Accurate prediction of noise propagation in gene network is also crucial for reverse engineering of natural networks and designing reliable synthetic genetic circuits [41].

Pedraza and van Oudenaarden [41] investigated how noise propagates through a gene regulatory cascade by constructing a synthetic network in single *E. coli* cells. Three genes were arranged so that the first gene regulates the expression of the second gene which, in turn, modulates the expression of the third gene. Fluorescent genes were fused and co-expressed with the genes of interest in order to quantify expression variability. The authors measured correlations between genes in the cascade and correlations with a constitutive gene and found that noise in a gene is determined by three sources: intrinsic fluctuations, transmitted noise from upstream genes and global noise affecting all genes. To interpret the results, a stochastic model was developed which demonstrated that expression variability of a target gene in the cascade was influenced most by transmitted noise from the upstream regulator. This finding means that variability of gene expression can be more strongly affected by the underlying network connectivity than by its own intrinsic noise. Austin *et al.* [42] lent more support to this hypothesis by showing that noise frequency content (spectra) is determined by the underlying gene circuit structure, establishing a mapping between the two. In a similar work to Pedraza's, gene regulatory cascade with varying lengths were constructed in *E. coli* to explore the effects of cascade length on the variability and sensitivity of the network's response [68]. It was shown that longer cascades, which are more sensitive, amplify cell-to-cell variability, particularly at intermediate input signal levels.

In addition to the experimental studies with simple synthetic gene regulatory networks, simulation studies have also been carried out. The obvious advantage of this type of studies over the synthetic methods is the large pool of *in silico* gene networks that can be tested at will with relatively low cost. For example, a variety of transcriptional networks using a stochastic model were simulated to examine the effect of delay and repression strength of negative feedback on intrinsic noise propagation [69]. Resulting analysis showed that addition of negative or positive feedback to transcriptional cascades does not reduce intrinsic noise. Moreover, for cascades with multigene negative feedback, increasing network length and tightening regulation at one of the repression stages can improve synchrony in oscillations among cell population [69].

2.3.7. *Noise and robustness*

Robustness in biological systems is defined as the ability of a system to maintain its functions despite external and internal perturbations. Robustness is present across many species, from transcriptional level to the level of systemic homoeostasis, assuring their survival during evolution. Many examples of robustness can be surveyed in different biological systems. Giaever *et al.* [70] have done a comprehensive study on yeast genome which showed that 80% of the genes were found dispensable under optimal laboratory conditions and 65% of the genes were dispensable under non-optimal conditions. Barkai and Leibler demonstrated using a computer model that the adaptation mechanism found in the chemotactic signalling pathway (responsible for bacterial chemotaxis) in *E. coli* is robust [71]. The bacterium is capable of chemotaxis over a wide range of chemo-attractant (a chemical that attracts bacteria) concentrations. A model of segmental polarity in *Drosophila melanogaster* has been shown to be insensitive to variations in initial values and rate constants, enabling stable pattern formation [72].

One major and fascinating question in studies of stochasticity is how biological systems can still function precisely with correct end results in the presence of noise. These systems seem to have mechanisms of noise-resistance or must be very robust against the influence of noise, if any. A remarkable example is of circadian rhythms. The network that controls circadian rhythms consists of multiple, complex, interlocking feedback loops. Deterministic as well as stochastic models have been constructed to investigate the mechanisms for noise resistance in circadian rhythms [73–75]. General models of chemical oscillators are sensitive to kinetic parameters. However, the proposed mechanisms for circadian rhythms produce regular oscillations in the presence of noise. Not only that, the stochastic model is able to produce regular oscillations whereas the deterministic models do not [76]. This suggests that the regulatory networks may even make use of molecular fluctuations.

Although robustness and stochasticity are two inherent features of cellular networks and are related, one is usually studied in isolation from the other [75]. Because kinetic parameters are subject to fluctuations, and robustness is often measured as how sensitive a system is to the changes in model parameters, studies of robustness need to explicitly take into account the presence of noise. Consideration of these two features together will offer good insights into the design and function of intracellular networks. One of the intriguing questions in this respect is what design features or control mechanisms are required for networks to function robustly given the occurrence of molecular noise.

2.3.8. *Noise controlling mechanisms*

To follow up the question at the end of the previous section, in this section we look at some of the main mechanisms used by cells to control noise. The most common mean of noise reduction is probably negative feedback. Negative feedback was shown to be an important control mechanism widely employed by cellular processes to regulate the level of signal molecules inside the cells [77]. A negative feedback loop occurs if a gene product directly or indirectly represses transcription of its own gene, aiming to stabilise its level and maintain constant conditions. Negative feedback control plays important roles in gene regulation: it helps to maintain homoeostasis of gene expression rate; it generates stability and therefore improves the robustness of developing cells [78]; it also facilitates sustaining oscillations of gene transcription rate [79]. Regarding the ability to attenuate noise, autorepression (a form of negative feedback where a transcriptional factor represses its own synthesis — for example, found in the regulation of approximately 40% of *Escherichia coli* genes) has been experimentally demonstrated to significantly reduce variability [80].

Integral feedback is another noise damping mechanism used in biological systems such as bacterial chemotaxis. It is a form of negative feedback often applied in engineering for ensuring the output of a system robustly tracks its desired value independent of noise or variation in system parameters [81].

The increase in gene number presents another way to control noise in gene expression. Many genes are present in identical copies on chromosome as results of events such as gene duplication or polyploidy. Compared to the case with only a single gene copy, intrinsic noise in gene expression is lowered as the gene dosage increases [82]. This is consistent with our expectation as a higher number of the same gene leads to an increased likelihood of gene expression which in turn, results in higher number of protein product and noise is therefore reduced. Intrinsic noise level is estimated to be inversely scaled with the square root of gene number [82].

Inefficient translation coupled with high transcription rate is also thought to be adopted by bacteria like the bacteriophage lambda as an effective mechanism for decreasing fluctuations in the concentration of the corresponding protein [33]. As mentioned earlier, this mechanism works in reverse to the translational bursting mechanism, which results in more noise in protein concentration. Other noise attenuating mechanism utilised by cells include regulatory checkpoints which ensure that a step in a pathway is only allowed to proceed after the preceding step finishes successfully [83].

2.4. Modelling Networks

2.4.1. *Importance of mathematical modelling*

Given the underlying complexity, mathematical models and computational simulation techniques are ideal tools for studying biochemical networks. They provide sophisticated frameworks for investigating the components of the networks and analysing the rules governing their interactions. The inherently complex structure of GRNs combined with the varying timescales on which different biological processes act, makes it particularly difficult to develop intuitions about how regulatory systems operate. Moreover, model simulation of known systems can provide validation of a particular modelling approach. Such simulations can also offer valuable guidance to target future studies by enabling experiments to be carried out *in silico* (on a computer) that would be expensive, time consuming or otherwise infeasible to perform *in vitro* and/or *in vivo*. In general, the role of mathematical models in systems biology is multi-faceted. First, while properly constructed mathematical models enable validation of current knowledge by comparing model predictions with experimental data when discrepancies are found in these types of comparisons, our knowledge of the underlying networks can be systematically expanded. Secondly, mathematical models can suggest novel experiments for testing hypotheses that are formulated from modelling experiences [84, 85]. Thirdly, they enable the study and analysis of system properties that are not accessible through *in vitro* experiments [86]. And, finally, mathematical models can also be used for designing desirable products based on existing biological networks [87].

Genetic regulatory pathways are only one type of networks within the complex web of intracellular interactions and interconnections. Metabolic pathways and signalling pathways are the other two major classes of cellular networks where significant modelling efforts are underway. A metabolic pathway is a series of chemical reactions occurring within a cell, resulting in either the formation of metabolic products or the initiation of another metabolic pathway. The dominant phenomenon in metabolism is enzymatic reactions. Scientists have characterised metabolism better than any other part of cellular behaviour due to more developed experimental techniques being available to quantify the network components. The typical mathematical modelling schemes are deterministic because, usually, a large number of molecules are involved in metabolic pathways. A signalling pathway is any process by which a cell converts one kind of signal or stimulus into another, where the dominant phenomenon is molecular binding. Signalling pathways normally have much fewer reactant molecules than metabolic systems, therefore, stochastic methods are more dominant in modelling signal pathways. As mentioned before, a GRN consists of a set of genes, proteins, metabolites (the intermediates and products of metabolism), and their mutual regulatory interactions. The dominant phenomena in GRNs are molecular binding, polymerization and degradation. Like signalling pathways, they tend to contain a small number of molecular entities, but typical modelling schemes are deterministic and/or stochastic depending on the purpose of the modelling. Ultimately, all three types of pathway have to be integrated into large networks to understand the functions of a cell.

2.4.2. *Nonlinearity issues in modelling*

Gene regulatory networks are inherently nonlinear. This is due to the complex, often not fully understood, interactions between cellular components of GNRs (genes, proteins, TFs, small molecules, etc.) manifested through intricate layers of feedback loops. By possessing this nonlinearity, GRNs are able to give birth to a diverse array of dynamical behaviours, including stable focus, sustained oscillations, hysteresis, toggle-switch, and multistability. However, nonlinearity also imposes great hurdles for GRN modelling. To obtain insights into these systems, nonlinear modelling frameworks must accordingly be developed, yet they are often analytically intractable. The problem gets even worse for large-scale networks.

To counter this problem, there have been attempts to simplify nonlinear models for ease of investigation while retaining their fundamental, important features. For instance, approaches based on linearisation have been developed in which systems are approximated by linearising around a fixed point, usually the steady state. Notably, stability analysis [88] has its foundation rest on local linearisation theory. S-systems approximation is another example where a nonlinear system is approximated with a power-law representation around a steady-state point. Although the resulting S-system is still nonlinear, it possesses the features that are more amicable for analysis [89, 90]. In stochastic context, Kampen [91] developed the Linear Noise Approximation method to approximate the Chemical Master Equation (CME) by meaningfully separating stochastic variables into a deterministic part and a random, fluctuating part (see Section 2.5.1.2 for a thorough discussion). The analysis of the CME now reduces to analysing change in the probability distribution of the fluctuations.

Linearisation, however, always comes with costs. Approximated models are often limited to locality and may fail to address global characteristics of the systems. Therefore, demand for tractable, efficient nonlinear modelling approaches is currently an important aspect of the field.

2.4.3. *Multi-scale issues in modelling biochemical systems*

In modelling systems, including biochemical systems, simplifications are unavoidable. Depending on the level of detail that a model intends to capture, certain assumptions should be made to ignore the effects of some unnecessarily detailed processes without a significant loss of higher level of knowledge that can be acquired in a system. In physics, there are some well-defined rules. For example, the macroscopic object obeys the laws of classical mechanics, whereas these laws no longer hold true in mesoscopic and microscopic physics, which obey the laws of quantum mechanics. Modelling a biochemical system is similar; modelling effort has to focus on a particular scale depending on the purpose of the exercise. We can define the scales involved in biochemical reactions in general as follows:

Macroscopic scale: In this scale, we assume that the system is a well-mixed solution or, equivalently, is homogeneous. The behaviour of every particle is assumed to be the average behaviour of its kind. Therefore, particles are treated as concentrations (the number of molecules per unit volume) and models at this level are normally expressed by differential equations. Because the chemical reaction is described by increasing or decreasing concentration levels, the changes in the state of the system are continuous.

Microscopic scale: This is the lowest level of reactions, where atom–atom, atom–molecule or molecule–molecule collisions take place. The Avogadro number is the number of formula units in a mole and it describes the fundamental quantitative relationship between macroscopic and microscopic levels: one mole of atoms or molecules $= 6.022 \times 10^{23}$ atoms or molecules. The system at the microscopic level is represented by single molecules, each with a position and a momentum. Hence, the dynamics are stochastic in contrast to macroscopic computation where the dynamics can be described through averaging theorems.

Mesoscopic scale: This intermediate description of chemical reactions incorporates the information between the microscopic and macroscopic scales in a suitable way. The boundaries are not sharp but can be roughly indicated. In the mesoscopic level, we eliminate some irrelevant or poorly understood variables. For example, we assume the solution is well mixed, therefore, we only count the molecules in a system, rather than keeping track of their individual properties. Because every particle is treated as an individual at this level, the dynamics of the system is stochastic with states changing discretely.

Inspired by the need to counter the multi-scale nature of biochemical systems, great efforts have been put into multi-scale modelling allowing adequate levels of description for different species and different reactions. The resulting models are sometimes referred to as hybrid models. Some researchers [92, 93], for example, have considered partitioning the system into subsets depending on the basis of propensity function values: fast and slow reaction subsets. They then constructed the Chemical Master Equation (see Section 2.5.1.2) which describes the joint probability density functions of the number and the occurrences of both subsets. Cao *et al.* [94] have proposed a virtual fast system which is Markovian to represent the real non-Markovian system. The authors used the stationary (asymptotic) properties of the virtual fast system to make a stochastic algorithm to construct the slow-scale reactions, which the authors argued, is more reliable and transparent. Takahashi [95], on the other hand, developed a modular, object-oriented meta-algorithm to integrate the multi-scaled system running on different types of algorithms and driven by different timescales. They applied the meta-algorithm to heat shock response model that combined Gillespie's and Gibson's Next Reaction algorithms and deterministic differential equations. The results showed a dramatic improvement in performance without the loss of significant accuracy. Overall, multi-scale stochastic modelling is certainly a worthwhile exercise in the future when the scale of modelled systems becomes large.

2.5. Approaches in Modelling

Currently, two theoretical approaches are used to analyse GRNs, reverse engineering (top-down) and forward modelling (bottom-up) [96]. Reverse engineering is used to analyse data that are not a priori known to contain any specific pathways [97]. It analyses expression changes of thousands of genes in parallel over time and attempts to determine regulatory interactions based on the gene expression profiles (expression values of different genes under different experimental conditions). By searching for clusters and motifs, and eventually deducing functional correlations, reverse engineering methods seek to reconstruct underlying regulatory networks. The advantage of reverse engineering is that the gene expression data themselves are used to identify meaningful or informative gene dynamical behaviours, and normally a large fraction, or almost all genes, of a cell can be covered. However, the difficulty associated with this approach is that the data derived from the current experimental tools, such as gene expression arrays or proteomic arrays, are normally too noisy to provide insights into the underlying relations between the genes.

Forward modelling is also known as "in silico cell", which tries to isolate some genetic pathways and build a detailed model that can be compared directly with experimental data [11, 98]. The basis of forward modelling is *a priori* knowledge or hypotheses about the processes of the interactions taking place during gene expression. It starts with building a conceptual model where elements and their interaction are extracted from the literature. The conceptual model is then converted into an appropriate computational model. Once the parameters are set, the model can produce the dynamics of the regulatory network. The advantage of forward modelling is that the models can be compared with experimental reality directly and testable hypotheses for further experiments can be obtained. The drawback of this approach is that its scope focuses only on local dynamics but the target pathways are frequently influenced by genes from other pathways. Moreover, it often lacks specific kinetic parameters for the individual processes under consideration.

Mathematical techniques used in forward modelling of GRNs will be introduced below, as they are used to investigate various GRNs.

2.5.1. *Bottom-up modelling*

The existing computational and mathematical methods for GRNs modelling can be generally divided into two groups: deterministic models, and stochastic models.

2.5.1.1. *Deterministic models*

Deterministic models are still the most common type of models used in gene regulatory network modelling. These include Boolean models, based on Boolean logic (true/false logic) and Differential Equation models. They take no account of any uncertainties (about network domain, or network components), and thus provide no information on prediction uncertainties. They are based purely on the principle of causality according to which there is a unique relationship between the causes and the resulting effects.

Boolean networks are the earliest models of regulatory networks developed by Kauffman [99, 100]. A model of Boolean network typically consists of N elements $\{x_1^k, x_2^k, \ldots, x_N^k\}$ at time t_k. These elements can be genes or encoded proteins. The state of elements is simply either active (ON), where the gene is expressed and its product is in an active conformation, or inactive (OFF). Each state of element is regulated by other elements that serve as inputs. The dynamic behaviour of each element is governed by the Boolean logical rules from its inputs. For example, Boolean OR function means output gene is ON when one of its inputs is ON, and Boolean AND function means output gene is ON when all of its inputs are ON. Let us assume:

$$x_i^k \in \{0, 1\} \quad \text{for } i = 1, \ldots, N. \tag{2.58}$$

As each element in the network is updated simultaneously, the system is deterministic. The state of a given element at time t_{k+1} can be determined by a Boolean logic function B (returning 0 or 1) whose input is the current state of the system at time t_k:

$$x_i^{k+1} = B_i(x_1^k, x_2^k, \ldots, x_N^k). \tag{2.59}$$

Because of its simplicity, analytical tractability and ease of implementation, Boolean networks serve as a useful conceptual tool for investigating network dynamics and organisation. Huang [101] pointed out Boolean model's advantage in observing qualitatively the generic features of large regulatory networks as a whole, even without the knowledge of particular values of kinetic parameters. Kauffman [102] showed that Boolean models, in which anonymous genes connect randomly, can qualitatively capture the typical collective behaviour of interconnected genes of a genome during evolution and development. Other applications with Boolean network used in observing global properties of large-scale regulatory systems are given by Kauffman [103], Somogyi and Sniegoski [104], Szallasi and Liang [105].

Gene expression, however, is not just a simple matter of being ON or OFF. In fact, expression levels often vary in a continuum [79]. This raises the issue of choosing threshold values for Boolean variables which characterise their states: expression level above (below) this threshold signifies ON (OFF) status. However, there is currently no stringent theoretical foundation to determine the value of the threshold *a priori* [106]. Most often, this choice requires direct intervention from the modeller with heuristics based on interpretation from experimental results, introducing subjectiveness into models.

Although Boolean networks have proven useful and enabled partial insights into large-scale networks, the simplifications of binary states and possible synchronous time update fail to capture multiple levels of activity and more complex temporal dynamics. As a result, products whose concentration varies in a continuous fashion may require a continuous function to accurately capture their dynamics [107]. A more common deterministic approach to modelling GRN is to model the interactions of elements in the GRN as chemical reactions and to include the cooperation of biochemical entities in the model. In this method, ordinary differential equations (ODE) are used to represent a GRN as a series of coupled chemical reactions where the states of the system are the concentration of molecules. ODE models have had a long history and have returned encouraging results.

However, under certain circumstances such as when the molecular populations of some reactant species are very small, or if the system's dynamic structure is susceptible to noise amplification, deterministic models no longer represent the physical system adequately and cannot be used to predict the concentration of chemical species. On the molecular scale of chemical reactions, individual molecules randomly bump into others. If the energies of the colliding molecules were sufficient, they might react, otherwise they might not. To convert this probabilistic system into a continuous, deterministic model, assumptions have to be made. These assumptions only hold within a defined regime in which, generally, the number of molecular species is large, and discrete, stochastic nature of chemical reactions is not apparent.

For simplicity, we first consider the case of binding one TF to a promoter with only one binding site. It can be formulated as

$$[T] + [P] \underset{k_2}{\overset{k_1}{\rightleftharpoons}} [TP], \tag{2.60}$$

where $[T]$ denotes the concentration of TFs which are not bound to a promoter. $[P]$ denotes the concentration of an unbound promoter, $[TP]$ is the concentration of a promoter bound with a TF, and k_1 and k_2 are the binding and unbinding rates at the unbound or bound site.

Assuming the binding and unbinding processes of TFs to a promoter are very fast, we can consider that the promoter is always in a specific state, either bound or unbound with the TF. We define v to be the fractional saturation of the TF,

$$v = \frac{[TP]}{[P_T]}, \tag{2.61}$$

where $[TP]$ is the total concentration of the promoter. We know that $[P_T] = [P] + [TP]$ and $k_2 = ([T][P])/[TP]$ according to the mass action rate law. In addition, if we define K_{dis} as the equilibrium dissociation constant following tradition, where $K_{\text{dis}} = k_2$, then v can be expressed in terms of free TFs (T) in the following way,

$$v = \frac{[T]}{K_{\text{dis}} + [T]}. \tag{2.62}$$

In some more complex cases where there are multiple binding sites in a promoter, there are three possible cooperativities existing between the binding sites, named non-cooperativity, positive cooperativity and negative cooperativity. For mathematically modelling the cooperativity, we consider two special cases first.

In the first case, if we assume n TFs bind to n binding sites independently of each other, then all the k_1 and k_2 values are the same. The expression of v can be derived by summing all the n binding sites,

$$v = \frac{n[T]}{K_{\text{dis}} + [T]}. \tag{2.63}$$

In the second case, if we assume n TFs bind to n binding sites with infinite cooperativity among the TFs, then there are only two possibilities for the binding of n TFs to the promoter, either all the binding sites in the promoter are bound or none of the binding sites in the promoter are bound. We can write the reaction,

$$[P] + n[T] \rightleftharpoons [PT_n], \tag{2.64}$$

where $[PT_n]$ is the concentration of a promoter where all the n binding sites are bound to n transcription factors. The expression for v can then be expressed as

$$v = \frac{n\,[T]^n}{K_{\text{dis}} + n\,[T]^n}, \tag{2.65}$$

where the dissociation constant $K_{\text{dis}} = [P][T]^n/[PT_n]$ according to the mass action rate law.

For modelling the cooperative binding to n equivalent sites over part of the saturation range (between the two special cases),

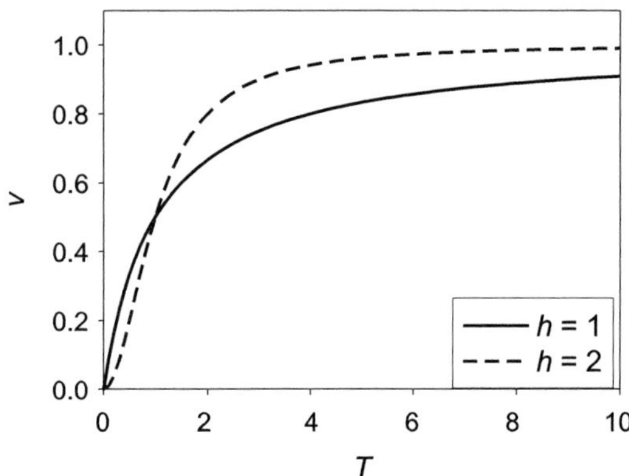

Fig. 2.7. The fractional saturation of reactions without cooperative binding ($h = 1$) and with positive cooperative binding ($h = 2$).

we can write an analogous equation, which is known as the Hill function [108],

$$v = \frac{n[T]^h}{K^h + n[T]^h},\qquad(2.66)$$

where h is the Hill constant or the Hill coefficient with $1 < h < n$.

To illustrate the effect of the Hill constant on the fractional saturation of reactant, v, in the reactions, we plotted v against the increasing concentration of reactant, T, from 0 to 10 using an h value of 1 (no cooperativity) and 2 (positive cooperativity), respectively (Fig. 2.7). For simplicity, a K value of 1 is taken. It is clearly shown that the fractional saturation of reactant reaches its maximum more rapidly with positive cooperativity.

2.5.1.2. *Stochastic models*

Stochastic models attempt to include the effects of uncertainties in the operation of components and their interactions in the gene regulatory network. They recognise the probabilistic nature of the outputs as opposed to the deterministic model, which suggests that there is only one "correct" set of outputs. It is evident that there are important stochastic elements inherent in cell and molecular processes as seen in the previous sections. Stochastic variation arises both from randomness of molecular diffusion and from effects of chance in the combinatorial assembly of transcription factor complexes at a DNA control sequences. The significance of stochastic variation in gene expression lies in the diversity it creates, explaining the intrinsic variability of biological systems and probabilistic uncertainty about specific outcomes, such as cell fates.

There are two main approaches to modelling stochastic events in gene expression: stochastic differential equations and stochastic simulation algorithms. Stochastic differential equations extend the standard differential equation description of the reaction dynamics to include a noise term

$$\frac{dx_i}{dt} = f_i(x_i) + v_i(t), \tag{2.67}$$

where $v_i(t)$ is an additive noise term. This equation, known as the Langevin Equation (LE), can be developed into an alternative formulation that describes the evolution of the probability density function [109]. This method works well when the number of molecules in the system is large as molecular fluctuations can be approximated in a continuous time scale. The limitation is that the equations are generally too complex to be solved using analytical techniques. When the number of molecules in the system is low, a discrete stochastic simulation algorithm (SSA) such as Gillespie's algorithm is generally used to effectively simulate reaction models numerically. The Gillespie algorithm makes timesteps of variable length; in each time step, based on the rate constant and population size of each chemical species, one random number is used to choose which reaction will occur, and another random number determines how long the step will last [110].

In this section, we give a survey of common approaches to stochastic modelling of coupled chemical reactions and their mathematical derivations. Two of them are exact and rigorous ways to predict the evolution of a biochemical system (such as a GRN): (1) Chemical Master Equation and (2) Numerical Simulation for constructing sample realizations through probability distribution, such as the Gillespie algorithm. While being advantageous with their exact description of systems, these methods however suffer intractability and expensive computational cost. The trade-off between precision and speed has to be made for modelling realistic systems. A number of approximated methods to the Chemical Master Equation are therefore also presented herein, including SDE — the Chemical Langevin equation, the Fokker–Planck equation and Linear Noise Approximation.

The Chemical Master Equation (CME): CME is used to describe the temporal variation in species of a biochemical reaction system [91, 111, 112]. Its formulation is based on the assumption that the content of the system is well mixed, i.e. there is a sufficiently long time between reactive collisions to ensure that each pair of molecules is equally likely to be the next to collide. CME demonstrates how the state of the system changes through the changes of probability distributions with time.

We generally have a reaction system consisting of n chemical species $S_1, S_2, \ldots, S_j, \ldots, S_n$ that take part in m possible types of reactions, or reaction channels $R_1, R_2, \ldots, R_j, \ldots, R_m$. The state vector of the system $X(t)$ is defined by $[X_1(t), X_2(t), \ldots, X_n(t)]$ with X_i ($i = 1, \ldots, n$) is the number of molecules of species S_i at time t. For $j = 1, \ldots, m$, the reaction channel R_j has an associated stoichiometric vector, $v_j \in \mathrm{R}^n$, whose ith component is the change in the number of molecules of species S_i as a result of that reaction. In other words, a reaction of channel R_j would change state vector from $X(t)$ to $X(t) + v_j$.

For illustrative purpose, consider a simple bimolecular reaction of reaction channel R_j:

$$S_{j1} + S_{j2} \xrightarrow{c_j} S_{j3}, \qquad (2.68)$$

where c_j is the rate constant.

Reaction (2.68) takes place whenever a molecule of species S_{j1} meets a molecule of species S_{j2}. Because there are $X_{j1}(t)$ molecules of S_{j1} and $X_{j2}(t)$ molecules of S_{j2} at time t, there are $X_{j1}(t)X_{j2}(t)$ different combinations of 2 molecules of species S_{j1} and S_{j2}. Therefore, if we define $c_j dt$ as the probability that one combination of reaction (2.68) reactants will meet in the next infinitesimal time dt, then the probability that a reaction of type (2.68) will occur in the time interval $[t, t + dt)$ is $X_{j1}(t)X_{j2}(t)c_j\,dt$. We then denote the propensity function of reaction channel j as,

$$a_j(X(t)) = X_{j1}(t)X_{j2}(t)c. \qquad (2.69)$$

Similarly, a unimolecular reaction and a dimerization reaction are given by

$$S_{j1} \xrightarrow{c_j} S_{j2} \quad \text{and} \quad 2S_{j1} \xrightarrow{c_j} S_{j2},$$

having propensity functions of

$$a_j(X(t)) = X_{j1}(t)\,c_j, \quad \text{and}$$
$$a_j(X(t)) = 1/2X_{j1}(t)\,[X_{j1}(t) - 1]\,c_j, \quad \text{respectively.}$$

Returning to our system, the propensity function of the reactions channel $R_j(j = 1, \ldots, m)$ is:

$$a_j(X(t)) = h_j(t)\,c_j, \qquad (2.70)$$

where c_j is the reaction rate constant and $h_j(t)$ is the number of different combinations of R_j reactants at time t. Therefore $a_j(X(t))dt$ indicates the probability that a reaction of type R_j will occur in the time interval $[t, t + dt)$ in the system.

Denote $P(x, t)$ as the probability that system state $X(t) = x$ at time t. The idea behind CME is how to calculate $P(x, t + dt)$ given that the probability of the system being in any possible states at time t and that dt is so small that at most one reaction can take place over $[t, t + dt)$.

Because at most one reaction can take place, during $[t, t + dt)$, there are a total of $m + 1$ potential state transitions from time t to $t + dt$: for $1 \leq j \leq m$, transition j indicates that the system was in state $x - v_j$ at time t and one reaction of type R_j fired over $[t, t+dt)$ bringing the system into state x with probability $a_j(x-v_j)dt$; transition $m + 1$ indicates that no reaction occurs during $[t, t + dt)$

and the system remains at state x, the probability of this event is

$$1 - \sum_{1}^{m} a_j(x)dt.$$

Therefore, $P(x, t + dt)$ can be expressed as follows:

$$P(x, t + dt) = \sum_{j=1}^{m} a_j(x - v_j)dt P(x - v_j, t)$$

$$+ \left(1 - \sum_{j=1}^{m} a_j(x)dt\right) P(x, t). \qquad (2.71)$$

Rearrange the equation and letting $dt \to 0$, we have the CME:

$$\frac{dP(x, t)}{dt} = \sum_{j=1}^{m} a_j(x - v_j)P(x - v_j, t) - \sum_{j=1}^{m} a_j(x)P(x, t). \qquad (2.72)$$

From a mathematical perspective, CME is a set of intuitive and simple linear ODEs with one ODE for each possible state of the system. However, the number of possible states is large when the number of chemical species n is high. It is therefore a daunting task just to write down all equations of the CME, let alone solve it. Analytical solution of CME is generally intractable as well. Two main approaches are usually used to tackle the intractability of the CME; the most common strategy is to apply a Monte-Carlo method, in this context the Stochastic Simulation Algorithm (SSA) of Gillespie [110], which serves as an exact numerical calculation of the CME, is a popular choice. The other approach is to approximate the CME so that the solution can be represented on a smaller space by using the Fokker–Planck equation or the Linear Noise Approximation method.

The Gillespie algorithm (SSA): SSA is derived based on quantity $P_{\text{no}}(\tau|x, t)$, defined as the probability that no reaction occurs in the time interval $[t, t + \tau)$ given that system state $X(t) = x$ at time t.

To calculate $P_{\text{no}}(\tau|x, t)$, consider the time interval $[t, t + \tau + d\tau)$. We assume that what happens over $[t, t + \tau)$ is independent of what happens over $[t + \tau, t + \tau + d\tau)$. Therefore,

$$P_{\mathrm{no}}(\tau + d\tau | x, t) = P_{\mathrm{no}}(\tau | x, t)^* P_{\mathrm{no}}(d\tau | x, t + \tau). \qquad (2.73)$$

Note, however, that:

$$P_{\mathrm{no}}(d\tau | x, t + \tau) = 1 - \sum_{j=1}^{m} a_j(x) d\tau.$$

Therefore,

$$P_{\mathrm{no}}(\tau + d\tau | x, t) = P_{\mathrm{no}}(\tau | x, t)^* \left(1 - \sum_{j=1}^{m} a_j(x) d\tau \right). \qquad (2.74)$$

This equation can be solved with:

$$P_{\mathrm{no}}(\tau | x, t) = e^{-a_{\mathrm{sum}}(x)\tau}, \quad \text{where } a_{\mathrm{sum}}(x) = \sum_{j=1}^{m} a_j(x). \qquad (2.75)$$

Now, given $X(t) = x$ we further define another quantity $P(\tau, j | x, t)$ such that $P(\tau, j | x, t) d\tau$ represents the probability that the next reaction will occur in infinitesimal time $[t + \tau, t + \tau + d\tau)$ and that reaction is of type R_j. This probability can be calculated

$$P(\tau, j | x, t) d\tau = [\text{Pro. no reaction occurs in } [t, t + \tau)]$$
$$\times [\text{Pro. reaction type } R_j \text{ occurs in}$$
$$[t + \tau, t + \tau + d\tau)].$$
$$\text{or} \quad P(\tau, j | x, t) d\tau = P_{\mathrm{no}}(\tau | x, t)\, a_j(x) d\tau. \qquad (2.76)$$

Recalling Eqs. (2.75) and (2.76) can be rewritten as

$$P(\tau, j | x, t) = \frac{a_j(x)}{a_{\mathrm{sum}}(x)} a_{\mathrm{sum}}(x) e^{-a_{\mathrm{sum}}(x)\tau}. \qquad (2.77)$$

This equation shows that $P(\tau, j | x, t)$ is a joint density function and can be expressed as product of two independent density functions: (1) corresponds to the reaction time to fire (a continuous random variable), and (2) corresponds to the reaction type to fire (a discrete random variable).

The SSA generates τ and j from values of $P(\tau|x,t)$ and $P(j|x,t)$ using the probability distribution inverse method via uniform (0,1) sample. The algorithm can be summarized in steps as follows:

(1) Set initial number of species $[X_1, X_2, \ldots, X_n]$, reaction rates $c_j (1 \leq j \leq m)$ and initial time $t = 0$.
(2) Evaluate $a_j(X(t))$ for $1 \leq j \leq m$ and the sum

$$a_{\mathrm{sum}}(X(t)) = \sum_{j=1}^{m} a_j(X(t)).$$

(3) Generate two independent uniform random variables r_1 and r_2. Set

$$\tau = \ln(1/r_2)/a_{\mathrm{sum}}(X(t)),$$

and set j as the index satisfying

$$\sum_{k=1}^{j-1} a_k(X(t)) \leq r_2 a_{\mathrm{sum}}(X(t)) \leq \sum_{k=1}^{j} a_k(X(t)).$$

(4) Update system state $X(t)$, set $t \to t + \tau$.
(5) Go to step 2 or Stop if $t = t_{\mathrm{max}}$.

SSA is considered as an exact calculation of CME, in the sense that the statistics from CME are produced precisely. However, this exactness is achieved at a high computational cost. A large amount of random numbers needs to be generated and kept track of if the number of molecules in the system is high and/or there are some fast reactions. Therefore, for systems possessing species with high numbers, simulations using SSA could be a long and expensive task. Nevertheless, SSA works well for small systems or systems of reasonable size. Because of its precise representation of the CME, SSA can be used as a useful tool to validate other approximate approaches of the CME.

Chemical Langevin Equation (CLE): In the SSA, the time interval length until next reaction τ is variable and can be very small which is the main reason leading to the high computational cost of SSA. It is therefore intuitive to consider a fixed τ and within this time, fire all reactions that would have occurred simultaneously. For each reaction channel R_j, suppose p_j is a random variable representing the number of reactions of this channel that fire within time interval τ. It is easy to see that p_j depends on propensity function $a_j(X(t))$ and on τ. System state at time $t+\tau$ can be simply updated from time t based on the stoichiometric vector v_j as follows:

$$X(t+\tau) = X(t) + \sum_{j=1}^{m} v_j p_j(a_j(X(t)), \tau). \qquad (2.78)$$

The task now is to derive values of $p_j(a_j(X(t)), \tau)$. Here, we introduce the first assumption: assume τ is small enough that the change in the state during $[t, t+\tau)$ will be so slight that no propensity function will suffer an appreciable change in its value. In other words, $a_j(X(t))$ is considered constant over $[t, t+\tau)$ for all j. Because of this, the probability of channel R_j firing in any infinitesimal interval $d\tau$ inside $[t, t+\tau)$ will be the same value, $a_j(X(t))d\tau$. This means that the number of type R_j reactions over $[t, t+\tau)$, p_j would be a Poisson random variable (counting variable) characterized by a Poisson distribution $P(k, \lambda)$ [113]:

$$P(k, \lambda) = \text{Prob}(p_j = k) = \frac{e^{-\lambda}\lambda^k}{k!}, \qquad (2.79)$$

where $\lambda = a_j(X(t))\tau$.

The mean and variance of this distribution are $\langle P(k, \lambda) \rangle = \text{var}\{P(k, \lambda)\} = a_j(X(t))\tau$. So the first assumption has allowed us to estimate random values of p_j from a Poisson distribution with a parameter that can be readily calculated. Probability theory states that a Poisson random variable with large mean is well approximated by a normal random variable with the same mean and variance. We thus introduce the second assumption: assume τ is large enough so that $a_j(X(t))\tau$ is large for all $1 \leq j \leq m$. This assumption permits us to approximate the Poisson random variables $p_j(a_j(X(t)), \tau)$ with normal random variables $N_j(a_j(X(t))\tau, a_j(X(t))\tau)$. The approximation converts p_j from an integer-valued, discrete random numbers to a real-valued, continuous random numbers.

Follow $N(m,\sigma^2) = m + \sigma N(0,1)$, Eq. (4.3) can then be rewritten as

$$X(t+\tau) = X(t) + \tau \sum_{j=1}^{m} v_j a_j(X(t)) + \sqrt{\tau} \sum_{j=1}^{m} v_j \sqrt{a_j(X(t))} N_j(0,1),$$

$$(2.80)$$

where $N_j(0,1)$ are independent normal random variables drawn from unit interval $(0,1)$. This equation is called the Chemical Langevin Equation and is referred to as its "second form".

Rearranging Eq. (2.84) and divide both sides by τ, we have

$$\frac{X(t+\tau) - X(t)}{\tau} = \sum_{j=1}^{m} v_j a_j(X(t)) + \sum_{j=1}^{m} v_j \sqrt{a_j(X(t))} \frac{N_j(0,1)}{\sqrt{\tau}},$$

or

$$\frac{X(t+\tau) - X(t)}{\tau} = \sum_{j=1}^{m} v_j a_j(X(t)) + \sum_{j=1}^{m} v_j \sqrt{a_j(X(t))} N_j(0,1/\tau).$$

If we pass to the limit $\tau \to dt \to 0$, we obtain

$$\frac{dX(t)}{dt} = \sum_{j=1}^{m} v_j a_j(X(t)) + \sum_{j=1}^{m} v_j \sqrt{a_j(X(t))} dW_j(t). \qquad (2.81)$$

This equation is usually referred to as the "white-noise" form of the CLE because of the so-called *white-noise* term

$$dW_j(t) = \lim_{dt \to \infty} N_j(0,1/dt).$$

It has t as a continuous variable. It is important to note that Eq. (2.81) is only a good approximation of Eq. (2.80) when we make the third assumption that the chosen fixed time τ has to be small enough.

To sum up, we see that the white-noise form of the CLE is obtained from the SSA (and so CME) based on three assumptions:

(1) τ is small enough so that $a_j(X(t))$ will be considered constant over $[t, t + \tau)$ for all j;
(2) τ is large enough so that $a_j(X(t))\,\tau$ will be large;
(3) τ is small enough for Eq. (2.80) to give a good approximation of Eq. (2.81).

These assumptions seem to contradict on the value of τ but are not impossible to achieve simultaneously. However, the satisfaction requires a large number of species molecules present in the system.

In modelling stochasticity, molecular fluctuations can be incorporated explicitly by including random variables (or rather stochastic processes) in the model parameters [75]. The model of ordinary differential equations is then converted into a model of stochastic differential equations (SDE). We have approximated the CLE from the CME, which is an example of SDE. However, it is usually set up by starting with the deterministic model of ODEs and appending a noise term to the end of the differential equations,

$$\frac{dX(t)}{dt} = vr(X(t)) + Noise(t), \tag{2.82}$$

where $Noise(t) = v\sqrt{r(X(t))}n(t)$ is the additive (white) noise term.

Fokker–Planck Equation: Another way to tackle the CME is to reduce its dimension by an approximation using Taylor expansion truncation. We rewrite here the CME formulated in the previous section:

$$\frac{\partial P(x,t)}{\partial t} = \sum_{j=1}^{m} a_j(x - v_j)P(x - v_j, t) - \sum_{j=1}^{m} a_j(x)P(x,t). \tag{2.83}$$

For each $1 \leq j \leq m$, denote a new function $F_j(x,t) = a_j(x)P(x,t)$. The CME now becomes:

$$\frac{\partial P(x,t)}{\partial t} = \sum_{j=1}^{m} F_j(x - v_j) - F_j(x). \tag{2.84}$$

Taylor expansion of a real function $f(x)$ about point $x = a$ is:
$f(x) = f(a) + f'(a)(x-a) + f''(a)(x-a)^2 + \cdots + f^{(n)}(a)(x-a)^n + \cdots$.
Apply the Taylor expansion for $F_j(x - v_j,\ t)$ about x point and truncate after the second-order terms, we obtain

$$F_j(x - v_j, t) = F(x) + ((x - v_j) - x)\frac{\partial}{\partial x}F(x) + ((x - v_j) - x)^2\frac{1}{2}\frac{\partial^2}{\partial x^2}F(x),$$

or

$$F_j(x - v_j, t) = F(x) - v_j\frac{\partial}{\partial x}F(x) + v_j^2\frac{1}{2}\frac{\partial^2}{\partial x^2}F(x). \tag{2.85}$$

Substituting this into Eq. (2.83) we have

$$\frac{\partial P(x,t)}{\partial t} = \sum_{j=1}^{m} -v_j\frac{\partial}{\partial x}F(x) + v_j^2\frac{1}{2}\frac{\partial^2}{\partial x^2}F(x). \tag{2.86}$$

Note that in this equation x and v_j are all in the form of vectors: $v_j = [v_{1j}, v_{2j}, \ldots, v_{nj}]$ and $x = [x_1, x_2, \ldots, x_n]$. It can be spelt out in details as follows:

$$\frac{\partial P(x,t)}{\partial t} = \sum_{j=1}^{m}\left\{ -\sum_{i=1}^{n} v_{ij}\frac{\partial(a_j(x)P(x,t))}{\partial x_i} \right.$$

$$\left. + \sum_{i=1}^{n}\sum_{k=1}^{n} v_{ij}v_{kj}\frac{\partial^2(a_j(x)P(x,t))}{\partial x_i\partial x_k} \right\}. \tag{2.87}$$

This is called the Fokker–Planck equation. One advantage of working with the Fokker–Planck equation is that it is possible to analyse the model. Tools such as sensitivity analysis and bifurcation theory are applicable. However, the downside of this approach is that it is impossible to solve the Fokker–Planck equation even numerically, when the number of molecules reaches beyond a few species [75]. To analyse these models, Monte-Carlo methods are used by most researchers, where one solves the Langevin equation many times and then use statistics to estimate the probability density function. Compared to deterministic equations, Monte-Carlo methods are more time consuming when simulating many molecules and reactions, although they currently are the only option for complex models.

Linear Noise Approximation (LNA): It is understood that as the number of molecules in a reaction system increases, the system's evolution becomes smoother and the deterministic framework is considered more appropriate for modelling. LNA [114] exploits this behavior and bases on the idea that the total number of molecules for each species can be meaningfully separated into two parts: a deterministic part, call it d_i, and a fluctuating part, call it f_i. The fluctuations scale roughly as the square-root of the number of molecules. Therefore, the molecules numbers can be written as follows:

$$X_i = \Omega d_i + \Omega^{1/2} f_i, \text{ where } \Omega \text{ is the system volume.} \tag{2.88}$$

We would like a set of equations that govern the change in the deterministic part d and an equation that governs the change in the probability distribution of the fluctuations, call it $\Pi(f, t)$, centred upon d. The probability distribution $P(X, t)$ in the CME for $X = [X_1, X_2, \ldots, X_n]^T = \Omega \chi$ is related to $\Pi(f, t)$ for $f = [f_1, f_2, \ldots, f_n]$ as follows:

$$P(X, t) = P(\Omega d + \Omega^{1/2} f, t) = \Pi(f, t). \tag{2.89}$$

Differentiate with respect to t of $\Pi(f, t)$ at constant molecule number, we have

$$\frac{\partial P(X, t)}{\partial t} = \frac{\partial \Pi(f, t)}{\partial t} + \sum_{i=1}^{n} \frac{\partial f_i}{\partial t} \frac{\partial \Pi(f, t)}{\partial f_i}. \tag{2.90}$$

Since $\frac{\partial X_i}{\partial t} = 0$ means $\frac{\partial f_i}{\partial t} = -\Omega^{1/2} \frac{\partial d_i}{\partial t}$, substitute into Eq. (2.90) gives

$$\frac{\partial P(X, t)}{\partial t} = \frac{\partial \Pi(f, t)}{\partial t} - \Omega^{1/2} \sum_{i=1}^{n} \frac{\partial d_i}{\partial t} \frac{\partial \Pi(f, t)}{\partial f_i}. \tag{2.91}$$

We define a step-operator E_i^k that increments the ith species by an integer k, acting on a function g as

$$E_i^k g(\ldots, X_i, \ldots) = g(\ldots, X_i + k, \ldots). \tag{2.92}$$

This step-operator can be approximated by using Taylor expansion around $k/\Omega^{1/2}$ as follows:

$$E_i^k g(\ldots, X_i, \ldots) = g(\ldots, X_i + k, \ldots)$$

$$= g\left(\ldots, \Omega d_i + \Omega^{1/2}\left(f_i + \frac{k}{\Omega^{1/2}}\right), \ldots\right)$$

$$\approx g(\ldots, X_i, \ldots) - \frac{k}{\Omega^{1/2}}\frac{\partial g}{\partial f_i} + \frac{k^2}{2\Omega}\frac{\partial^2 g}{\partial f_i^2} + \cdots, \qquad (2.93)$$

so

$$E_i^k g(\ldots, X_i, \ldots) \approx \left[1 - \Omega^{-1/2}k\frac{\partial}{\partial f_i} + \Omega^{-1}\frac{k^2}{2}\frac{\partial^2}{\partial f_i \partial f_i} + O(\Omega^{-3/2})\right]$$

$$\times g(\ldots, X_i, \ldots).$$

When this operator is applied many times, for example, when the system state is updated after a reaction of type R_j, we have

$$E^{-v_j}g(X) = E_1^{-v_{1j}} \ldots E_n^{-v_{nj}}g(X)$$

$$\approx \left[1 - \Omega^{-1/2}\sum_{i=1}^{n}v_{ij}\frac{\partial}{\partial f_i} + \frac{\Omega^{-1}}{2}\sum_{i=1}^{n}\sum_{l=1}^{n}v_{ij}v_{lj}\frac{\partial^2}{\partial f_i \partial f_l}\right.$$

$$\left. + O(\Omega^{-3/2})\right]g(X). \qquad (2.94)$$

Taylor expansion of the propensity function $a_j(\chi)$ around the macroscopic value $a_j(d)$ gives

$$a_j(\chi) = a_j(d + \Omega^{1/2}f) = a_j(d) + \Omega^{-1/2}\sum_{i=1}^{n}\frac{\partial a_j(d)}{\partial d_i}f_i + O(\Omega^{-1}).$$

$$(2.95)$$

The CME can be represented with this operator as this:

$$\frac{\partial P(X,t)}{\partial t} = \Omega\sum_{j=1}^{m}(E^{-v_j} - 1)a_j(\chi, \Omega)P(X,t). \qquad (2.96)$$

Substitute equations above into the CME, we obtain

$$\frac{\partial \Pi(f,t)}{\partial t} - \Omega^{1/2} \sum_{i=1}^{n} \frac{\partial d_i}{\partial t} \frac{\partial \Pi(f,t)}{\partial f_i}$$

$$= \Omega \sum_{j=1}^{m} \left[-\Omega^{-1/2} \sum_{i=1}^{n} v_{ij} \frac{\partial}{\partial f_i} + \frac{\Omega^{-1}}{2} \sum_{i=1}^{n} \sum_{l=1}^{n} v_{ij} v_{lj} \frac{\partial^2}{\partial f_i \partial f_l} \right.$$

$$\left. + O(\Omega^{-3/2}) \right] \left(a_j(d) + \Omega^{-1/2} \sum_{i=1}^{n} \frac{\partial a_j(d)}{\partial d_i} f_i + O(\Omega^{-1}) \right) \Pi(f,t).$$

$$(2.97)$$

On both sides of this equation, we identify terms of order $\Omega^{1/2}$ and compare them

$$\Omega^{1/2} : \sum_{i=1}^{n} \frac{\partial d_i}{\partial t} \frac{\partial \Pi(f,t)}{\partial f_i} = \sum_{j=1}^{m} \sum_{i=1}^{n} v_{ij} a_j(d) \frac{\partial \Pi(f,t)}{\partial f_i}. \qquad (2.98)$$

We now have a coupled set of nonlinear differential equations that govern the deterministic evolution of the system, which happens to coincide with the macroscopic reaction rate equations because at macroscopic level

$$\frac{\partial d_i}{\partial t} = \sum_{j=1}^{m} v_{ij} a_j(d).$$

Identify terms of order Ω^0 and compare them we have

$$\Omega^0 : \frac{\partial \Pi(f,t)}{\partial t} = \sum_{j=1}^{m} \left[\sum_{i=1}^{n} \sum_{l=1}^{n} -v_{ij} \frac{\partial a_j(d)}{\partial d_l} \frac{\partial \left(f_l \Pi(f,t) \right)}{\partial f_i} \right.$$

$$\left. + \frac{1}{2} a_j(d) \sum_{i=1}^{n} \sum_{l=1}^{n} v_{ij} v_{lj} \frac{\partial^2 \Pi(f,t)}{\partial f_i \partial f_l} \right]$$

$$= - \sum_{i,k=1}^{n} A_{ik} \frac{\partial \left(f_k \Pi \right)}{\partial f_i} + \frac{1}{2} \sum_{i,k=1}^{n} B_{ik} \frac{\partial^2 \Pi}{\partial f_i \partial f_l}, \qquad (2.99)$$

with $A_{ik} = \sum_{j=1}^{m} v_{ij}\frac{\partial a_j(d)}{\partial d_k}$ and $B_{ik} = \sum_{j=1}^{m} v_{ij}v_{kj}a_j(d) = [v * diag(a_j(d)) * v^T]_{ik}$.

We now also have a partial differential equation that characterizes the probability distribution of the fluctuations part. This is referred to as a linear Fokker–Planck equation with coefficient matrix A and B that depend on time through the deterministic rates. Despite being referred to with the same name, this equation is essentially different to the Fokker–Planck equation governing the evolution of the probability distribution of the system state described in the previous section.

Higher orders than Ω^0 are not considered in the linear noise approximation. The stationary solution of the fluctuation probability distribution can be solved by setting:

$$-\sum_{i,k=1}^{n} A_{ik}\frac{\partial (f_k \Pi)}{\partial f_i} + \frac{1}{2}\sum_{i,k=1}^{n} B_{ik}\frac{\partial^2 \Pi}{\partial f_i \partial f_k} = 0, \qquad (2.100)$$

where matrix A and B are evaluated in the macroscopic stationary state d_{ss}. It is given as the multi-dimensional Normal distribution

$$\Pi(f,t) = \frac{1}{\sqrt{(2\pi)^n Det[C(t)]}} \exp\left(-\frac{1}{2}f^T \cdot C(t) \cdot f\right), \qquad (2.101)$$

with C is the covariance matrix $C_{ik} = \langle f_i f_k \rangle - \langle f_i \rangle \langle f_k t \rangle$ which satisfies $A \cdot C + C \cdot A^T + B = 0$ at steady state.

The linear Fokker–Planck provides insights into the physics of the system. Notice matrix A is simply the Jacobian matrix of the deterministic system, evaluated pointwise along the deterministic trajectory. It therefore represents the local damping or dissipation of the fluctuations. While matrix B indicates how much the microscopic system is changing at each point along the deterministic trajectory, hence represents the local fluctuations. The equation above which is evaluated at steady state, describes the balance of two effects coming from dissipation and fluctuations. It is thus usually called the dissipation–fluctuation relation.

Noise measurements: Noise can be quantified in a number of ways. Autocorrelation has been used to summarize both the magnitude and frequency of fluctuations [115]. However, most models so far have focused on exploring the steady-state statistics of gene regulatory networks. Two most important characteristics are the mean and the variance of the number of molecules of each species within the networks. The advantages of these system properties are that they are fundamental and simple to understand, provide clear interpretations and more importantly, they are easily accessible experimentally [116].

Two major measurements have been used to compute noise strength (NS). The first measurement is formulated as the variance over the squared mean (Eq. (2.102)), which allows for a clean separation of different noise sources as long as the models are weakly nonlinear [115]:

$$NS_1 = \frac{var}{mean^2} = \frac{\sigma^2}{\langle m \rangle^2}. \tag{2.102}$$

The second measurement is the Fano factor, formulated as the variance over the mean (Eq. (2.103)). The Fano factor is equal to one for Poisson distribution because var $=$ mean, it therefore can measure how close to Poisson distributed a given process may be

$$NS_1 = \frac{var}{mean} = \frac{\sigma^2}{\langle m \rangle}. \tag{2.103}$$

However, the comparison with the Poissonian only works well for uni-variate discrete random processes since the variance is often proportional to the mean, but not relevant for multi-variate random processes. The use of the Fano factor therefore can be misleading for these latter, more realistic processes [115].

Choice of measurement of noise is largely due to which aspects of cell-to-cell variability that need to emphasize in the study, which are different from experiment to experiment [117]. However, the variance over squared mean provide better physical insights than the Fano factor [118]. Moreover, the variance over squared mean is also more suitable for experimental interpretations [115].

2.5.2. *Top-down: Reverse engineering of networks*

The advent of microarray technology, which allows us to measure expression levels of tens of thousands of genes simultaneously, has opened up a new opportunity for identifying biologically plausible regulatory interaction between genes. Gene expression data from microarray is used to construct expression profiles, which is often done by putting together expression levels from different experimental conditions, or time intervals. Similarities and differences between the expression profiles, as well as changes in the expression levels, provide important insights into regulatory relationships. Therefore, it is possible to infer the interactions between components of a network from gene expression data, this process is known as reverse engineering of GRNs. It relies on the following assumption: given enough data on gene expression levels, it is possible to infer how genes are regulated [119, 120]. A number of methods and associated tools have been developed and used for the purpose of reverse engineering. Bayesian networks [121–124], ordinary differential equation [125–127] and information-theoretic approach can be applied to infer the network topology depending on the type and form of available data as well as the results that one wants to achieve. Other methods aiming to find correlations between expression levels and consequent regulatory relations are also utilized, including methods to detect gene relations from relevant databases; analytical modelling methods to derive formulas from gene data; correlation analysis methods to discover correlations between gene expression over time, etc. More recent techniques such as artificial neural networks are also being used for cluster analysis in which genes are clustered based on their expressions.

References

[1] Fleischmann R. *et al.* (1995). Whole-genome random sequencing and assembly of Haemophilus influenzae Rd. *Science*, 269, pp. 496–512.

[2] Venter J.C. *et al.* (2001). The sequence of the human genome. *Science*, 291, pp. 1304–1351.

[3] Baldi P. and Hatfield G.W. (2002). *DNA Microarrays and Gene Expression: From Experiments to Data Analysis and Modeling* (Cambridge University Press, Cambridge).

[4] Zhu H. *et al.* (2001). Global analysis of protein activities using proteome chips. *Science*, 293, pp. 2101–2105.

[5] Uetz P. *et al.* (2000). A comprehensive analysis of protein-protein interactions in Saccharomyces cerevisiae. *Nature*, 403, pp. 623–637.

[6] Kitano H. (2002). Systems biology: a brief overview. *Science*, 295, pp. 1662–1664.

[7] Wiener N. (1961). *Cybernetics or Control and Communication in the Animal and the Machine* (MIT Press, Cambridge, MA).

[8] Lewin B. (2004). *Genes 8* (Pearson Prentice-Hall, Upper Saddle River).

[9] von Hippel P.H. (2004). Biochemistry: completing the view of transcriptional regulation. *Science*, 305, pp. 350–352.

[10] Stamatoyannopoulos J.A. (2004). The genomics of gene expression. *Genomics*, 84, pp. 449–457.

[11] Bower J.M. and Bolouri H. (2001). *Computational Modeling of Genetic and Biochemical Networks* (MIT Press, Cambridge, MA).

[12] Goldbeter A. (1995). A model for circadian oscillations in the Drosophila period protein (PER). *Proc R Soc London B*, 261, pp. 319–324.

[13] Xie Z. and Kulasiri D. (2007). Modelling of circadian rhythms in Drosophila incorporating the interlocked PER/TIM and VRI/PDP1 feedback loops. *J Theor Biol*, 245, pp. 290–304.

[14] Novak B., Pataki Z., Ciliberto A. and Tyson J.J. (2001). Mathematical model of the cell division cycle of fission yeast. *Chaos*, 11, pp. 277–286.

[15] Bootman M.D., Lipp P. and Berridge M.J. (2001). The organisation and functions of local Ca^{2+} signals. *J Cell Sci*, 114, pp. 2213–2222.

[16] Becskei A., Séraphin B. and Serrano L. (2001). Positive feedback in eukaryotic gene networks: cell differentiation by graded to binary response conversion. *EMBO J*, 20, pp. 2528–2535.

[17] Smolen P., Baxter D.A. and Byrne J.H. (2000). Mathematical modeling of gene networks. *Neuron*, 26, pp. 567–580.

[18] Elowitz M.B. and Leibler S. (2000). A synthetic oscillatory network of transcriptional regulators. *Nature*, 403, pp. 335–338.

[19] Angeli D., Ferrell J.E. and Sontag E.D. (2004). Detection of multistability, bifurcations, and hysteresis in a large class of biological positive-feedback systems. *Proc Natl Acad Sci USA*, 101, pp. 1822–1827.

[20] Pomerening J.R., Sontag E.D. and Ferrell J.E. (2003). Building a cell cycle oscillator: hysteresis and bistability in the activation of Cdc2. *Nat Cell Biol*, 5, pp. 346–351.

[21] Waage P. and Gulberg C.M. (1986). Studies concerning affinity. *J Chem Educ*, 63, p. 1044.

[22] Michaelis L. and Menten M. (1913). Die kinetik der invertinwirkung. *Biochem*, 49, pp. 333–369.

[23] Hill A. (1910). The possible effects of the aggregation of the molecules of haemoglobin on its dissociation curves. *J. Physiol*, 40, pp. 1115–1121.

[24] Briggs G.E. and Haldane J.B. (1925). A note on the kinetics of enzyme action. *Biochem J*, 19, pp. 338–339.

[25] Weiss J. (1997). The Hill equation revisited: uses and misuses. *FASEB J*, 11, pp. 835–841.

[26] Andrieu C., Freitas N., de Doucet A. and Jordan M.I. (2003). An introduction to MCMC for machine learning. *Mach Learn*, 50, pp. 5–43.

[27] Hastings W.K. (1970). Monte Carlo sampling methods using Markov chains and their applications. *Biometrika*, 57, pp. 97–109.

[28] Metropolis N. *et al.* (1953). Equation of State Calculations by Fast Computing Machines. *J Chem Phys*, 21, p. 1087.

[29] Haario H., Laine M., Mira A. and Saksman E. (2006). DRAM: efficient adaptive MCMC. *Stat Comput*, 16, pp. 339–354.

[30] Kitano H. (2007). Towards a theory of biological robustness. *Mol Syst Biol* 3, p. 137.

[31] McKay M.D., Beckman R.J. and Conover W.J. (1979). A Comparison of three methods for selecting values of input variables in the analysis of output from a computer code. *Technometrics*, 21, pp. 239–245.

[32] Delbrück M. (1945). The burst size distribution in the growth of bacterial viruses (bacteriophages). *J Bacteriol*, 50, pp. 131–135.

[33] Powell E.O. (1958). An outline of the pattern of bacterial generation times. *J Gen Microbiol*, 18, pp. 382–417.

[34] Singh U.N. and Gupta R.S. (1971). Polyribosomes and unstable messenger RNA II. Some further implications of the tape theory of protein synthesis. *J Theor Biol*, 30, pp. 603–619.

[35] Singh U.N. (1969). Polyribosomes and unstable messenger RNA: a stochastic model of protein synthesis. *J Theor Biol*, 25, pp. 444–460.

[36] Rigney D.R. and Schieve W.C. (1977). Stochastic model of linear, continuous protein synthesis in bacterial populations. *J Theor Biol*, 69, pp. 761–766.

[37] Ozbudak E.M. *et al.* (2002). Regulation of noise in the expression of a single gene. *Nat Genet*, 31, pp. 69–73.

[38] Elowitz M.B., Levine A.J., Siggia E.D. and Swain P.S. (2002). Stochastic gene expression in a single cell. *Science*, 297, pp. 1183–1186.

[39] Blake W.J. *et al.* (2003). Noise in eukaryotic gene expression. *Nature*, 422, pp. 633–637.

[40] Raser J.M. and O'Shea E.K. (2004). Control of stochasticity in eukaryotic gene expression. *Science*, 304, pp. 1811–1814.

[41] Pedraza J.M. and van Oudenaarden A. (2005). Noise propagation in gene networks. *Science*, 307, pp. 1965–1969.

[42] Austin D.W. *et al.* (2006). Gene network shaping of inherent noise spectra. *Nature*, 439, pp. 608–611.

[43] Dublanche Y. *et al.* (2006). Noise in transcription negative feedback loops: simulation and experimental analysis. *Mol Syst Biol*, 2, p. 41.

[44] Kepler T.B. and Elston T.C. (2001). Stochasticity in transcriptional regulation: origins, consequences, and mathematical representations. *Biophys J*, 81, pp. 3116–3136.

[45] Swain P.S., Elowitz M.B. and Siggia E.D. (2002). Intrinsic and extrinsic contributions to stochasticity in gene expression. *Proc Natl Acad Sci USA*, 99, pp. 12795–12800.

[46] Paulsson J. (2004). Summing up the noise in gene networks. *Nature*, 427, pp. 415–418.

[47] Tao Y. (2004). Intrinsic and external noise in an auto-regulatory genetic network. *J Theor Biol*, 229, pp. 147–156.

[48] Longo D. and Hasty J. (2006). Dynamics of single-cell gene expression. *Mol Syst Biol*, 2, p. 64.

[49] McAdams H.H. and Arkin A. (1999). It's a noisy business! Genetic regulation at the nanomolar scale. *Trends Genet*, 15, pp. 65–69.

[50] Benzer S. (1953). Induced synthesis of enzymes in bacteria analyzed at the cellular level. *Biochim Biophys Acta*, 11, pp. 383–395.

[51] Maloney P.C. and Rotman B. (1973). Distribution of suboptimally induced β-d-galactosidase in Escherichia coli. *J Mol Biol*, 73, pp. 77–91.

[52] Novick A. and Weiner M. (1957). Enzyme induction as an all-or-none phenomenon. *Proc Natl Acad Sci USA*, 43, pp. 553–566.

[53] Spudich J.L. and Koshland D.E. (1976). Non-genetic individuality: chance in the single cell. *Nature*, 262, pp. 467–471.

[54] Hasty J., Pradines J., Dolnik M. and Collins J.J. (2000). Noise-based switches and amplifiers for gene expression. *Proc Natl Acad Sci USA*, 97, pp. 2075–2080.

[55] McAdams H.H. and Arkin A. (1997). Stochastic mechanisms in gene expression. *Proc Natl Acad Sci USA*, 94, pp. 814–819.

[56] Rosenfeld N. *et al.* (2005). Gene regulation at the single-cell level. *Science*, 307, pp. 1962–1965.

[57] Volfson D. *et al.* (2006). Origins of extrinsic variability in eukaryotic gene expression. *Nature*, 439, pp. 861–864.

[58] Colman-Lerner A. *et al.* (2005). Regulated cell-to-cell variation in a cell-fate decision system. *Nature*, 437, pp. 699–706.

[59] Kaern M., Elston T.C., Blake W.J. and Collins J.J. (2005). Stochasticity in gene expression: from theories to phenotypes. *Nat Rev Genet*, 6, pp. 451–464.

[60] Bar-Even A. *et al.* (2006). Noise in protein expression scales with natural protein abundance. *Nat Genet*, 38, pp. 636–643.

[61] Becskei A., Kaufmann B.B. and van Oudenaarden A. (2005). Contributions of low molecule number and chromosomal positioning to stochastic gene expression. *Nat Genet*, 37, pp. 937–944.

[62] Lewis K. (2000). Programmed death in bacteria. *Microbiol Mol Biol Rev*, 64, pp. 503–514.

[63] Maughan H. and Nicholson W.L. (2004). Stochastic processes influence stationary-phase decisions in Bacillus subtilis. *J Bacteriol*, 186, pp. 2212–2214.

[64] Thattai M. and van Oudenaarden A. (2004). Stochastic gene expression in fluctuating environments. *Genetics*, 167, pp. 523–530.

[65] Fraser H.B. *et al.* (2004). Noise minimization in eukaryotic gene expression. *PLoS Biol*, 2, p. e137.

[66] Kurakin A. (2005). Self-organization vs Watchmaker: stochastic gene expression and cell differentiation. *Dev Genes Evol*, 215, pp. 46–52.

[67] Paldi A. (2003). Stochastic gene expression during cell differentiation: order from disorder? *Cell Mol Life Sci*, 60, pp. 1775–1778.

[68] Hooshangi S., Thiberge S. and Weiss R. (2005). Ultrasensitivity and noise propagation in a synthetic transcriptional cascade. *Proc Natl Acad Sci USA*, 102, pp. 3581–3586.

[69] Hooshangi S. and Weiss R. (2006). The effect of negative feedback on noise propagation in transcriptional gene networks. *Chaos*, 16, p. 26108.

[70] Giaever G. *et al.* (2002). Functional profiling of the Saccharomyces cerevisiae genome. *Nature*, 418, pp. 387–391.

[71] Barkai N. and Leibler S. (1997). Robustness in simple biochemical networks. *Nature*, 387, pp. 913–917.

[72] von Dassow G., Meir E., Munro E.M. and Odell G.M. (2000). The segment polarity network is a robust developmental module. *Nature*, 406, pp. 188–192.

[73] Dunlap J.C. (1999). Molecular bases for Circadian clocks. *Cell*, 96, pp. 271–290.

[74] Gonze D., Halloy J. and Goldbeter A. (2002). Robustness of circadian rhythms with respect to molecular noise. *Proc Natl Acad Sci USA*, 99, pp. 673–678.

[75] Rao C.V., Wolf D.M. and Arkin A.P. (2002). Control, exploitation and tolerance of intracellular noise. *Nature*, 420, pp. 231–237.

[76] Vilar J.M.G., Kueh H.Y., Barkai N. and Leibler S. (2002). Mechanisms of noise-resistance in genetic oscillators. *Proc Natl Acad Sci USA*, 99, pp. 5988–5992.

[77] Krishna S., Andersson A.M.C., Semsey S. and Sneppen K. (2006). Structure and function of negative feedback loops at the interface of genetic and metabolic networks. *Nucleic Acids Res*, 34, pp. 2455–2462.

[78] Freeman M. (2000). Feedback control of intercellular signalling in development. *Nature*, 408, pp. 313–319.

[79] Smolen P., Baxter D.A. and Byrne J.H. (2000). Modeling transcriptional control in gene networks–methods, recent results, and future directions. *Bull Math Biol*, 62, pp. 247–292.

[80] Becskei A. and Serrano L. (2000). Engineering stability in gene networks by autoregulation. *Nature*, 405, pp. 590–593.

[81] Yi T.-M., Huang Y., Simon M.I. and Doyle J. (2000). Robust perfect adaptation in bacterial chemotaxis through integral feedback control. *Proc Natl Acad Sci USA*, 97, pp. 4649–4653.

[82] Raser J.M. and O'Shea E.K. (2005). Noise in gene expression: origins, consequences, and control. *Science*, 309, pp. 2010–2013.

[83] Doncic A., Ben-Jacob E. and Barkai N. (2006). Noise resistance in the spindle assembly checkpoint. *Mol Syst Biol*, 2, p. 2006.0027.

[84] Yuh C.H., Bolouri H. and Davidson E.H. (2001). Cis-regulatory logic in the endo16 gene: switching from a specification to a differentiation mode of control. *Development*, 128, pp. 617–629.

[85] McAdams H. and Shapiro L. (1995). Circuit simulation of genetic networks. *Science*, 269, pp. 650–656.

[86] Pritchard L. and Kell D.B. (2002). Schemes of flux control in a model of Saccharomyces cerevisiae glycolysis. *Eur J Biochem*, 269, pp. 3894–3904.

[87] Arkin A.P. (2001). Synthetic cell biology. *Curr Opin Biotechnol*, 12, pp. 638–644.

[88] Heinrich R., Rapoport S.M. and Rapoport T.A. (1977). Metabolic regulation and mathematical models. *Prog Biophys Mol Biol*, 32, pp. 1–82.

[89] Savageau M.A. and Voit E.O. (1987). Recasting nonlinear differential equations as S-systems: a canonical nonlinear form. *Math Biosci*, 87, pp. 83–115.

[90] Savageau M.A. (1969). Biochemical systems analysis. *J Theor Biol*, 25, pp. 365–369.

[91] Kampen V.N.G. (2007). *Stochastic Processes in Physics and Chemistry*, 3rd edn. (Elesvier, Oxford, UK).

[92] Haseltine E.L. and Rawlings J.B. (2002). Approximate simulation of coupled fast and slow reactions for stochastic chemical kinetics. *J Chem Phys*, 117, p. 6959.

[93] Rao C.V. and Arkin A.P. (2003). Stochastic chemical kinetics and the quasi-steady-state assumption: application to the Gillespie algorithm. *J Chem Phys*, 118, p. 4999.

[94] Cao Y., Gillespie D.T. and Petzold L.R. (2005). The slow-scale stochastic simulation algorithm. *J Chem Phys*, 122, p. 14116.

[95] Takahashi K., Kaizu K., Hu B. and Tomita M. (2004). A multi-algorithm, multi-timescale method for cell simulation. *Bioinformatics*, 20, pp. 538–546.

[96] Kauffman S. (2004). A proposal for using the ensemble approach to understand genetic regulatory networks. *J Theor Biol*, 230, pp. 581–590.

[97] Armstrong N.J. and van de Wiel M.A. (2004). Microarray data analysis: from hypotheses to conclusions using gene expression data. *Cell Oncol*, 26, pp. 279–290.

[98] Endy D. and Brent R. (2001). Modelling cellular behaviour. *Nature*, 409, pp. 391–395.

[99] Kauffman S. (1969). Homeostasis and differentiation in random genetic control networks. *Nature*, 224, pp. 177–178.

[100] Kauffman S. (1974). The large scale structure and dynamics of gene control circuits. *J Theor Biol*, 44, pp. 167–190.

[101] Huang S. (1999). Gene expression profiling, genetic networks, and cellular states: an integrating concept for tumorigenesis and drug discovery. *J Mol Med (Berl)*, 77, pp. 469–480.

[102] Kauffman S.A. (1993). *The Origins of Order: Self Organization and Selection in Evolution* (Oxford University Press, New York).

[103] Kauffman S.A. (1991). Antichaos and adaptation: biological evolution may have been shaped by more than just natural selection, computer models suggest that certain complex systems tend toward self-organization. *Sci Am*, 265, p. 78.

[104] Somogyi R. and Sniegoski C.A. (1996). Modeling the complexity of genetic networks: Understanding multigenic and pleiotropic regulation. *Complexity*, 1, pp. 45–63.

[105] Szallasi Z. and Liang S. (1998). Modeling the normal and neoplastic cell cycle with "realistic Boolean genetic networks": their application for understanding carcinogenesis and assessing therapeutic strategies. *Pac Symp Biocomput*, pp. 66–76.

[106] Serra R., Villani M., Graudenzi A. and Kauffman S.A. (2007). Why a simple model of genetic regulatory networks describes the distribution of avalanches in gene expression data. *J Theor Biol*, 246, pp. 449–460.

[107] Bolouri H. and Davidson E.H. (2002). Modeling transcriptional regulatory networks. *Bioessays*, 24, pp. 1118–1129.

[108] Hill A.V. (1910). The possible effects of the aggregation of the molecules of haemoglobin on its oxygen dissociation curve. *J Physiol*, 40, pp. 4–7.

[109] de Jong H. (2002). Modeling and simulation of genetic regulatory systems: a literature review. *J Comput Biol*, 9, pp. 67–103.

[110] Gillespie D.T. (1977). Exact stochastic simulation of coupled chemical reactions. *J Phys Chem*, 81, pp. 2340–2361.

[111] Gillespie D.T. (1992). *Markov Processes: An Introduction for Physical Scientists* (Academic Press, Boston).

[112] McQuarrie D.A. (1967). Stochastic approach to chemical kinetics. *J Appl Probab*, 4, pp. 417–478.

[113] Gillespie D.T. (2001). Approximate accelerated stochastic simulation of chemically reacting systems. *J Chem Phys*, 115, p. 1716.

[114] Kampen V. (1976). The expansion of the Master equation. *Adv Chem Phys*, 34, p. 245.

[115] Paulsson J. (2005). Models of stochastic gene expression. *Phys Life Rev*, 2, pp. 157–175.

[116] Thattai M. and van Oudenaarden A. (2001). Intrinsic noise in gene regulatory networks. *Proc Natl Acad Sci USA*, 98, pp. 8614–8619.

[117] Kaufmann B.B. and van Oudenaarden A. (2007). Stochastic gene expression: from single molecules to the proteome. *Curr Opin Genet Dev*, 17, pp. 107–112.

[118] Scott M., Ingalls B. and Kaern M. (2006). Estimations of intrinsic and extrinsic noise in models of nonlinear genetic networks. *Chaos*, 16, p. 26107.

[119] Vohradsky J. (2001). Neural model of the genetic network. *J Biol Chem*, 276, pp. 36168–36173.

[120] Wahde M. and Hertz J. (2000). Coarse-grained reverse engineering of genetic regulatory networks. *Biosystems*, 55, pp. 129–136.

[121] Friedman N., Linial M., Nachman I. and Pe'er D. (2000). Using Bayesian networks to analyze expression data. *J Comput Biol*, 7, pp. 601–620.

[122] Hartemink A.J., Gifford D.K., Jaakkola T.S. and Young R.A. (2001). Using graphical models and genomic expression data to statistically validate models of genetic regulatory networks. *Pac Symp Biocomput*, pp. 422–433.

[123] Nagarajan R., Aubin J.E. and Peterson C.A. (2004). Modeling genetic networks from clonal analysis. *J Theor Biol*, 230, pp. 359–373.

[124] Sachs K. *et al.* (2005). Causal protein-signaling networks derived from multiparameter single-cell data. *Science*, 308, pp. 523–529.

[125] Tegner J., Yeung M.K.S., Hasty J. and Collins J.J. (2003). Reverse engineering gene networks: integrating genetic perturbations with dynamical modeling. *Proc Natl Acad Sci USA*, 100, pp. 5944–5949.

[126] Yeung M.K.S., Tegnér J. and Collins J.J. (2002). Reverse engineering gene networks using singular value decomposition and robust regression. *Proc Natl Acad Sci USA*, 99, pp. 6163–6168.

[127] Chen T., He H.L. and Church G.M. (1999). Modeling gene expression with differential equations. *Pac Symp Biocomput*, pp. 29–40.

Chapter 3

Proteins, Mechanisms and Networks in NMDAR-dependent Synaptic Plasticity

3.1. Introduction

In Chapter 1, we discuss the structure of neurons and synapses as seen in Fig. 1.1, where pre- and postsynaptic terminals and synaptic cleft are displayed as major components of a synapse. The end of the axon of the sending neuron is the presynaptic terminal; the synaptic cleft is 20–40 nm in length, situated between the two terminals containing a matrix of fibrous extracellular proteins to adhere the pre- and postsynaptic membranes. An action potential travels through the axon to the presynaptic terminal and it must pass through the cleft to reach postsynaptic membrane to transmit the information contained in the action potential. However, the biochemical processes and time-scales involved in the information transmission have been fined tuned to be robust through millions of years of evolution; we are going to explore the marvels of some of these interactions in this chapter.

Neurotransmitters, as the name suggests, transmit the information from one terminal to the other by travelling through the protein matrix in the cleft. Neurotransmitters reside in synaptic vesicles as seen in Fig. 1.2 and these spherical shaped vesicles are packed in the presynaptic terminal. The membrane enclosing the vesicles are embedded with vesicular transporters, which pump the neurotransmitters into the vesicle. The active zone of the presynaptic cell contains the docked synaptic vesicles filled with

87

neurotransmitters ready to be emitted into the cleft in response to an action potential.

The postsynaptic terminal is the dendrite or soma of the receiving neuron. A very small region of 30–50 nm in length called postsynaptic density (PSD) is directly exposed to the synaptic cleft [1] at the distal tip of 200–800 nm in diameter and 0.5–2 μm in length on the dendritic spine. PSD is a potentially very active area with approximately 10,000 copies of a variety of more than 100 proteins, including adhesion molecules receptors, ion channels, tyrosine kinases, and G-protein-coupled receptors [1]. These diverse groups of proteins orchestrate the postsynaptic response to the presynaptic reactions induced by the environmental stimulus by forming biochemical pathways and adapting the topology of synapses structurally. The functionalities of these proteins currently are a very active area of research and the most of our discussion in this chapter is focussed on these functionalities. Synaptic strength is positively correlated to the dendritic spine and PSD volumes. The spine volume increase allows more trafficking of proteins into PSD and these proteins contribute to the structural changes within PSD and postsynaptic cell and also to increase the synaptic strength.

Ligand-gated receptors in the postsynaptic PSD act as sensors for neurotransmitters in the synaptic cleft to sense them and the post-synaptic potential is generated. The α-amino-3-hydroxy-5-methyl-4-isoxazolepropionic acid receptor (AMPAR) is such a receptor of the neurons in the hippocampus. Neurotransmitters bind to AMPAR anchored in PSD (PSD AMPAR) to open its embedded ion channel allowing the cations, mostly Na^+ and K^+, to enter and depolarise the postsynaptic membrane. This depolarisation is crucial to induce a potential difference across the membrane known as the excitatory postsynaptic potential (EPSP) and the AMPAR mediated current known as the excitatory postsynaptic current (EPSC) determines the magnitude of EPSP [2].

The number of PSD AMPAR and its single channel conductance are the major factors influencing the AMPAR mediated EPSC [3]. AMPAR, a mobile receptor, shuttles between the cytoplasm and PSD of the synapse [4], and the number of AMPAR among various synapses varies from 0 to 200 [4] and therefore we can expect a large variation in EPSP or EPSC as well. AMPARs residing in the cytoplasmic pool are transported to extrasynaptic sites by the process called exocytosis and then they laterally diffuse into the PSD [5–9] where they bind to the PSD scaffolding proteins, such as PSD-95 [10–12] mediated by transmembrane AMPAR regulatory protein (TARP), the binding partner of AMPAR [13, 14]. The alterations of the properties of proteins involved in mobility of AMPAR can change the number of PSD AMPAR and influence the overall receptor conductance without affecting the unitary channel conductance, which is a measure of the conductance of a single AMPAR ion channel and its alteration would change the rate at which cations pass through the ion channel within a unit time.

3.2. NMDAR-Dependent Synaptic Plasticity and Memory

NMDAR, blocked by a Mg^{2+} ion [15, 16], requires the postsynaptic membrane to be heavily depolarised to repel the Mg^{2+} ion and to expose the embedded Ca^{2+} channel of NMDAR. The exposed iron channel opens when a neurotransmitter, often a glutamate molecule, binds to NMDAR allowing Ca^{2+} influx to diffuse into the postsynaptic cell along the membrane. The Ca^{2+} gradient is very high across the postsynaptic membrane as its concentrations are $0.1\,\mu M$ inside the postsynaptic cell and $1800\,\mu M$ outside, respectively; therefore, the induced Ca^{2+} influx is very rapid leading to a quick elevation in the intracellular Ca^{2+} level. This elevation of Ca^{2+} is crucial in activating the subsequent synaptic plasticity pathways.

The rapid Ca^{2+} elevation triggers postsynaptic cascades which modulate the phosphorylation levels on serine 831 residue (S831) of AMPAR [17], serine 845 residue (S845) of AMPAR [5–9], and TARP [13, 14]. These are important processes because S831 regulates the unitary channel conductance, S845 controls the AMPAR translocation into extrasynaptic sites and TARP regulates the binding of AMPAR to PSD-95. Figure 3.1 shows the specific changes caused by the postsynaptic cascades and we delve into some details of the cascade next.

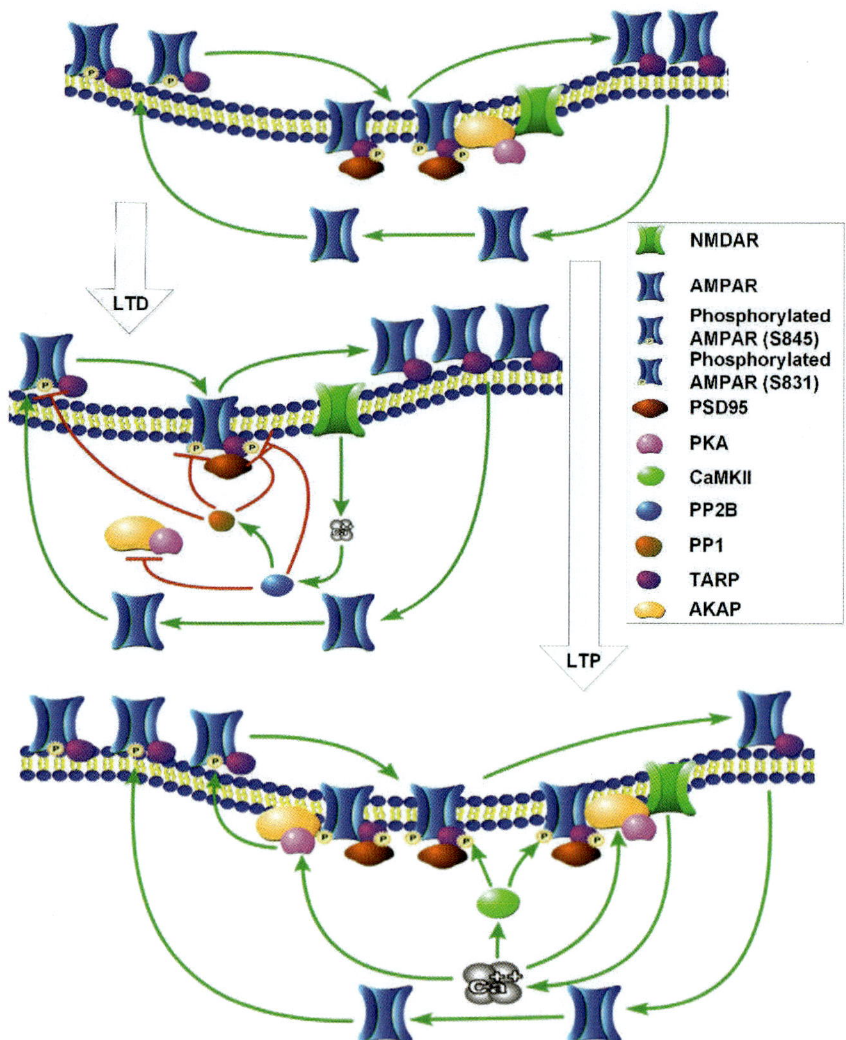

Fig. 3.1. Specific changes of the postsynaptic neuron during LTP and LTD. Three states of the postsynaptic neuron are shown here: basal (top), LTD (middle) and LTP (bottom). The details of the specific changes during LTP and LTD are summarised in Table 3.1.

Calmodulin (CaM), a protein with a high concentration in the postsynaptic cell, is a very important actor in synaptic plasticity and we discuss in later chapters its multifunctional aspects. In its inactive form, CaM binds neurogranin, which inhibits and holds postsynaptic CaM. Ca^{2+} at the high concentration activates protein kinase C (PKC) and PKC phosphorylates neurogranin to relinquish the bound CaM [17, 18]. The released CaM associates Ca^{2+} to form Ca^{2+}/CaM complex to expose the hydrophobic domain, which is the main binding site of CaM and increases the affinity of CaM to other proteins [19]. Ca^{2+}/CaM complex then selectively activates long-term potentiation (LTP) pathway or long-term depression (LTD) pathway (Table 3.1).

Transient high Ca^{2+} elevation activates the LTP pathway triggering the activations of kinases, including CaMKII and protein kinase A (PKA) (Fig. 3.1). CaMKII is the key protein that phosphorylates S831 of AMPAR to increase the unitary channel conductance [20–22] and TARP to increase the surface expression of AMPAR [13, 14]. Meanwhile, PKA phosphorylates S845 of AMPAR to move more AMPAR into extrasynaptic sites through exocytosis [23–25].

The LTD pathway is activated by a prolonged moderate Ca^{2+} elevation. This pattern of Ca^{2+} usually activates phosphatases, including calcineurin (PP2B) and protein phosphatase 1 (PP1) [26] (Fig. 3.1). PP1 dephosphorylates both S831 and S845 of AMPAR to decrease the unitary channel conductance and to enhance the internalisation of AMPAR into the cytoplasm by endocytosis [27–30]. Moreover, both PP2B and PP1 dephosphorylate TARP to reduce the surface expression of AMPAR [14, 30]. In addition, the latest

Table 3.1. Specific changes of the postsynaptic neuron during LTP and LTD.

States	Ca^{2+} (μM)	Induction frequency	Phosphorylation level			AMPAR number and conductance	PKA removal
			S831	S845	TARP		
Basal	0.1	NA	Low	Basal	Low	Basal	NA
LTD	<1	1 Hz	/	−	−	−	Yes
LTP	>10	100–200 Hz	+	+	+	+	NA

+ increase; − decrease; / insignificant change; NA not applicable or no report found.

evidence suggests that PP2B removes PKA out of dendritic spine or PSD to prevent the rephosphorylation on AMPAR [31, 32].

All the biochemical processes that are discussed so far show the flexibility or "plasticity" of the neuron to adapt to different stimulations and respond differentially. Because the processes discussed are primarily dependent on the activation of NMDAR, we call this plasticity, NMDAR-dependent synaptic plasticity. As evidence shows that sophisticated and complex functionalities of the brain, such as memory formation and retrieval, are facilitated by synaptic plasticity by enabling the modulation of synaptic strength: for example, memory is impaired by the related antagonists or gene mutations [33–39]. Synaptic plasticity and memory formation are closely linked but because of the complexity involved, their relationship is arduous to unravel; but we now understand that the experience-dependent memory formation is strongly associated with synaptic plasticity [39, 40]. The adaptation of synapses to the environmental stimulus, if highly selective, can be considered as memory, and brain can "remember" according to Herbbian theoretical perspective [40]. A particular cell assembly consisting of interconnected neurons could recognise the environment stimulus at the next occurrence [41]. The hippocampal neurons associated with spatial memory, for example, fire restrictively at a particular place for the organism to orient itself spatially [41]. The connections within a cluster of neurons are strongly related to the functions of the cell assembly as seen in the experiments, and the diversity of these connections may explain different types of memories.

LTP and LTD, the well-studied sub-forms of *NMDAR*-dependent synaptic plasticity, do not diminish after the vanishing of the Ca^{2+} signal and last for hours to days. Furthermore, during the later stages of LTP and LTD, protein transcription is a necessary function to induce structural changes of the synapse [49], i.e. gene expressions are required in these later stages. Therefore, it is necessary to distinguish early phase LTP, without gene expression, and late phase LTP, with gene expression [42–44]. We discuss the mechanisms mostly associated with the early phase of *NMDAR*-dependent synaptic plasticity.

In addition to the bidirectional behaviour in which two antithetic pathways share a single Ca^{2+} upstream, the modulators of synaptic plasticity have fundamental importance in facilitating the dynamics of synaptic plasticity, including the bistability in which the modulations are maintained much longer than the signal duration; and the robustness within the context of the low protein copy numbers. Synaptic plasticity depends on the stimulation frequency and the pattern of Ca^{2+} elevation. The prolonged low-frequency stimulations (LFS, 1 Hz for 900 s) produce the prolonged moderate intracellular Ca^{2+} rise and LTD. The transient high-frequency stimulations (HFS, 100–200 Hz for 1 s) cause the brief high intracellular Ca^{2+} rise and produce LTP [2]. We next explore the relationships between the proteins (modulators) and their dynamics related to synaptic plasticity in terms of their unique structures or functionalities.

3.2.1. *NMDAR*

NMDAR is a tetramer having two NR1 subunits and two NR2 subunits and the NR2 subunits are selected from either NR2A or NR2B subunits. By binding to the PDZ domain of membrane-associated guanylate kinase protein family (MAGUKs), NMDAR is contained in PSD. PSD-95 is the most known member of MAGUK family [45]. NMDAR is one of the major postsynaptic Ca^{2+} sources and is critical in the emergence of synaptic plasticity despite the small number of NMDAR in PSD of a mature synapse — the average number is only 20 [4, 54]. The dysfunctional NMDARs impair the formation of spatial memory and synaptic plasticity [35, 37, 46, 47].

The difference in the current induction between NR2A-containing and NR2B-containing NMDARs is important: the current induced by NR2A is shorter in duration and higher in amplitude than that induced by NR2B [48]. Therefore, the patterns of the Ca^{2+} influx passing through the receptor are different between NR2A- and NR2B-NMDAR subunits. LTP becomes dominant over LTD during the neuronal developmental period [49] and the NR2A/NR2B ratio of NMDARs increases during this period [50]. As the number of

NR2B-NMDAR subunits is synaptic-size independent [51], the bi-directional behaviour of synaptic plasticity may be influenced by the NR2A/NR2B ratio.

3.2.2. Ca^{2+} and CaM interaction

The selectivity of Ca^{2+}/CaM complex stems from the unique Ca^{2+} sensitivities of the CaM-binding proteins, some of which are key modulators of synaptic plasticity. In this chapter, "Ca^{2+}/CaM complex" refers to the fully Ca^{2+} bound CaM at all of its four Ca^{2+} binding sites (N terminus and C terminus of CaM contain four Ca^{2+} binding sites, two each). Filling these Ca^{2+} binding sites the following should be noted: (1) N terminus and C terminus have different binding rates; (2) at the presence of CaM binding proteins, the affinity of CaM for Ca^{2+} increases. The enhancement is in a heterotrophic positive cooperative manner in relation to the numbers of unfilled Ca^{2+} binding sites [52, 53]; and (3) the states of CaM binding proteins induce distinct degrees of the enhancement [54]. As a result, CaM binding proteins display unique Ca^{2+} sensitivities for the CaM binding and activation. For example, PP2B has a much higher Ca^{2+} sensitivity for CaM binding [55]. Having different Ca^{2+} sensitivities for CaM binding is physiologically important to distinguish the groups of proteins involved in LTP and LTD, and thus to facilitate the bidirectional behaviour of synaptic plasticity.

CaM is diffusive as shown *in vivo* [56]. It is proposed that neurogranin targets CaM to ideal locations to form CaM pools [17, 56, 57], given the highly mobile nature of CaM molecules in neuron [56]. In response to Ca^{2+} elevation through NMDAR, CaM is released when neurogranin is phosphorylated by Ca^{2+} activated PKC and is captured by Ca^{2+} to Ca^{2+}/CaM complex.

3.2.3. CaMKII

A CaMKII subunit has multiple isoforms and the dominant isoforms are expressed from α and β genes in the hippocampus. Both isoforms are synthesised in the same location and they may form CaMKII hetero-oligomers [58]. A CaMKII subunit is composed of a kinase domain, a regulatory domain, and a hub domain with a linker sitting in between the regulatory and hub domains (Fig. 3.2A). Twelve subunits form two hexametric rings and each ring contains six subunits where their hub domains are linked to form a central hub (Fig. 3.2B) [59]. The functionality of CaMKII relies on a few critical sites (Fig. 3.2A): (1) substrate binding site (S site) in the kinase domain for the catalytic activity; (2) pseudosubstrate segment in the regulatory domain containing CaM footprint which is critical for Ca^{2+}/CaM complex binding and CaMKII activation; (3) T site in the kinase domain for NMDAR binding; (4) threonine 286 site (abbreviation: T286 for α isoform or T287 for β isoform) in the regulatory domain for autonomous activity, the autophosphorylation; and (5) threonine 305/306 sites (T305/T306) in the pseudosubstrate segment for regulation of CaMKII activity but the exact mechanism as to how this regulation happens is unknown.

The inhibited subunits of CaMKII have the regulatory domain attached to the kinase domain; the S site binds to the pseudosubstrate segment and the T site binds to the T286 site so that all the critical sites are blocked and the kinase is inactive (Fig. 3.2C(a)) [59]. However, inactive CaMKII subunits are switching between autoinhibited compact and autoinhibited extended states (Fig. 3.2B). The critical sites are unavailable in the autoinhibited compact state and they are exposed in the autoinhibited extended state [60]. The major factor determining the rate of switching is the length of the linker, which varies among CaMKII isoforms [59].

To activate a CaMKII subunit, a Ca^{2+}/CaM complex must bind to the CaM footprint triggering a conformational change, which opens the S site for the catalytic processes (Fig. 3.2C(b)) [59, 61]. Once the S site is exposed, the adjacent active subunits trans-autophosphorylate the exposed T286 site (autophosphorylation) which makes the rebinding between the kinase and the regulatory domains impossible (Fig. 3.2C(c)) [59, 62–65]. As a result, the autophosphorylated CaMKII subunit may remain active partially even with the dissociation of CaM. Moreover, the autophosphorylation enhances the affinity of a subunit for the binding to Ca^{2+}/CaM complex [54]. Hence, the autophosphorylation prevents

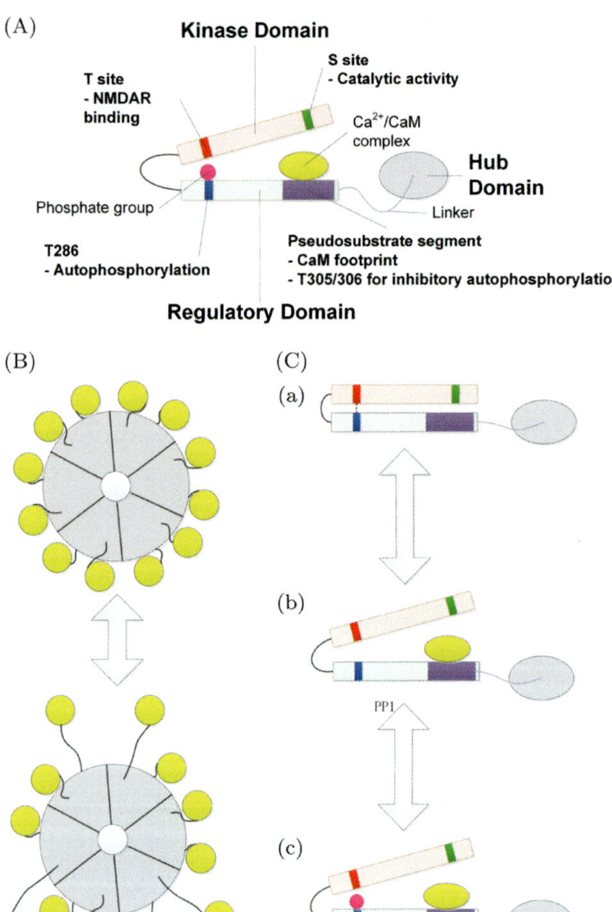

Fig. 3.2. Structure and conformational changes of CaMKII. (A) The structure and functional domains of a CaMKII subunit. (B) The structure of a CaMKII holoenzyme. Inhibited CaMKII is switching between autoinhibited compact (top) and autoinhibited extended (bottom) states. The rate of the switching is dependent on the length of the linker. (C) The activation of a CaMKII subunit: (a) Inhibited subunit: the S site binds to the pseudosegment and the T site binds to the T286 site; (b) Ca^{2+}/CaM complex attaches to the CaM footprint to cause a conformational change where the kinase domain detaches from the regulatory domain. At this state, the subunit is active; and (c) the activated subunit is autophosphorylated at the T286 site to retain the opening between the kinase and regulatory domains. Phosphorylation can be reverted by protein phosphatase 1 (PP1)-dependent dephosphorylation [60].

the reversing the active CaMKII subunits into the inhibited state upon the decrease of intracellular Ca^{2+} level [66–68]. After CaM dissociation, T305/T306 becomes autophosphorylated (inhibitory autophosphorylation) to prevent the rebinding of Ca^{2+}/CaM complex [69, 70].

F-actin holds CaMKII away from PSD [58] and Ca^{2+}/CaM-dependent activation releases CaMKII allowing its translocation into PSD [71]. CaMKII binds to the NR2B subunit of NMDAR and the binding anchors CaMKII in PSD [72–74]. The binding captures CaMKII in an active conformation independent of the autophosphorylation [75].

The difference in CaMKII isoforms may be important for synaptic plasticity. β isoform translocates much slower than α isoform [71]. The slower translocation rate may be caused by two factors: (1) F-actin binds to β isoform, but not to α isoform [58, 71]; and (2) the rates of the switching between the autoinhibition states are different among CaMKII isoforms [59]. As a result, the slower translocation rate of β isoform subunit may be unable to keep up with the transient signal of the LTP induction. On the other hand, β isoform has a higher catalytic property than that of α isoform [76], which may enhance the phosphorylation of the substrates. Moreover, the binding between β isoform and F-actin may facilitate the structural changes in the late phase [77]. Further evidence shows that the α/β isoforms ratio of a CaMKII is proactively regulated in hippocampal neurons which could be an important factor of LTP induction [58, 78].

The prolonged activity of CaMKII may be critical for the long-lasting activity of LTP. Two processes maintain the activity of CaMKII: (1) the autophosphorylation maintains the activity of CaMKII for up to one minute after the Ca^{2+} level diminishes [79], and (2) the NMDAR binding maintains the staying of CaMKII in PSD for at least 30 minutes [80], and potentially maintains CaMKII in an active conformation [75]. The prolonged activity of CaMKII potentially maintains the phosphorylation level of AMPAR and TARP, both of which are critical for LTP.

3.2.4. *Cyclic adenosine monophosphate (cAMP) and PKA*

PKA holoenzyme has two catalytic subunits and a regulatory dimer; the regulatory dimer binds to and inhibits the catalytic subunits [81]. RIIβ is the dominant PKA isoform expressed in the brain. The regulatory dimer contains two tandem cAMP-binding domains, each has two cAMP binding sites, and cAMP binding to the domains are required to trigger the release of the catalytic subunits. The binding affinity (or called dissociation constant, K_d) between the regulatory dimmer and catalytic subunits decreases from 0.6 nM to 9 nM, by more than 10 folds, when cAMP binds to the domains in regulatory dimer [81]. The released catalytic subunits are allowed to phosphorylate the target substrates. Furthermore, cAMP-PKA signalling pathway may be cross-talking to the activation pathway of the gene expression for further maintenance of LTP [82].

The regulation of PKA may facilitate the bi-directional behaviour of synaptic plasticity. The regulation involves interactions among adenylyl cyclase (AC), cAMP and phosphodiesterase (PDE) (Fig. 3.3). AC catalyses the conversion of adenosine triphosphate (ATP) into cAMP in response to a few signals, i.e. Ca^{2+}/CaM complex, dopamine, G-proteins. Among eight types of ACs, two are Ca^{2+}/CaM complex dependent in the neuron. Meanwhile, PDE converts cAMP into adenosine monophosphate (AMP) and both Ca^{2+}/CaM complex and PKA promote the activity of PDE [83, 84]. As a result, cAMP may be dual inhibited by (1) Ca^{2+}/CaM complex through a direct inhibitory circuit; and (2) PKA through an inhibitory feedback loop. Such inhibition loop forms a switch on the activation of PKA with respect to the intracellular Ca^{2+} level: the modest elevation in the Ca^{2+} level triggers a stronger PDE-inhibition than the AC-dependent activation leading to the low cAMP levels to prevent the activation of PKA and LTP; and the high elevation in the Ca^{2+} level saturates the PDE-dependent inhibition so that the AC-dependent activation dominate the promotion of the activation of PKA and LTP.

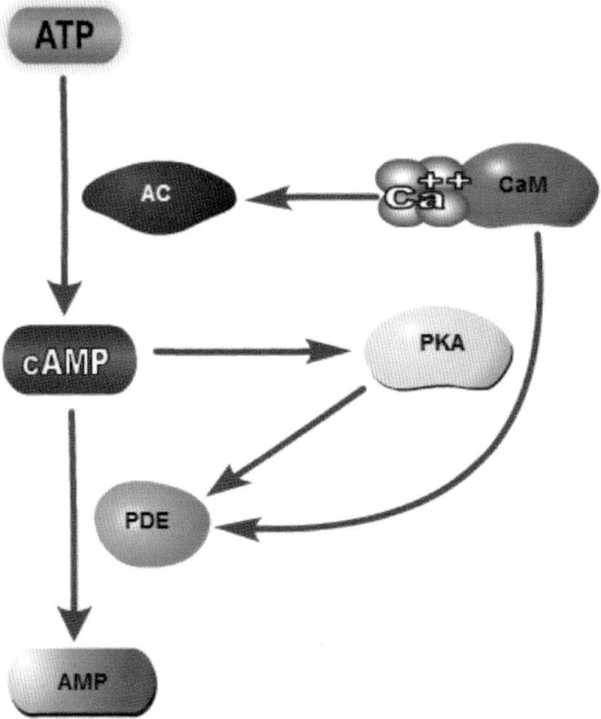

Fig. 3.3. The regulation of PKA through PDE-dependent cAMP inhibition. Ca^{2+}/CaM complex stimulates AC, which converts ATP into cAMP. cAMP further activates PKA. Meanwhile, cAMP is inhibited by PDE by converting into AMP. The activity of PDE is promoted by both PKA and Ca^{2+}/CaM complex forming a dual inhibition loop of cAMP.

3.2.5. *A-kinase anchoring proteins*

While a lot of synaptic proteins undergo spatial movements between synaptic compartments, the targeting and the spatial specificity of proteins become important for synaptic plasticity. One protein family involved in the processes is A-kinase anchoring proteins (AKAPs) (see [85]), which bind and translocate PKA into the dendritic spine and then into PSD [86]. Moreover, AKAPs can bind to many other proteins, some of which are critical to synaptic plasticity.

The most studied type of AKAPs is AKAP150 in mouse, and its homologue, AKAP79, in human. The translocation of AKAP79/150 into dendritic spine and PSD is enabled by its target domain near the N-terminus, through binding to proteins in the specific synaptic compartment, including F-actin [31], acidic phospholipid phosphatidylinositol-4,5-bisphosphate (PIP$_2$) [87], and cadherin cell adhesion molecule [32]. AKAP79/150 binds to PKC, AC, PP2B, and MAGUKs other than PKA [88–97]. The diverse binding ability of AKAP79/150 establishes links among the modulators and facilitates critical interactions among them in the emergence of synaptic plasticity. Many other types of AKAPs may contribute to synaptic plasticity as well. For example, Yotiao binds to PP1 instead of PP2B, and regulates NMDAR activity [98, 99]. AKAP allows synaptic proteins to be co-localised for their optimal functioning. Such an arrangement promotes the targeting of modulators of synaptic plasticity to critical sites of the synaptic transmission. The exact co-localisations of synaptic proteins are important and require further investigations.

The removal of AKAP from dendritic spine or PSD, caused by PP2B-dependent actin depolymerisation, is critical for LTD (Fig. 3.1) [31, 32]. Phospholipase C cleavage of PIP2 diminishes the activity of PKA [31, 100, 101] suggesting that the functionality of PKA relies on the binding to AKAP79/150 to translocate into dendritic spine and PSD [102]. Moreover, the removal slightly lags behind the occurring of the dephosphorylation of target substrate [102, 103], which may indicate that the removal of AKAP is a potential mechanism to obstruct the rephosphorylation of substrates by PKA. Particularly, the rephosphorylation may cause AMPAR to re-enter extrasynaptic sites. There is no evidence showing the involvement of the removal of AKAP in LTP.

3.2.6. *Interactions among modulators*

The interaction between kinases and phosphatases are necessary for the differentiation of LTP from LTD and vice versa. Current understanding indicates two places of the interaction: (1) the regulation of PP1 and (2) potentially the removal of AKAP [26, 85].

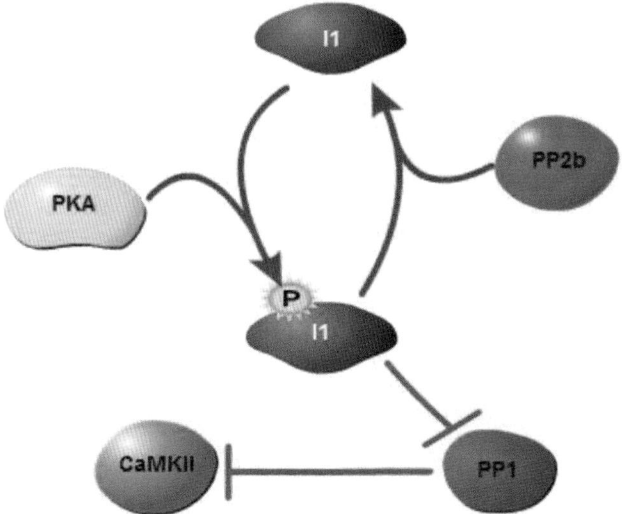

Fig. 3.4. The regulation of PP1. The phosphorylated I1 binds to and inhibits PP1. The phosphorylation of I1 is driven by PKA and some other kinases. PP2B activates PP1 by dephosphorylating I1 to release PP1 from the I1 binding.

PP1 strongly prevents the LTP induction; it dephosphorylates all the critical sites of synaptic proteins related to LTP: S831, S845, TARP, and CaMKII. The phosphorylated inhibitor 1 (I1) binds to and inhibits PP1. It is understood that I1 is phosphorylated by PKA [21], while PP2B dephosphorylates I1 to release PP1 [26] (Fig. 3.4). As a result, the activation of PP2B triggers the activation of PP1 to dephosphorylate these critical sites. The activation of PKA, on the other hand, prevents the PP1-dependent dephosphorylation of these critical sites, which may be another reason for the removal of PKA during LTD discussed previously.

However, the mechanism which prevents the PP2B-dependent removal of AKAPs, as well as PKA, is not well understood. The following mechanism is plausible: kinases phosphorylate ADF/cofilin and the phosphorylated ADF/cofilin suppresses the F-actin depolymerisation [104–106], which increases the level of F-actin and thus stabilises AKAPs in the dendritic spine. Moreover, evidence shows the elevation in the actin level during LTP [107].

3.2.7. *Summary of the alterations on AMPAR*

AMPAR is a tetramer consisting of two heterodimers, each having subunits from GluR1 to GluR4. Hippocampal AMPAR has mainly GluR1/GluR2 or GluR2/GluR3 heterodimers [108]. Cations influx through AMPAR depolarises the postsynaptic membrane when a glutamate-gated ion channel within it opens. This makes AMPAR the most important receptor in synaptic transmission. The peak AMPAR current (or AMPAR channel conductance), which influences EPSC, is affected by two important properties of AMPAR [109]: number of functional receptors and unitary channel conductance (the product of them is AMPAR channel conductance) [2, 22]. In this section, the potential mechanisms underlying the alterations on AMPAR channel conductance are discussed with respect to: (1) the number of functional AMPAR; and (2) the unitary channel conductance of AMPAR. Functional AMPAR is contained in PSD through the binding to MAGUKs [12]. Synapses with high numbers of functional AMPAR carry strong synaptic strength (the number is regulated by the Ca^{2+}-driven processes [110] ranging from 0 to 200 [111]). Further evidence shows that a large number of AMPARs are moved into or away from PSD during LTP and LTD, respectively [4, 27, 112–115].

PKA phosphorylates S845 to promote the externalisation of AMPAR into the extrasynaptic sites by exocytosis during LTP [5, 23, 25, 116]. In contrast, phosphatases dephosphorylate S845 to decrease the number of AMPAR in the extrasynaptic sites. Both PP2B and PP1 dephosphorylate S845 to promote the internalisation of extrasynaptic AMPAR into the cytoplasm by endocytosis [27, 29, 117–119]. Evidence shows that the level of S845 phosphorylation is rapidly decreased while the activities of phosphatases are increased during LTD [103].

TARP binds to the PDZ domain of PSD-95 through its cytoplasmic tail in a phosphorylated basis [13, 14]. Since TARP is tightly associated with AMPAR in the dendritic spine, TARP helps to anchor extrasynaptic AMPAR into PSD [120]. In the absence of TARP, surface expression of AMPAR is reduced by 75% [120]. CaMKII and PKC phosphorylate TARP to increase surface expression of AMPAR [12, 13, 121] while PP2B and PP1 dephosphorylate TARP to remove AMPAR away from PSD [14]. These synaptic proteins (PKC, CaMKII, PP2B, and PP1) are Ca^{2+} dependent; therefore, the regulation of TARP phosphorylation is also Ca^{2+} driven. TARP regulation may play a major role in the emergence of synaptic plasticity.

As mentioned before, an increase in the unitary AMPAR conductance results in a raise in depolarisation of the postsynaptic membrane. Phosphorylating S831 increases the unitary AMPAR conductance [22] by increasing the probability of AMPAR to switch to a higher unitary channel conductance, among possible four levels (9, 14, 20, and 28 ρS) [22]. The conductance values with higher probabilities are 9 and 14 ρS for an unphosphorylated S831 AMPAR. Overall, phosphorylating S831 by CaMKII causes an average of 50% increase in the unitary AMPAR conductance.

3.3. Mathematical Models of Synaptic Plasticity

Synaptic plasticity is modelled using different approaches to gain insights into its core components and processes (Fig. 3.5). We categorise them into single component models (SCMs) (Fig. 3.5A), simplified models (SMs) (Fig. 3.5B) and complete pathway models (CPMs) (Fig. 3.5C). SCMs are formulated for a single synaptic component to understand its potential structural and functional properties which are important to the dynamic behaviour of synaptic plasticity. SMs are theoretical frameworks to form a simplified synaptic circuit by encapsulating relationships among synaptic proteins that are claimed to be critical for synaptic plasticity (often selected from Ca^{2+}, CaM, cAMP, CaMKII, PKA, PP2B, PP1, and AMPAR). The relationships formulated in SMs may not have direct biochemical relationships to experimentally discovered interactions, but are plausible from evidence in the literature or are simplified from a multi-step cascades. The results of SMs show the behaviours of synaptic proteins within the interacting synaptic protein networks and provide insights into the potential roles of a synaptic protein plays in the networks contributing to

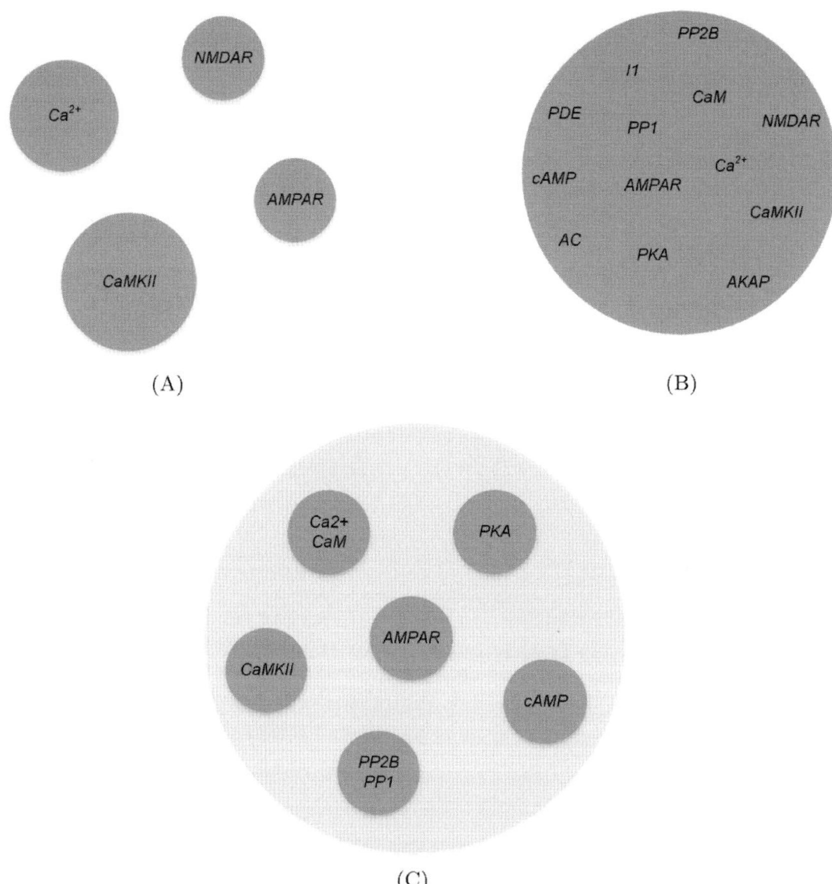

Fig. 3.5.   The scope of models related to synaptic plasticity. A circle represents a model related to synaptic plasticity and the text inside represents the scope of the model. Three parts of the figures represent (A) single component models; (B) simplified models; and (C) complete pathway models.

synaptic plasticity. Moreover, SMs reveal the emergent properties of the synaptic circuits contributing to the emergence of synaptic plasticity. SMs provide valuable insights into the complexity of synaptic interactions despite the limitations to the experimentally established data due to the simplification. Although CPMs and SMs share similarities in their usages to understand the emergent properties of the synaptic circuits, they are different in many ways: (1) CPMs contain the full spectrum of relationships among the synaptic proteins of a (or a part of a) synapse. These relationships are formulated based on experimentally revealed biochemical reactions or plausible biochemical reactions to reflect the realistic operation of the synapse; (2) due to the details included in CPMs, CPMs tend to interpret the experimental data better so that its predictions made have higher believability than SMs; and (3) also due to the details included in CPMs, the parameter estimation of CPMs is very challenging.

3.3.1. *Generating Ca^{2+} signals*

It is very important to generate Ca^{2+} signals as input to test model behaviours. Zhabotinsky [122] uses a framework of exponential build-up and decay to describe the Ca^{2+} elevation triggered by tetanus and our studies are all based on this theoretical framework. For example, assume that a tetanus of n pulses at frequency f stimulates the hippocampal neuron, each pulse increases the intracellular Ca^{2+} concentration by A, and the relaxation time of Ca^{2+} decay is τ. Then the resulting intracellular Ca^{2+} concentration following this tetanus can be described [122] as follows:

$$[Ca^{2+}] = [Ca^{2+}]_{rest} + A \sum_{i=1}^{n} \exp(-i/(f\tau)), \qquad (3.1)$$

where $[Ca^{2+}]_{rest}$ is the rest Ca^{2+} concentration. The parameter values used by Zhabotinsky [122] are $0.4\,\mu M$ ($1\,Hz$), or $1\,\mu M$ ($10\,Hz$ and $100\,Hz$) for A and $0.2\,s$ for τ.

We present four patterns of Ca^{2+} stimulation used in our studies: (1) tetanus contains 100 pulses at $100\,Hz$ (HFS), (2) tetanus contains 100 pulses at $10\,Hz$, (3) tetanus contains $100/900$ pulses at $1\,Hz$ (LFS), and (4) theta burst stimulation (TBS) (Fig. 3.6). The intracellular Ca^{2+} concentrations following these patterns of stimulations are shown in Fig. 3.7.

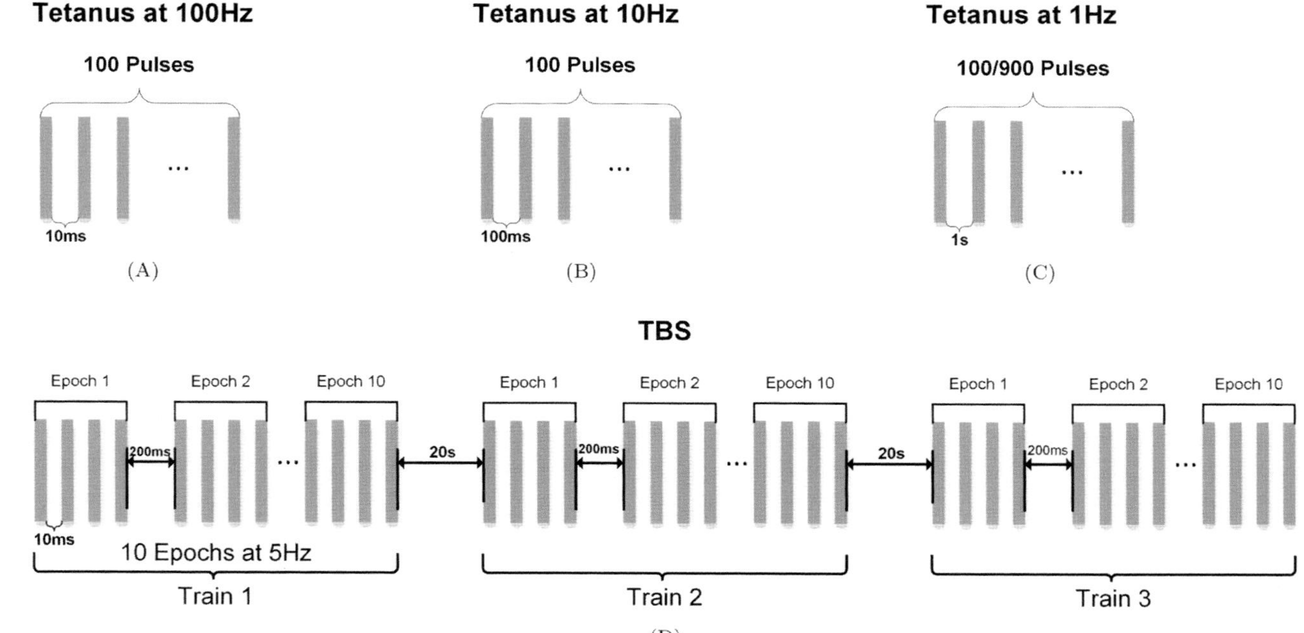

Fig. 3.6.   Stimulation patterns. Each pillar represents one pulse of the stimulation. (A) Tetanus at 100 Hz contains 100 pulses delivered with 10 ms separation. This pattern of stimulation is also referred as HFS for LTP induction. (B) Tetanus at 10 Hz contains 100 pulses delivered with 100 ms separation. (C) Tetanus at 1 Hz contains 100 or 900 pulses delivered with 1 s separation. This pattern of stimulation is also referred as LFS for LTD induction. (D) Theta burst stimulation contains three trains delivered with 20 s separation. Each train contains 10 epochs delivery at 5 Hz (separate by 200 ms). Each epoch contains four pulses at 100 Hz (separated by 10 ms).

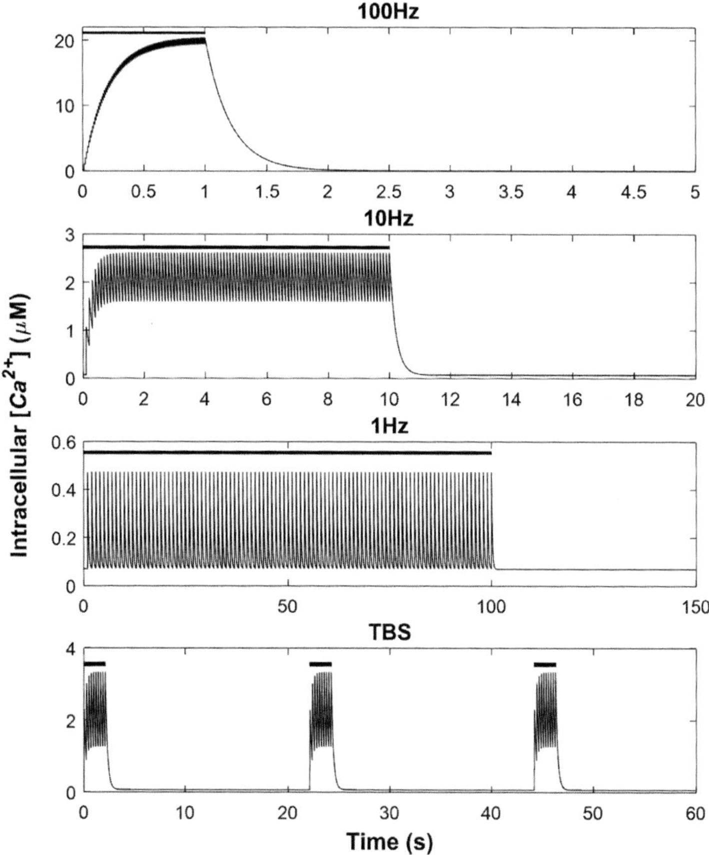

Fig. 3.7.   Intracellular Ca^{2+} elevations in response to the stimulation patterns. Each stimulation pattern in Fig. 3.6 is applied as indicated by the black bar(s). From top to bottom, figures show the simulated Ca^{2+} elevation following 100 Hz tetanus, 10 Hz tetanus, 1 Hz tetanus, and TBS, respectively.

3.3.2. *Single component models*

Ca^{2+} dynamics, CaMKII autophosphorylation, and AMPAR trafficking are often investigated using SCMs.

3.3.2.1. *Ca^{2+} dynamics*

As discussed previously, intracellular Ca^{2+} level is the key element for the emergence of synaptic plasticity. Hence, the propagating pattern of Ca^{2+} around the synaptic compartments and the relationship between the Ca^{2+} level and the activation of synaptic proteins are the central questions which SCMs try to answer [123–125]. Ca^{2+} displays distinct behaviours between the local (close to Ca^{2+} sources) and global (far away from Ca^{2+} sources) areas [126], although it has a fast diffusion rate (0.3–1 $\mu m^2/ms$) and should be instantaneously spread over and evenly distributed across the synapse given the small size of the dendritic spine (0.03–0.8 μm^3). Therefore, it is important to understand the propagating pattern of Ca^{2+} ions across postsynaptic compartments as well as whether the propagating pattern contains motifs of the Ca^{2+} selectivity.

Naoki *et al.* [124] use a Ca^{2+} SCM to investigate the potential relationship between Ca^{2+} dynamics and the bi-directional behaviour of synaptic plasticity. This research deals with three questions: (1) the diffusive processes cause the difference between the local and global Ca^{2+} dynamics; (2) the implication of the diffusible CaM to the formation of Ca^{2+}/CaM complex; and (3) the characteristics of Ca^{2+} pattern through NR2A- and NR2B-NMDAR subunits. The results suggest a dual decoding system that the information encoded in the frequency of the stimulus and the influx rate of NMDAR subunits are decoded by the local/global Ca^{2+} dynamics and the diffusible CaM. The direction of synaptic plasticity (LTP or LTD) is decided by the dual decoding system and errors are allowed to some extent. At the end, they propose a hypothesis that LTP or LTD are decided by the pairing between frequency and Ca^{2+} influx through NMDAR subunits; HFS paired NR2A induces LTP while LFS paired with NR2B induces LTD. This interpretation is consistent with the experimental observations [50].

3.3.2.2. *CaMKII autophosphorylation*

The existing SCMs of CaMKII investigate the CaMKII subunit state transitions using a number of approaches. In these models, the main objectives are to understand the complicated dynamics of the twelve CaMKII subunits as their critical sites undergo structural changes following Ca^{2+} signals and the linkages among the Ca^{2+}

levels, the formation of Ca^{2+}/CaM complex, and the state transitions of CaMKII subunits (particularly the T286 autophosphorylation) during LTP.

The dynamical changes of the autophosphorylation were extensively studied using the deterministic approach [127–129]. In these studies, the behaviours of the autophosphorylation were investigated under different frequencies of stimulations and multiple-pulse stimulations to understand the expression of the autophosphorylation in relation to the pattern of the stimulation signal [127–129]. The conclusion drawn is that the autophosphorylation occurs significantly only when the frequency of the stimulation is high, hence the autophosphorylation is an ideal decoder for the frequency of the stimulation [127, 128, 130]. Moreover, LTP is induced by HFS, hence the significant expression of the autophosphorylated CaMKII subunits following HFS may be an important factor for both induction and maintenance of LTP [131–133].

The robustness and time-span of the autophosphorylation against the low number of subunits and the turnover (averaged at one per 30 h for CaMKII) are investigated using stochastic approaches [134]. The major conclusion of the study is that the stability of the autophosphorylation increases as the number of the subunits of CaMKII increases and the persistent activation time span could reach from a few years to the whole lifetime.

3.3.2.3. *AMPAR trafficking*

As discussed previously, AMPAR undergoes continuous trafficking between cytoplasm and PSD and the trafficking is central to the operation of synaptic plasticity. The trafficking involves a number of synaptic proteins that drive the exocytosis/endocytosis of AMPAR between cytoplasm and extrasynaptic sites, and the anchoring/removal of AMPAR into and out of PSD. The pace of these processes at which AMPAR is travelling between the cytoplasm and PSD is important to explain the experimental observation. The Markovian approach was applied to look at the stochastic spatial movement of AMPAR in the PSD [135]. The study provides us with some clues for the timescales of trapping AMPAR in the PSD and the lateral movement of AMPAR. In addition, PSD scaffolding proteins (i.e. PSD-95) are suggested as an important factor in the trapping of AMPAR to retain its function.

3.3.3. *Simplified models*

The protein interacting network that is responsible for synaptic plasticity may contain many small networks, each provides a set of functions contributing to the dynamical behaviours of synaptic plasticity. These small networks include but not limited to: the cAMP-dependent PKA regulation network; the AKAPs dependent co-localisation of PKA; the PP1 regulation network; the CaMKII/PP1 interacting network; and the phosphorylation states of AMPAR controlled by kinases and phosphatases. Modelling these small interacting networks using SMs helps us to understand the functions of the small networks that may contribute significantly to synaptic plasticity. This knowledge allows us to build-up the relation of synaptic proteins to the emergence of synaptic plasticity by interpreting them with regard to the functions of small networks which contain them. Moreover, the complex interacting networks of synaptic plasticity can possibly be described through the integration of the small networks.

The phosphorylation states of AMPAR in the hippocampus is interpreted as the target of synaptic plasticity and is modelled using SMs by integrating the complex protein interacting networks of modulators, or kinases and phosphatases. The contribution of the modulators to synaptic plasticity is interpreted by their activation levels under different stimulation patterns [136] while the effects of synaptic plasticity are reflected by a "U-shape" in the quantities relating to AMPAR phosphorylation states, such as the AMPAR conductance [137], usually having Ca^{2+} level as the control parameter. The "U-shape" is caused by the shifting in the balance between kinases and phosphatase activities since they carry different Ca^{2+} affinities for the activation [136]. In addition, some criteria in basal conditions need to be obeyed to have the "U-shape" [136]. Moreover, the stimulus frequency is thought to encode the functional direction of synaptic plasticity. The information of functional direction is transduced by the Ca^{2+}/CaM interacting network as the Ca^{2+}/CaM complex level and is finally decoded by the autophosphorylation [130, 137–139].

SMs are developed for many areas of the brain. For example, a model for cerebellum [140] suggests that the threshold for inhibitory synaptic plasticity is set by the PDE1-dependent cAMP regulation network. A similar protein interacting network exists in hippocampus so that the finding may be applicable in the hippocampus.

3.3.4. *Complete pathway models*

There are a large number synaptic proteins included in CPMs. In addition to the similar set of essential modulators in SMs, CPMs also take account of the proteins in the production, maintenance, and signalling among the essential modulators. The relationship between these proteins and synaptic plasticity is mediated through the expression of the modulators so that they do not have direct implications to synaptic plasticity. The addition of these proteins enhance the models with regard to the believability and the accuracy against experiments, but they add more complexity into the model so that the model is challenging to develop and even more challenging to estimate its parameters. In the following sections, we discuss two examples of CMPs: the model of kinase and phosphatase interactions developed by Hayer and Bhalla [141], and the model of PKA and cAMP interactions developed by Kim *et al.* [142].

3.3.4.1. *Model of kinase and phosphatase interactions*

Before the work by Hayer and Bhalla [141], a model published in 1999 [143] includes all the biochemical reactions of the complex interactions between kinases, phosphatases, and transcriptional factors. Hayer and Bhalla [141] expand on the 1999 model to include two compartments representing the PSD and postsynaptic cytoplasm and the trafficking of AMPAR between the two compartments controlled by its phosphorylation states. The new model is bi-stable which expresses both LTP and LTD and robustness of the bi-stability is analysed in the study using stochastic simulation approaches.

The model is developed based on mass action rate laws and include the detailed reaction cascade initialised from Ca^{2+} influx through NMDAR to the resulting changes in phosphorylation states of AMPAR. For this complex protein interacting network, the model has 256 variables divided into many sub-models of the small networks, each describing the activation and inactivation of a modulator. These modulators, as described previous, include CaMKII, CaM, cAMP, PKA, PP2B, PP1, and AMPAR. In a small network, the modulator contains at least one active state and many inactive states, and the sub-model of the small network is formulated based on the transition among these states. The linkages between the small networks, usually the interactions between the active states of modulators, form the larger protein interacting network for the emergence of synaptic plasticity.

The model aims to investigate the bi-stability of synaptic plasticity under a number of hypothetical environments which are likely to be existing in the synaptic system as networks 0–5. Network 0 is formulated for parameter estimation and validation and networks 1–5 are modified upon network 0 to test the hypothetical environments in order to answer specific research questions. A number of insights are drawn from the analyses: (1) networks 1 and 2 show a new bi-stable switch of the functional AMPAR (AMPAR switch) which may underlie the mechanism required for the structural changes that activate a silent synapse; (2) network 3 shows a bi-stable switch of the CaMKII autophosphorylation (CaMKII switch), which is consistent with many models [122, 127, 128] and the literature [144]; and (3) The coupling between the AMPAR switch and the CaMKII switch under tight coupling (network 4), where AMPAR and CaMKII are dephosphorylated by a common pool of PP1, and weak coupling (network 5), where the processes in network 4 are accomplished by independent PP1 pools. Both tight and weak couplings are possible because AKAPs may hold synaptic proteins in separating clusters having local interactions within the same cluster (network 4) or global interactions to a different cluster (network 5). The conclusion is that LTP is induced predominantly by the AMPAR switch by driving more AMPAR into PSD while the CaMKII switch fine tunes the AMPAR conductance as indicated by the frequency of the stimulus. However, AMPAR switch can be slowly turned on by first turning on the CaMKII switch along. The underlying mechanism may be the saturation of the PP1 dephosphorylation, which is consistent with the experimental observations [14]. cAMP-PKA pathway suppresses the PP1 activity, we may hypothesise cAMP-PKA pathway has the following role: (1) PKA facilitates the CaMKII switch by restricting the PP1 dephosphorylation which would accelerate the turning on of AMPAR switch, and (2) drives more AMPAR into extrasynaptic sites. Interestingly, a single train of HFS is unable to induce LTP without PKA [145, 146]. Furthermore, turning on both switches are very difficult, but the systems are very robust against stochasticity when both are on. In addition, when both are on, the duration of high states can last even to lifetime that is unlikely with just one switch on.

3.3.4.2. *Model of PKA and cAMP interactions*

Kim *et al.* [142] advance their 2010 work [147] using multi-compartmental stochastic reaction-diffusion approach to investigate the temporal pattern of PKA activity with an emphasis on the spatial specificity of PKA in the emergence of synaptic plasticity. As discussed previously, the spatial targeting of synaptic proteins is essential elements of synaptic plasticity, i.e. the targeting of PKA by AKAPs, hence it is an extraordinarily important direction, but this direction lacks investigation [148]. Later, the same approach is applied in striatal synaptic plasticity [149].

Kim *et al.* [142] model the detailed synaptic interactions related to synaptic plasticity, including a Ca^{2+} and dopamine sub-model for the dynamics of Ca^{2+} and dopamine; a cAMP sub-model for the regulation of cAMP and the production of PKA having inputs feeding from the Ca^{2+} and dopamine sub-model; an interaction sub-model integrating CaMKII activation and phosphorylation, and phosphatases activations (PP2B and PP1) involving I1. These processes involve a complex interaction among PKA, CaMKII, PP2B, and PP1; and an AMPAR phosphorylation states sub-model.

As a result, the interactions among the sub-models form the signalling pathway connecting Ca^{2+} influx to phosphorylation states of AMPAR.

Different from the model of kinases and phosphatase interactions, most of the molecules are randomly diffused among many predefined sub-volumes towards their target locations in a synapse. In Kim *et al.* model, the synapse is divided into two large volumes: dendrite and dendritic spine. The dendrite is defined to contain many rectangular sub-volumes and molecules in dendrite are allowed for two-dimensional diffusion across the dendrite. The dendritic spine is a bulge on the dendrite and is defined to contain a few cylindrical or conical sub-volumes. Molecules in the dendritic spine are allowed for one-dimensional diffusion, i.e. either towards or away from PSD. The model considers the following proteins to be diffusible: cAMP, ATP, CaM, CaMKII, I1, and PKA catalytic subunit. Simulation of the diffusion is based on the modelling-morphology described above and the diffusion constants of the proteins.

The location of PKA significantly influences its activity. Results show the activity of PKA is enhanced when it is anchored near either its targets, AMPAR and I1, or near AC, the source of cAMP; however, anchored near AC gains larger enhancement than anchored near its targets. Moreover, the larger enhancement is independent of the location of AC. Although dendrite is far from PSD, anchoring PKA near AC in dendrite still gains larger enhancement, although its targets are in PSD. Further analysis shows the less diffusible nature of cAMP compared to that of PKA due to the strong PDE inactivation, although the diffusibility of cAMP itself is greater than PKA. As a result, microdomains of cAMP are formed near AC. The PKA anchored near the microdomains gains greater activation. The inactivation of PKA catalytic subunits is low, which allows PKA to diffuse across a wider space to reach its targets. Hence, the co-localisation of PKA by AKAPs is important to achieve the greater activity which may be crucial for the expression of synaptic plasticity. As a confirmation, LTP is impaired in mice expressing Ht31 peptide, which disrupts the PKA anchoring [142].

References

[1] Sheng M. and Hoogenraad C.C. (2007). The postsynaptic architecture of excitatory synapses: a more quantitative view. *Annu Rev Biochem*, 76, pp. 823–847.

[2] Citri A. and Malenka R.C. (2007). Synaptic plasticity: multiple forms, functions, and mechanisms. *Neuropsychopharmacology*, 33, pp. 18–41.

[3] He Y., Kulasiri D. and Samarasinghe S. (2014). Systems biology of synaptic plasticity: a review on N-methyl-D-aspartate receptor mediated biochemical pathways and related mathematical models. *Biosystems*, 122, pp. 7–18.

[4] Lüscher C. *et al.* (1999). Role of AMPA receptor cycling in synaptic transmission and plasticity. *Neuron*, 24, pp. 649–658.

[5] Banke T.G. *et al.* (2000). Control of GluR1 AMPA receptor function by cAMP-dependent protein kinase. *J Neurosci*, 20, pp. 89–102.

[6] Ehlers M.D. (2000). Reinsertion or degradation of AMPA receptors determined by activity-dependent endocytic sorting. *Neuron*, 28, pp. 511–525.

[7] He K. *et al.* (2009). Stabilization of Ca^{2+}-permeable AMPA receptors at perisynaptic sites by GluR1-S845 phosphorylation. *Proc Natl Acad Sci USA*, 106, pp. 20033–20038.

[8] Oh M.C., Derkach V.A., Guire E.S. and Soderling T.R. (2006). Extrasynaptic membrane trafficking regulated by GluR1 serine 845 phosphorylation primes AMPA receptors for long-term potentiation. *J Biol Chem*, 281, pp. 752–758.

[9] Yang Y., Wang X., Frerking M. and Zhou Q. (2008). Delivery of AMPA receptors to perisynaptic sites precedes the full expression of long-term potentiation. *Proc Natl Acad Sci USA*, 105, pp. 11388–11393.

[10] Ehrlich I., Klein M., Rumpel S. and Malinow R. (2007). PSD-95 is required for activity-driven synapse stabilization. *Proc Natl Acad Sci USA*, 104, pp. 4176–4181.

[11] Garner C.C., Nash J. and Huganir R.L. (2000). PDZ domains in synapse assembly and signalling. *Trends Cell Biol*, 10, pp. 274–280.

[12] Hayashi Y. *et al.* (2000). Driving AMPA receptors into synapses by LTP and CaMKII: requirement for GluR1 and PDZ domain interaction. *Sci Signal*, 287, p. 2262.

[13] Opazo P. *et al.* (2010). CaMKII triggers the diffusional trapping of surface AMPARs through phosphorylation of stargazin. *Neuron*, 67, pp. 239–252.

[14] Tomita S. *et al.* (2005). Bidirectional synaptic plasticity regulated by phosphorylation of stargazin-like TARPs. *Neuron*, 45, pp. 269–277.

[15] Mayer M.L., Westbrook G.L. and Guthrie P.B. (1984). Voltage-dependent block by Mg^{2+} of NMDA responses in spinal cord neurones. *Nature*, 309, pp. 261–263.

[16] Nowak L.L. *et al.* (1984). Magnesium gates glutamate-activated channels in mouse central neurones. *Nature*, 307, pp. 462–465.

[17] Baudier J. *et al.* (1991). Purification and characterization of a brain-specific protein kinase C substrate, neurogranin (p17). Identification of a consensus

amino acid sequence between neurogranin and neuromodulin (GAP43) that corresponds to the protein kinase C phosphorylation si. *J Biol Chem*, 266, pp. 229–237.

[18] Prichard L., Deloulme J.C. and Storm D.R. (1999). Interactions between neurogranin and calmodulin *in vivo*. *J Biol Chem*, 274, pp. 7689–7694.

[19] LaPorte D.C., Wierman B.M. and Storm D.R. (1980). Calcium-induced exposure of a hydrophobic surface on calmodulin. *Biochemistry*, 19, pp. 3814–3819.

[20] Barria A. *et al.* (1997). Regulatory phosphorylation of AMPA-type glutamate receptors by CaMKII during long-term potentiation. *Science*, 276, pp. 2042–2045.

[21] Blitzer R.D. *et al.* (1998). Gating of CaMKII by cAMP-regulated protein phosphatase activity during LTP. *Science*, 280, pp. 1940–1943.

[22] Derkach V., Barria A. and Soderling T.R. (1999). Ca^{2+}/calmodulin-kinase II enhances channel conductance of α-amino-3-hydroxy-5-methyl-4-isoxazolepropionate type glutamate receptors. *Proc Natl Acad Sci USA*, 96, pp. 3269–3274.

[23] Esteban J.A. *et al.* (2003). PKA phosphorylation of AMPA receptor subunits controls synaptic trafficking underlying plasticity. *Nat Neurosci*, 6, pp. 136–143.

[24] Lee H.-K. *et al.* (2010). Specific roles of AMPA receptor subunit GluR1 (GluA1) phosphorylation sites in regulating synaptic plasticity in the CA1 region of hippocampus. *J Neurophysiol*, 103, pp. 479–489.

[25] Zheng Z. and Keifer J. (2009). PKA has a critical role in synaptic delivery of GluR1-and GluR4-containing AMPARs during initial stages of acquisition of *in vitro* classical conditioning. *J Neurophysiol*, 101, pp. 2539–2549.

[26] Mulkey R.M., Endo S., Shenolikar S. and Malenka R.C. (1994). Involvement of a calcineurin/inhibitor-1 phosphatase cascade in hippocampal long-term depression. *Nature*, 369, pp. 486–488.

[27] Beattie E.C. *et al.* (2000). Regulation of AMPA receptor endocytosis by a signaling mechanism shared with LTD. *Nat Neurosci*, 3, pp. 1291–1300.

[28] Carroll R.C., Beattie E.C., von Zastrow M. and Malenka R.C. (2001). Role of AMPA receptor endocytosis in synaptic plasticity. *Nat Rev Neurosci*, 2, pp. 315–324.

[29] Mulkey R.M., Herron C.E. and Malenka R.C. (1993). An essential role for protein phosphatases in hippocampal long-term depression. *Science*, 261, pp. 1051–1055.

[30] Snyder G.L. *et al.* (2003). Regulation of AMPA receptor dephosphorylation by glutamate receptor agonists. *Neuropharmacology*, 45, pp. 703–713.

[31] Gomez L.L. *et al.* (2002). Regulation of A-kinase anchoring protein 79/150–cAMP-dependent protein kinase postsynaptic targeting by NMDA receptor activation of calcineurin and remodeling of dendritic actin. *J Neurosci*, 22, pp. 7027–7044.

[32] Gorski J.A., Gomez L.L., Scott J.D. and Dell'Acqua M.L. (2005). Association of an A-kinase-anchoring protein signaling scaffold with cadherin

adhesion molecules in neurons and epithelial cells. *Mol Biol Cell*, 16, pp. 3574–3590.

[33] Bourtchuladze R. *et al.* (1994). Deficient long-term memory in mice with a targeted mutation of the cAMP-responsive element-binding protein. *Cell*, 79, pp. 59–68.

[34] Chang H.P. *et al.* (1999). Impaired memory retention and decreased long-term potentiation in integrin-associated protein-deficient mice. *Learn Mem*, 6, pp. 448–457.

[35] Davis S., Butcher S.P. and Morris R.G. (1992). The NMDA receptor antagonist D-2-amino-5-phosphonopentanoate (D-AP5) impairs spatial learning and LTP *in vivo* at intracerebral concentrations comparable to those that block LTP *in vitro*. *J Neurosci*, 12, pp. 21–34.

[36] Giese K.P., Fedorov N.B., Filipkowski R.K. and Silva A.J. (1998). Autophosphorylation at Thr286 of the α calcium-calmodulin kinase II in LTP and learning. *Science*, 279, pp. 870–873.

[37] Morris R.G. (1989). Synaptic plasticity and learning: selective impairment of learning rats and blockade of long-term potentiation *in vivo* by the *N*-methyl-*D*-aspartate receptor antagonist AP5. *J Neurosci*, 9, pp. 3040–3057.

[38] Silva A.J., Paylor R., Wehner J.M. and Tonegawa S. (1992). Impaired spatial learning in alpha-calcium-calmodulin kinase II mutant mice. *Science*, 257, pp. 206–211.

[39] Silva A.J., Stevens C.F., Tonegawa S. and Wang Y. (1992). Deficient hippocampal long-term potentiation in alpha-calcium calmodulin kinase II mutant mice. *Science*, 257, pp. 201–206.

[40] Lisman J.E. (1989). A mechanism for the Hebb and the anti-Hebb processes underlying learning and memory. *Proc Natl Acad Sci USA*, 86, pp. 9574–9578.

[41] Bear M.F., Connor B.W. and Paradiso M.A. (2007). *Neuroscience: Exploring the Brain*, 3rd edn. (Lippincott Williams & Wilkins, Baltimore).

[42] Abel T. *et al.* (1997). Genetic demonstration of a role for PKA in the late phase of LTP and in hippocampus-based long-term memory. *Cell*, 88, pp. 615–626.

[43] Bliss T.V.P. and Collingridge G.L. (1993). A synaptic model of memory: long-term potentiation in the hippocampus. *Nature*, 361, pp. 31–39.

[44] Frey U., Huang Y.Y. and Kandel E.R. (1993). Effects of cAMP simulate a late stage of LTP in hippocampal CA1 neurons. *Science*, 260, pp. 1661–1664.

[45] Cho K.-O., Hunt C.A. and Kennedy M.B. (1992). The rat brain postsynaptic density fraction contains a homolog of the Drosophila discs-large tumor suppressor protein. *Neuron*, 9, pp. 929–942.

[46] Morris R.G.M., Garrud P., Rawlins J.N.P. and O'Keefe J. (1982). Place navigation impaired in rats with hippocampal lesions. *Nature*, 297, pp. 681–683.

[47] Tsien J.Z. *et al.* (1996). Subregion- and cell type-restricted gene knockout in mouse brain. *Cell*, 87, pp. 1317–1326.

[48] Erreger K. *et al.* (2005). Subunit-specific gating controls rat NR1/NR2A and NR1/NR2B NMDA channel kinetics and synaptic signalling profiles. *J Physiol*, 563, pp. 345–358.

[49] Puyal J. *et al.* (2003). Developmental shift from long-term depression to long-term potentiation in the rat medial vestibular nuclei: role of group I metabotropic glutamate receptors. *J Physiol*, 553, pp. 427–443.

[50] Liu L. *et al.* (2004). Role of NMDA receptor subtypes in governing the direction of hippocampal synaptic plasticity. *Science*, 304, pp. 1021–1024.

[51] Shinohara Y. *et al.* (2008). Left-right asymmetry of the hippocampal synapses with differential subunit allocation of glutamate receptors. *Proc Natl Acad Sci USA*, 105, pp. 19498–19503.

[52] Keller C.H., Olwin B.B., LaPorte D.C. and Storm D.R. (1982). Determination of the free-energy coupling for binding of calcium ion and troponin I to calmodulin. *Biochemistry*, 21, pp. 156–162.

[53] Olwin B.B. and Storm D.R. (1985). Calcium binding to complexes of calmodulin and calmodulin binding proteins. *Biochemistry*, 24, pp. 8081–8086.

[54] Meyer T., Hanson P.I., Stryer L. and Schulman H. (1992). Calmodulin trapping by calcium-calmodulin-dependent protein kinase. *Science*, 256, pp. 1199–1202.

[55] Quintana A.R., Wang D., Forbes J.E. and Waxham M.N. (2005). Kinetics of calmodulin binding to calcineurin. *Biochem Biophys Res Commun*, 334, pp. 674–680.

[56] Petersen A. and Gerges N.Z. (2015). Neurogranin regulates CaM dynamics at dendritic spines. *Sci Rep*, 5, p. 11135.

[57] Mons N., Enderlin V., Jaffard R. and Higueret P. (2001). Selective age-related changes in the PKC-sensitive, calmodulin-binding protein, neurogranin, in the mouse brain. *J Neurochem*, 79, pp. 859–867.

[58] Shen K., Teruel M.N., Subramanian K. and Meyer T. (1998). CaMKIIβ functions as an F-actin targeting module that localizes CaMKIIα/β heterooligomers to dendritic spines. *Neuron*, 21, pp. 593–606.

[59] Chao L.H. *et al.* (2011). A mechanism for tunable autoinhibition in the structure of a human Ca^{2+}/calmodulin-dependent kinase II holoenzyme. *Cell*, 146, pp. 732–745.

[60] Stratton M.M., Chao L.H., Schulman H. and Kuriyan J. (2013). Structural studies on the regulation of Ca^{2+}/calmodulin dependent protein kinase II. *Curr Opin Struct Biol*, 23, pp. 292–301.

[61] Lisman J.E., Schulman H. and Cline H. (2002). The molecular basis of CaMKII function in synaptic and behavioural memory. *Nat Rev Neurosci*, 3, pp. 175–190.

[62] Lou L.L., Lloyd S.J. and Schulman H. (1986). Activation of the multifunctional Ca^{2+}/calmodulin-dependent protein kinase by autophosphorylation: ATP modulates production of an autonomous enzyme. *Proc Natl Acad Sci USA*, 83, pp. 9497–9501.

[63] Miller S.G. and Kennedy M.B. (1986). Regulation of brain Type II Ca^{2+} calmodulin-dependent protein kinase by autophosphorylation: A Ca^{2+}-triggered molecular switch. *Cell*, 44, pp. 861–870.

[64] Saitoh T. and Schwartz J.H. (1985). Phosphorylation-dependent subcellular translocation of a Ca^{2+}/calmodulin-dependent protein kinase produces an autonomous enzyme in Aplysia neurons. *J Cell Biol*, 100, pp. 835–842.

[65] Yang E. and Schulman H. (1999). Structural examination of autoregulation of multifunctional calcium/calmodulin-dependent protein kinase II. *J Biol Chem*, 274, pp. 26199–26208.

[66] Fong Y.L., Taylor W.L., Means A.R. and Soderling T.R. (1989). Studies of the regulatory mechanism of Ca2+/calmodulin-dependent protein kinase II. Mutation of threonine 286 to alanine and aspartate. *J Biol Chem*, 264, pp. 16759–16763.

[67] Miller S.G., Patton B.L. and Kennedy M.B. (1988). Sequences of autophosphorylation sites in neuronal type II CaM kinase that control Ca^{2+}-independent activity. *Neuron*, 1, pp. 593–604.

[68] Waxham M.N., Aronowski J., Westgate S.A. and Kelly P.T. (1990). Mutagenesis of Thr-286 in monomeric Ca^{2+}/calmodulin-dependent protein kinase II eliminates Ca^{2+}/calmodulin-independent activity. *Proc Natl Acad Sci USA*, 87, pp. 1273–1277.

[69] Elgersma Y. *et al.* (2002). Inhibitory autophosphorylation of CaMKII controls PSD association, plasticity, and learning. *Neuron*, 36, pp. 493–505.

[70] Goh J.J. and Manahan-Vaughan D. (2015). Role of inhibitory autophosphorylation of calcium/calmodulin-dependent kinase II (αCAMKII) in persistent (>24 h) hippocampal LTP and in LTD facilitated by novel object-place learning and recognition in mice. *Behav Brain Res*, 285, pp. 79–88.

[71] Shen K. and Meyer T. (1999). Dynamic control of CaMKII translocation and localization in hippocampal neurons by NMDA receptor stimulation. *Science*, 284, pp. 162–167.

[72] Gardoni F. *et al.* (1998). Calcium/calmodulin-dependent protein kinase II is associated with NR2A/B subunits of NMDA receptor in postsynaptic densities. *J Neurochem*, 71, pp. 1733–1741.

[73] Leonard A.S. *et al.* (1999). Calcium/calmodulin-dependent protein kinase II is associated with the *N*-methyl-*D*-aspartate receptor. *Proc Natl Acad Sci USA*, 96, pp. 3239–3244.

[74] Strack S. and Colbran R.J. (1998). Autophosphorylation-dependent targeting of calcium/calmodulin-dependent protein kinase II by the NR2B subunit of the *N*-methyl-*D*-aspartate receptor. *J Biol Chem*, 273, pp. 20689–20692.

[75] Bayer K.U., Paul De Koninck A., Hell J.W. and Schulman H. (2001). Interaction with the NMDA receptor locks CaMKII in an active conformation. *Nature*, 411, pp. 801–805.

[76] Huynh Q.K. and Pagratis N. (2011). Kinetic mechanisms of Ca^{2+}/calmodulin dependent protein kinases. *Arch Biochem Biophys*, 506, pp. 130–136.

[77] Okamoto K., Bosch M. and Hayashi Y. (2009). The roles of CaMKII and F-actin in the structural plasticity of dendritic spines: a potential molecular identity of a synaptic tag? *Physiology (Bethesda)*, 24, pp. 357–66.

[78] Thiagarajan T.C., Piedras-Renteria E.S. and Tsien R.W. (2002). α-and βCaMKII: inverse regulation by neuronal activity and opposing effects on synaptic strength. *Neuron*, 36, pp. 1103–1114.

[79] Lee S.J., Escobedo-Lozoya Y., Szatmari E.M. and Yasuda R. (2009). Activation of CaMKII in single dendritic spines during long-term potentiation. *Nature*, 458, pp. 299–304.

[80] Bayer K.U. *et al.* (2006). Transition from reversible to persistent binding of CaMKII to postsynaptic sites and NR2B. *J Neurosci*, 26, pp. 1164–1174.

[81] Zawadzki K.M. and Taylor S.S. (2004). cAMP-dependent protein kinase regulatory subunit type IIbeta: active site mutations define an isoform-specific network for allosteric signaling by cAMP. *J Biol Chem*, 279, pp. 7029–7036.

[82] Mayford M., Siegelbaum S.A. and Kandel E.R. (2012). Synapses and memory storage. *Cold Spring Harb Perspect Biol*, 4, p. a005751.

[83] Lugnier C. (2006). Cyclic nucleotide phosphodiesterase (PDE) superfamily: a new target for the development of specific therapeutic agents. *Pharmacol Ther*, 109, pp. 366–398.

[84] Stanton P.K., Bramham C. and Scharfman H.E. (2005). *Synaptic Plasticity and Transsynaptic Signaling* (Springer Science & Business Media, New York).

[85] Sanderson J.L. and Dell'Acqua M.L. (2011). AKAP signaling complexes in regulation of excitatory synaptic plasticity. *Neurosci*, 17, pp. 321–336.

[86] Carr D.W. *et al.* (1992). Localization of the cAMP-dependent protein kinase to the postsynaptic densities by A-kinase anchoring proteins. Characterization of AKAP 79. *J Biol Chem*, 267, pp. 16816–16823.

[87] Dell'Acqua M.L. *et al.* (1998). Membrane-targeting sequences on AKAP79 bind phosphatidylinositol-4, 5-bisphosphate. *Embo J*, 17, pp. 2246–2260.

[88] Bhattacharyya S. *et al.* (2009). A critical role for PSD-95/AKAP interactions in endocytosis of synaptic AMPA receptors. *Nat Neurosci*, 12, pp. 172–181.

[89] Coghlan V.M. *et al.* (1995). Association of protein kinase A and protein phosphatase 2B with a common anchoring protein. *Science*, 267, pp. 108–111.

[90] Colledge M. *et al.* (2000). Targeting of PKA to glutamate receptors through a MAGUK-AKAP complex. *Neuron*, 27, pp. 107–119.

[91] Dell'Acqua M.L., Dodge K.L., Tavalin S.J. and Scott J.D. (2002). Mapping the protein phosphatase-2B anchoring site on AKAP79 binding and inhibition of phosphatase activity are mediated by residues 315–360. *J Biol Chem*, 277, pp. 48796–48802.

[92] Efendiev R. *et al.* (2010). AKAP79 interacts with multiple adenylyl cyclase (AC) isoforms and scaffolds AC5 and-6 to α-amino-3-hydroxyl-5-methyl-4-isoxazole-propionate (AMPA) receptors. *J Biol Chem*, 285, pp. 14450–14458.

[93] Klauck T.M. *et al.* (1996). Coordination of three signaling enzymes by AKAP79, a mammalian scaffold protein. *Science*, 271, pp. 1589–1592.

[94] Oliveria S.F., Gomez L.L. and Dell'Acqua M.L. (2003). Imaging kinase–AKAP79–phosphatase scaffold complexes at the plasma membrane in living cells using FRET microscopy. *J Cell Biol*, 160, pp. 101–112.

[95] Robertson H.R., Gibson E.S., Benke T.A. and Dell'Acqua M.L. (2009). Regulation of postsynaptic structure and function by an A-kinase anchoring protein–membrane-associated guanylate kinase scaffolding complex. *J Neurosci*, 29, pp. 7929–7943.

[96] Tavalin S.J. *et al.* (2002). Regulation of GluR1 by the A-kinase anchoring protein 79 (AKAP79) signaling complex shares properties with long-term depression. *J Neurosci*, 22, pp. 3044–3051.

[97] Willoughby D. *et al.* (2010). AKAP79/150 interacts with AC8 and regulates Ca^{2+}-dependent cAMP synthesis in pancreatic and neuronal systems. *J Biol Chem*, 285, pp. 20328–20342.

[98] Lin J.W. *et al.* (1998). Yotiao, a novel protein of neuromuscular junction and brain that interacts with specific splice variants of NMDA receptor subunit NR1. *J Neurosci*, 18, pp. 2017–2027.

[99] Westphal R.S. *et al.* (1999). Regulation of NMDA receptors by an associated phosphatase–kinase signaling complex. *Science*, 285, pp. 93–96.

[100] Horne E.A. and Dell'Acqua M.L. (2007). Phospholipase C is required for changes in postsynaptic structure and function associated with NMDA receptor-dependent long-term depression. *J Neurosci*, 27, pp. 3523–3534.

[101] Keith D.J. *et al.* (2012). Palmitoylation of A-kinase anchoring protein 79/150 regulates dendritic endosomal targeting and synaptic plasticity mechanisms. *J Neurosci*, 32, pp. 7119–7136.

[102] Smith K.E., Gibson E.S. and Dell'Acqua M.L. (2006). cAMP-dependent protein kinase postsynaptic localization regulated by NMDA receptor activation through translocation of an A-kinase anchoring protein scaffold protein. *J Neurosci*, 26, pp. 2391–2402.

[103] Sanderson J.L. *et al.* (2012). AKAP150-anchored calcineurin regulates synaptic plasticity by limiting synaptic incorporation of Ca2+-permeable AMPA receptors. *J Neurosci*, 32, pp. 15036–15052.

[104] Carlier M.-F. and Pantaloni D. (1997). Control of actin dynamics in cell motility. *J Mol Biol*, 269, pp. 459–467.

[105] Chen H., Bernstein B.W. and Bamburg J.R. (2000). Regulating actin-filament dynamics *in vivo*. *Trends Biochem Sci*, 25, pp. 19–23.

[106] Pantaloni D., Le Clainche C. and Carlier M.-F. (2001). Mechanism of actin-based motility. *Science*, 292, pp. 1502–1506.

[107] Fukazawa Y. *et al.* (2003). Hippocampal LTP is accompanied by enhanced F-actin content within the dendritic spine that is essential for late LTP maintenance *in vivo*. *Neuron*, 38, pp. 447–460.

[108] Wenthold R.J., Petralia R.S. and Niedzielski A.S. (1996). Evidence for multiple AMPA receptor complexes in hippocampal CA1/CA2 neurons. *J Neurosci*, 16, pp. 1982–1989.

[109] Benke T.A. *et al.* (1998). Modulation of AMPA receptor unitary conductance by synaptic activity. *Nature*, 393, pp. 793–797.

[110] Isaac J.T.R., Nicoll R.A. and Malenka R.C. (1995). Evidence for silent synapses: implications for the expression of LTP. *Neuron*, 15, pp. 427–434.

[111] Ribrault C., Sekimoto K. and Triller A. (2011). From the stochasticity of molecular processes to the variability of synaptic transmission. *Nat Rev Neurosci*, 12, pp. 375–387.

[112] Ashby M.C. *et al.* (2004). Removal of AMPA receptors (AMPARs) from synapses is preceded by transient endocytosis of extrasynaptic AMPARs. *J Neurosci*, 24, pp. 5172–5176.

[113] Lu W.-Y. *et al.* (2001). Activation of synaptic NMDA receptors induces membrane insertion of new AMPA receptors and LTP in cultured hippocampal neurons. *Neuron*, 29, pp. 243–254.

[114] Shi S.-H. *et al.* (1999). Rapid spine delivery and redistribution of AMPA receptors after synaptic NMDA receptor activation. *Science*, 284, pp. 1811–1816.

[115] Wang Y.T. and Linden D.J. (2000). Expression of cerebellar long-term depression requires postsynaptic clathrin-mediated endocytosis. *Neuron*, 25, pp. 635–647.

[116] Man H.-Y., Sekine-Aizawa Y. and Huganir R.L. (2007). Regulation of α-amino-3-hydroxy-5-methyl-4-isoxazolepropionic acid receptor trafficking through PKA phosphorylation of the Glu receptor 1 subunit. *Proc Natl Acad Sci USA*, 104, pp. 3579–3584.

[117] Kameyama K., Lee H.-K., Bear M.F. and Huganir R.L. (1998). Involvement of a postsynaptic protein kinase A substrate in the expression of homosynaptic long-term depression. *Neuron*, 21, pp. 1163–1175.

[118] Lee H.-K., Kameyama K., Huganir R.L. and Bear M.F. (1998). NMDA induces long-term synaptic depression and dephosphorylation of the GluR1 subunit of AMPA receptors in hippocampus. *Neuron*, 21, pp. 1151–1162.

[119] Lee H.-K. *et al.* (2000). Regulation of distinct AMPA receptor phosphorylation sites during bidirectional synaptic plasticity. *Nature*, 405, pp. 955–959.

[120] Tomita S. *et al.* (2003). Functional studies and distribution define a family of transmembrane AMPA receptor regulatory proteins. *J Cell Biol*, 161, pp. 805–816.

[121] Groc L. *et al.* (2004). Differential activity-dependent regulation of the lateral mobilities of AMPA and NMDA receptors. *Nat Neurosci*, 7, pp. 695–696.

[122] Zhabotinsky A.M. (2000). Bistability in the Ca^{2+}/Calmodulin-dependent protein kinase-phosphatase system. *Biophys J*, 79, pp. 2211–2221.

[123] Holmes W.R. and Levy W.B. (1990). Insights into associative long-term potentiation from computational models of NMDA receptor-mediated

calcium influx and intracellular calcium concentration changes. *J Neurophysiol*, 63, pp. 1148–1168.

[124] Naoki H., Sakumura Y. and Ishii S. (2005). Local signaling with molecular diffusion as a decoder of Ca2$^+$; signals in synaptic plasticity. *Mol Syst Biol.* 1, 2005.0027.

[125] Zador A., Koch C. and Brown T.H. (1990). Biophysical model of a Hebbian synapse. *Proc Natl Acad Sci USA*, 87, pp. 6718–6722.

[126] Augustine G.J., Santamaria F. and Tanaka K. (2003). Local calcium signaling in neurons. *Neuron*, 40, pp. 331–346.

[127] Dosemeci A. and Albers R.W. (1996). A mechanism for synaptic frequency detection through autophosphorylation of CaM kinase II. *Biophys J*, 70, pp. 2493–2501.

[128] Kubota Y. and Bower J.M. (2001). Transient versus asymptotic dynamics of CaM kinase II: possible roles of phosphatase. *J Comput Neurosci*, 11, pp. 263–279.

[129] Pepke S., Kinzer-Ursem T., Mihalas S. and Kennedy M.B. (2010). A dynamic model of interactions of Ca^{2+}, calmodulin, and catalytic subunits of Ca^{2+}/calmodulin-dependent protein kinase II. *PLoS Comput Biol*, 6, p. e1000675.

[130] Chiba H., Schneider N.S., Matsuoka S. and Noma A. (2008). A simulation study on the activation of cardiac CaMKII δ-isoform and its regulation by phosphatases. *Biophys J*, 95, pp. 2139–2149.

[131] Aslam N., Kubota Y., Wells D. and Shouval H.Z. (2009). Translational switch for long-term maintenance of synaptic plasticity. *Mol Syst Biol*, 5, p. 284.

[132] Graupner M. and Brunel N. (2007). STDP in a bistable synapse model based on CaMKII and associated signaling pathways. *PLoS Comput Biol*, 3, p. e221.

[133] Lisman J.E. and Zhabotinsky A.M. (2001). A model of synaptic memory: a CaMKII/PP1 switch that potentiates transmission by organizing an AMPA receptor anchoring assembly. *Neuron*, 31, pp. 191–201.

[134] Miller P., Zhabotinsky A.M., Lisman J.E. and Wang X.-J. (2005). The stability of a stochastic CaMKII switch: dependence on the number of enzyme molecules and protein turnover. *PLoS Biol*, 3, p. e107.

[135] Holcman D. and Triller A. (2006). Modeling synaptic dynamics driven by receptor lateral diffusion. *Biophys J*, 91, pp. 2405–2415.

[136] d'Alcantara P., Schiffmann S.N. and Swillens S. (2003). Bidirectional synaptic plasticity as a consequence of interdependent Ca^{2+}-controlled phosphorylation and dephosphorylation pathways. *Eur J Neurosci*, 17, pp. 2521–2528.

[137] Castellani G.C. *et al.* (2005). A model of bidirectional synaptic plasticity: from signaling network to channel conductance. *Learn Mem*, 12, pp. 423–432.

[138] Bradshaw J.M., Kubota Y., Meyer T. and Schulman H. (2003). An ultrasensitive Ca^{2+}/calmodulin-dependent protein kinase II-protein phosphatase 1

switch facilitates specificity in postsynaptic calcium signaling. *Proc Natl Acad Sci USA*, 100, pp. 10512–10517.

[139] Dupont G., Houart G. and De Koninck P. (2003). Sensitivity of CaM kinase II to the frequency of Ca^{2+} oscillations: a simple model. *Cell Calcium*, 34, pp. 485–497.

[140] Kitagawa Y., Hirano T. and Kawaguchi S.Y. (2009). Prediction and validation of a mechanism to control the threshold for inhibitory synaptic plasticity. *Mol Syst Biol*, 5, p. 280.

[141] Hayer A. and Bhalla U.S. (2005). Molecular switches at the synapse emerge from receptor and kinase traffic. *PLoS Comput Biol*, 1, p. e20.

[142] Kim M. *et al.* (2011). Colocalization of protein kinase A with adenylyl cyclase enhances protein kinase A activity during induction of long-lasting long-term-potentiation. *PLoS Comput Biol*, 7, p. e1002084.

[143] Bhalla U.S. and Iyengar R. (1999). Emergent properties of networks of biological signaling pathways. *Science*, 283, pp. 381–387.

[144] Lisman J.E. and McIntyre C.C. (2001). Synaptic plasticity: a molecular memory switch. *Curr Biol*, 11, pp. R788–R791.

[145] Otmakhova N.A., Otmakhov N., Mortenson L.H. and Lisman J.E. (2000). Inhibition of the cAMP pathway decreases early long-term potentiation at CA1 hippocampal synapses. *J Neurosci*, 20, pp. 4446–4451.

[146] Zhang M., Storm D.R. and Wang H. (2011). Bidirectional synaptic plasticity and spatial memory flexibility require Ca^{2+}-stimulated adenylyl cyclases. *J Neurosci*, 31, pp. 10174–10183.

[147] Kim M., Huang T., Abel T. and Blackwell K.T. (2010). Temporal sensitivity of protein kinase a activation in late-phase long term potentiation. *PLoS Comput Biol*, 6, p. e1000691.

[148] Kotaleski J.H. and Blackwell K.T. (2010). Modelling the molecular mechanisms of synaptic plasticity using systems biology approaches. *Nat Rev Neurosci*, 11, pp. 239–251.

[149] Oliveira R.F., Kim M. and Blackwell K.T. (2012). Subcellular location of PKA controls striatal plasticity: stochastic simulations in spiny dendrites. *PLoS Comput Biol*, 8, p. e1002383.

Chapter 4

A Theoretical Model
of State Transition of CaMKII

4.1. Introduction

CaMKII has a significant role in the induction of synaptic plasticity as well as the associated phenomena, such as memory formation. Two states of CaMKII, the autophosphorylation and the CaMKII-NMDAR binding, show strong correlation to synaptic plasticity and memory formation [1 10]. The CaMKII-NMDAR complex, formed when T site of CaMKII is attached to NMDAR, especially provides major contributions to synaptic plasticity: (1) the complex maintains the staying of CaMKII in PSD for more than 30 minutes [11] that may be a potential mechanism underlying the connection between E-LTP and late-phase LTP (L-LTP); (2) the formation of complex displays high associativity to the induction of LTP [6, 9, 12]; and (3) inhibiting the formation of the complex impairs LTP [1, 4, 10].

Modelling the state transition (ST) of CaMKII has been a predominant path to understand the dynamic behaviour of synaptic plasticity. Inner state transitions of CaMKII (ISTs) are the state transitions of a single subunit, which has no direct impact on the state of other subunits in the same holoenzyme, for example, the Ca^{2+}/CaM complex binding and the autophosphorylation; holoenzyme state transitions of CaMKII (HSTs) are the state transitions impacting the state of a holoenzyme as well as all its constituting subunits, for example, the translocation and the NMDAR binding. Many models successfully describe ISTs and

125

investigate the significance of the autophosphorylation in decoding the frequency of the stimulation. However, they neither include the HSTs nor provide insight into the relationships among the autophosphorylation, CaMKII-NMDAR complex, and LTP.

HSTs are major steps in the formation of CaMKII-NMDAR complex; an experiment reveals the holoenzyme structure of CaMKII as the essential component for the attachment to NMDAR [13]. We develop a dynamic model of the HSTs (MoHST) of CaMKII to include both ISTs based on the conformation of CaMKII subunits, and HSTs based on the translocation of the CaMKII holoenzyme between dendritic spine and PSD as well as a probabilistic framework for the binding between NMDAR and the CaMKII holoenzyme [14]. Using MoHST, we hope to understand the relationships among ISTs, particularly the autophosphorylation, the CaMKII-NMDAR complex and LTP through studying the dynamic behaviours of CaMKII in the process of CaMKII-NMDAR complex formation following Ca^{2+} signals. To the best of our knowledge, this is the first model of the dynamics of HSTs and the model provides insights into the factors of CaMKII-NMDAR complex formation.

In this chapter, we present the detailed information of MoHST. The development of MoHST is described in Sec. 4.2. Section 4.3 gives the details related to the simulation with MoHST, including generation of Ca^{2+} signals, parameter estimation, and computational experiments. Section 4.4 gives the details of parameter perturbation and the associated results. Finally, a brief discussion/summary is given in Sec. 4.5.

4.2. Model Development

This section elaborates the development of MoHST: the assumptions of MoHST and the details of MoHST, including the probabilistic framework, the conceptual models, and the mathematical equations.

4.2.1. *Model assumptions*

(1) Two compartments, PSD and dendritic spine, are considered. This consideration is rooted on the translocation of CaMKII into PSD, which is prior to the formation of CaMKII-NMDAR complex, following the LTP induction signal. Both PSD and dendritic spine are considered as cylinder with mean volumes 0.01 fL (200–800 nm diameter and 30–50 nm height) and 0.1 fL (200–800 nm diameter and 0.5–2 μm height), respectively ($1\,\text{fL} = 10^{-15}\,\text{L} = 10^{-18}\,\text{m}^3$). The volumes are used for the unit conversion.

(2) The switching between the autoinhibited states of the CaMKII subunit to expose critical sites is ignored. The switching rate is dependent on the linker length, which is different among isoforms of the CaMKII subunit. But, isoforms are not a consideration of MoHST. Hence, the switching is assumed to instantaneously reach an equilibrium ratio in comparison to the slower activation process, which involves binding between Ca^{2+}/CaM complex and CaMKII subunits through spatial movement and targeting. Hence, the equilibrium ratio can be merged into the binding rate.

(3) Gene expression is unaltered during E-LTP. The total number of CaMKII, both in subunits and holoenzymes, is assumed to remain constant to reflect the situation that new CaMKII synthesis remains at a rate which exactly offsets the degradation of CaMKII.

(4) Phosphorylation and dephosphorylation are reactions catalyzed by enzymes. Michaelis–Menten kinetics provides a simplified framework for catalytic reactions. All other reactions are assumed to be elementary which obeys the conditions of the mass action kinetics.

(5) For simplicity, all molecules, except CaMKII, are assumed to evenly distribute across the whole synapse, although some of them display local dynamics, for example, Ca^{2+} has local dynamics around Ca^{2+} sources [15]. We assume that their local dynamics are insignificant for the dynamics of CaMKII.

(6) The autophosphorylated or the NR2B bound CaMKII subunits always have a CaM attached. The underlying reason is that both the autophosphorylation and the NR2B binding prevent the CaM dissociation — the CaM dissociation rate is decreased by 1000 folds and 20 folds, respectively, in autophosphorylated and NR2B bound CaMKII subunits [16, 17]. The resulting dissociation rate of CaM is negligible in these CaMKII subunits.

(7) The translocation of CaMKII is controlled by the CaM binding [18]. Hence, the translocation is assumed to be dependent on the levels of the CaM-bound and the autophosphorylated CaMKII subunits; both have a CaM bound.

(8) We consider the formation of CaMKII-NMDAR complex through S site mediated T site binding mechanism only (Fig. 4.3A). We also assume that T site binding is irreversible, at least for the duration of E-LTP.

(9) We assume one-to-one binding between NR2B of NMDAR and CaMKII subunit. Given that a CaMKII subunit is 475 amino acids long, it binds to and covers the binding region, amino acids 839–1303, of NR2B that blocks the NR2B from other CaMKII subunit. We assume that NMDAR has only one NR2B subunit according to the work reported by Shinohara *et al.* [19].

(10) We assume that PP1 does not dephosphorylate the NMDAR bound CaMKII in PSD. This assumption is based on the experimental observation that PP1 is inaccessible to CaMKII in PSD [20], which applies to the NMDAR bound case of anchoring CaMKII in PSD.

4.2.2. *The states of CaMKII*

We consider eight states of CaMKII (summarised in Table 4.1, we use variable names to denote state for brevity). ISTs establish the linkage among three inner states (IS) of a subunit (from Chiba *et al.* [21]): (1) iCaMKII (number of subunits in this state), (2) CaMCaMKII, and (3) CaMCaMKII*. HSTs establish the linkage among five holoenzyme states (HS) of CaMKII: (1) EoPSD (the number of holoenzymes in this state), (2) EiPSD, (3) CaMKIIS, (4) CaMKII*S,

Table 4.1. States of CaMKII and their definitions.

	Variable	Definition
IS	iCaMKII	Inhibited CaMKII subunit
	CaMCaMKII	CaM bound CaMKII subunit
	CaMCaMKII*	Autophosphorylated CaMKII subunit
HS	EoPSD	CaMKII outside of PSD
	EiPSD	NMDAR unbound CaMKII inside of PSD
	CaMKIIS	CaMKII bound to NMDAR at S site of a CaM bound subunit
	CaMKII*S	CaMKII bound to NMDAR at S site of an autophosphorylated subunit
	CaMKII*T	CaMKII bound to NMDAR at T site of any active subunit

and (5) CaMKII*T. (An active subunit could be either a CaM-CaMKII or a CaMCaMKII*).

HSs are composed of different ISs: (1) EoPSD and EiPSD have all of their subunits (12 subunits) belonging to the three ISs, and (2) CaMKIIS, CaMKII*S, and CaMKII*T have exactly one subunit bound to an NR2B according to assumption (9) (the bound subunit is denoted by the same variable name as the holoenzyme: CaMKIIS, CaMKII*S, or CaMKII*T) and others (11 subunits) belonging to the three ISs. As the total number of CaMKII subunits, $CaMKII_{\text{Total}}$, is conserved (assumption (3)) during E-LTP, and the following constraints are given:

$$CaMKII_{\text{Total}} = iCaMKII + CaMCaMKII + CaMCaMKII^*$$
$$+ CaMKIIS + CaMKII^*S + CaMKII^*T, \quad \text{or}$$
$$CaMKII_{\text{Total}} = 12\,(EoPSD + EiPSD + CaMKIIS$$
$$+ CaMKII^*S + CaMKII^*T).$$

4.2.3. *A probabilistic framework of the CaMKII-NMDAR binding*

Multiple CaMKII subunits may be available to bind to NMDAR (we call these subunits potential binding subunit (pBS); see

Sec. 4.2.3.2 for the explanation), but at most one of them can bind to NMDAR (assumption (9)). The conformations of the 12 subunits constituting the CaMKII holoenzyme determine the propensity of the binding between the holoenzyme and NMDAR. Given the 91 possible conformations of the CaMKII holoenzyme: select 12 subunits from 3 possible ISs,

$$C_2^{12+3-1} = \frac{14!}{2!(14-12)!} = 91 \quad [22];$$

modelling the exact binding dynamics is very difficult since 91 reaction rate constants are required. To avoid having 91 reaction rate constants, we develop a probabilistic framework to consider an average rate constant for the binding. The average rate constant is calculated by the product of a basal rate constant multiplied by the modifiers given by this framework (Fig. 4.1). Table 4.2 summaries the symbols used in the framework.

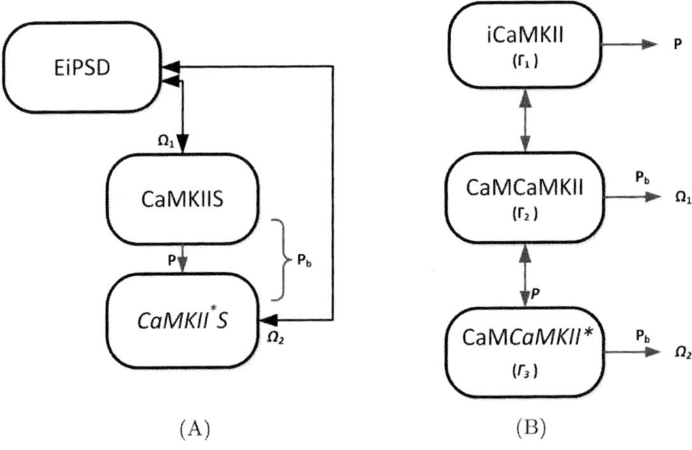

Fig. 4.1.   The schematic diagram of probabilistic framework. (A) The framework of CaMKII-NMDAR binding: NMDAR binds to the S site of an active subunit of EiPSD to form CaMKIIS or CaMKII*S. The rates of the bindings are modified by Ω_1 or Ω_2, respectively, to reflect the average binding rate with respect to 91 possible conformations of CaMKII holoenzyme. (B) ISs of CaMKII determine Γ_s ($s = 1, 2, 3$) which represent the fractions of the three potential ISs which constitute a CaMKII holoenzyme. Ωs depend on Γs and P_b (see Sec. 4.2.3.3).

Table 4.2. Definition of the probabilities and modifiers used in the model.

	Symbols	Definition
IS	Γ_1	Fraction of iCaMKII before binding to NMDAR
	Γ_2	Fraction of CaMCaMKII before binding to NMDAR
	Γ_3	Fraction of CaMCaMKII* before binding to NMDAR
	P	Probability that either or both sides of a neighbouring subunit are active
HS	P_b	Probability of S site mediated binding between CaMKII and NMDAR
	Ω_1	Modifier for the average rate constant of transition to CaMKIIS
	Ω_2	Modifier for the average rate constant of transition to CaMKII*S

4.2.3.1. *Composition of CaMKII*

We need to know the probabilities of the 91 possible conformations given, the fractions of the three ISs of CaMKII subunits. We use Γ_1, Γ_2, and Γ_3 to represent the fractions of iCaMKII, CaMCaMKII, and CaMCaMKII* ISs, respectively (Fig. 4.1B). The bound subunits (CaMKIIS, CaMKII*S, or CaMKII*T) lack PP1-dephosphoyrlation (assumption (10)) and carry different fractions. Hence, Γs are calculated excluding the bound subunits:

$$\Gamma_1 = \frac{iCaMKII}{CaMKII_{\text{Total}} - (CaMKIIS + CaMKII^*S + CaMKII^*T)}, \tag{4.1}$$

$$\Gamma_2 = \frac{CaMCaMKII}{CaMKII_{\text{Total}} - (CaMKIIS + CaMKII^*S + CaMKII^*T)}, \tag{4.2}$$

$$\Gamma_3 = \frac{CaMCaMKII^*}{CaMKII_{\text{Total}} - (CaMKIIS + CaMKII^*S + CaMKII^*T)}, \quad \text{and} \tag{4.3}$$

$$\Gamma_1 + \Gamma_2 + \Gamma_3 = 1.$$

The 91 conformations are distributed according to the multinomial distribution and their probabilities are calculated based on Γs; for example, the probability that CaMKII contains exactly n CaMCaMKII and mCaMCaMKII* is given by

$$\beta \cdot \Gamma_2^n \Gamma_3^m \Gamma_1^{12-n-m}, \quad (n = 0, 1, \ldots, 12), \tag{4.4}$$

where β is the multinomial coefficient:

$$\binom{12}{n}\binom{12-n}{m} = \frac{12!}{n!(12-n)!} \cdot \frac{12-n}{m!(12-n-m)!} \quad [23].$$

P_b is given by the fraction of bound NMDAR at S site of a CaMKII subunit (Fig. 4.1A):

$$P_b = \frac{CaMKIIS + CaMKII^*S}{NMDAR_{\text{Total}}}. \tag{4.5}$$

4.2.3.2. *Derivation of probabilistic binding between CaMKII and NMDAR*

pBS is a CaMKII subunit that: (1) belongs to an unbound holoenzyme in PSD (subunits of EiPSD), and (2) is active. In terms of IS defined, pBS can be either CaMCaMKII pBS (CaM-pBS), or CaMCaMKII* pBS (T286-pBS). The two types of pBS are considered separately because of their different dissociation rates from NMDAR binding [11].

Assume firstly that NMDAR can bind to up to n pBSs. Let X be the number of pBSs which NMDAR binds to. Let P_b be the probability that NMDAR binds to a single pBS. Assuming the n pBSs are independent, we have

$$X \sim \text{Binomial}(n, P_b),$$

with probability function:

$$P(X = x) = \binom{n}{x} P_b^x (1 - P_b)^{n-x} \quad \text{for } x = 0, 1, \ldots, n. \tag{4.6}$$

From (4.6) we have

$$P(X = 0) = (1 - P_b)^n, \quad \text{and} \tag{4.7}$$

$$P(X = 1) = nP_b(1 - P_b)^{n-1}. \tag{4.8}$$

Given by assumption (9), NMDAR can bind to only one pBS so that two events are possible: (1) NMDAR binds to one of the pBSs (E_1); and (2) NMDAR does not bind to any of the pBSs (E_2). We want to express the probabilities (4.7) and (4.8) conditional on $X \leq 1$ to give the probabilities of E_1 and E_2:

$$P(E_1) = P(X = 0 | X \leq 1)$$

$$= \frac{P(X = 1)}{P(X \leq 1)}$$

$$= \frac{nP_b (1 - P_b)^{n-1}}{(1 - P_b)^n + nP_b (1 - P_b)^{n-1}}$$

$$= \frac{nP_b}{(n - 1) P_b + 1}, \quad \text{and} \tag{4.9}$$

$$P(E_2) = 1 - P(E_1) = \frac{1 - P_b}{(n - 1)P_b + 1}. \tag{4.10}$$

4.2.3.3. *Calculation of Ω*

The transition from EiPSD to CaMKIIS occurs when NMDAR binds to one of the n CaM-pBSs of EiPSD, but none of other subunits at the same time (EiPSD has n CaM-pBS and m T286-pBS). We know that there are three ISs for an unbound CaMKII and iCaMKII subunit is not a pBS, so that we only need to know the probability of which NMDAR binds to one of the n CaM-pBSs of EiPSD, but none of the m T286-pBSs at the same time. Given the assumption of independent pBSs stated in the previous section, the probability of the transition to CaMKIIS, $P_{CaMKIIS}$, is given by

$$P_{CaMKIIS} = (P_{\text{CaM}-\text{pBS}} \cap P_{\overline{\text{T286}-\text{pBS}}})$$

$$= \frac{1 - P_b}{(m-1)\,P_b + 1} \cdot \frac{nP_b}{(n-1)\,P_b + 1}. \qquad (4.11)$$

Considering an elementary EiPSD with 1 CaM-pBS and 0 T286-pBS, the probability of which the elementary EiPSD transits to CaMKIIS is P_b according to Eq. (4.11). If the basal transition rate constant of the elementary EiPSD is k_{3f}, then the rate constant of a given conformation of CaMKII can be expressed by the product of k_{3f} and a coefficient. The coefficient reflects the magnitude of the difference between the rate constant of the elementary EiPSD and that of the given conformation. The coefficient can be expressed as the ratio of their probabilities to transit to CaMKIIS as follows:

$$\frac{P_{CaMKIIS}}{P_b} = \frac{1 - P_b}{(m-1)P_b + 1} \cdot \frac{n}{(n-1)P_b + 1}. \qquad (4.12)$$

A comprehensive model with 91 kinetic parameters may be used to model the exact dynamics of transition to CaMKIIS. As discussed previously, we use the average transition rate constant calculated by taking the mean of the 91 transition rate constants. We define Ω_1 as the modifier governing the average rate constant given by

$$\Omega_1 = \sum_{n=1}^{12} \sum_{m=0}^{12-n} \binom{12}{n} \binom{12-n}{m} \Gamma_2^n \Gamma_3^m \Gamma_1^{12-m-n}$$

$$\times \frac{1 - P_b}{(m-1)P_b + 1} \cdot \frac{n}{(n-1)P_b + 1}, \qquad (4.13)$$

where $\binom{12}{n}\binom{12-n}{m}\Gamma_2^n \Gamma_3^m \Gamma_1^{12-m-n}$ is the probability of a conformation (Eq. (4.4)), $\frac{1-P_b}{(m-1)P_b+1} \cdot \frac{n}{(n-1)P_b+1}$ is the coefficient of the conformation for the rate constant (Eq. (4.12)) and sum of their products gives the average coefficient. The average rate constant of the 91 conformations is $k_{3f} \cdot \Omega_1$.

Similarly, we define Ω_2 as the coefficient governing the average rate constant of transition to CaMKII*S. But, the elementary CaMKII in this case is the conformation with 1 T286-pBS and 0 CaM-pBS. Ω_2 is given by

$$\Omega_2 = \sum_{m=1}^{12} \sum_{n=0}^{12-m} \binom{12}{m} \binom{12-m}{n} \Gamma_2^n \Gamma_3^m \Gamma_1^{12-m-n}$$

$$\times \frac{1-P_b}{(n-1)P_b+1} \cdot \frac{m}{(m-1)P_b+1}. \tag{4.14}$$

4.2.4. *IST of CaMKII*

ISTs of CaMKII are modelled based on Chiba *et al.*'s work [21], which is developed based on Holmes's work of the CaM dynamics [24]. The advantage of Holmes's approach is the excellent agreement to the experimental data [21] based on a simple structure. Holmes considers four mechanistic steps governing the formation of Ca^{2+}/CaM complex where each step has one Ca^{2+} ion attached to the CaM molecule (Fig. 4.2A), and priority of the binding is given to two Ca^{2+} sites on C-terminus of CaM [24]. The variables associated with the formation of Ca^{2+}/CaM complex are listed in Table 4.3. Using the level of Ca^{2+}/CaM complex as the input, Chiba *et al.* [21] model ISTs of CaMKII among four states, with CaM dissociated autophosphorylated subunits in addition to the ISs defined in MoHST. CaM dissociated autophosphorylated state is ignored according to assumption (6) (Fig. 4.2B).

Table 4.3. Definition of the states/variables associated with IST of CaMKII.

States (#)	Definition
Ca^{2+}	Intracellular calcium ion
CaCaM	One Ca^{2+} ion binds to CaM
Ca_2CaM	Two Ca^{2+} ions bind to CaM
Ca_3CaM	Three Ca^{2+} ions bind to CaM
Ca_4CaM	Ca^{2+}/CaM complex

The reaction rates, denoted by R_1–R_7 (unit: molecular number (#) per sec, abbreviated as $\#\,s^{-1}$), are given by Holmes [24] and Chiba *et al.* [21] and we reproduce them in the following equations:

(1) Ca^{2+} attaches to CaM to form CaCaM [24],

$$R_1 = k_{6f} Ca^{2+} \cdot CaM - k_{6b} CaCaM. \qquad (4.15)$$

(2) Ca^{2+} attaches to CaCaM to form Ca_2CaM [24],

$$R_2 = k_{7f} Ca^{2+} \cdot CaCaM - k_{7b} Ca_2CaM. \qquad (4.16)$$

(3) Ca^{2+} attaches to Ca_2CaM to form Ca_3CaM [24],

$$R_3 = k_{8f} Ca^{2+} \cdot Ca_2CaM - k_{8b} Ca_3CaM. \qquad (4.17)$$

(4) Ca^{2+} attaches to Ca_3CaM to form Ca^{2+}/CaM complex [24],

$$R_4 = k_{9f} Ca^{2+} \cdot Ca_3CaM - k_{9b} Ca_4CaM. \qquad (4.18)$$

(5) We modified the equation for the rate of the binding of Ca^{2+}/CaM complex to iCaMKII. The original attempt is based on two pathways of the Ca^{2+}/CaM complex dissociation from the CaMKII subunit [21]. The pathway one has the Ca^{2+}/CaM complex to dissociate as a whole, and the pathway two has Ca^{2+} to dissociate first from the Ca^{2+}/CaM complex following by the dissociate of the remaining CaM from the CaMKII subunit [25]. The Ca^{2+} level is the deciding factor for the percentages of the dissociate by either of the two pathways and the rise of Ca^{2+} level causes more dissociation to undergo the pathway one. Therefore, the peak percentage of pathway two occurs at the rest low, which is around 0.07–0.11 μM in a normal hippocampal neuron. Using the parameter values of the original attempt, we find that pathway two takes account for only 1% of the dissociation at rest. We remove the pathway two and the rate

of the binding of Ca^{2+}/CaM complex to iCaMKII is given by

$$R_5 = k_{1f}Ca_4CaM \cdot iCaMKII - k_{1b}CaMCaMKII. \quad (4.19)$$

(6) Autophosphorylation of CaMCaMKII to form CaMCaMKII* [21],

$$R_6 = \frac{K_{\text{cat1}}CaMCaMKII \cdot P \cdot ATP}{K_{m1} + ATP}, \quad (4.20)$$

where P is defined by Chiba *et al.* [21] as "the probability that either or both sides of a neighbouring subunit are active" and is given by the expression, $1 - \Gamma_1^2$. ATP is a constant (the number of ATP in the synapse).

(7) Dephosphorylation of CaMKII* into CaMKIICaM by PP1 [21],

$$R_7 = \frac{K_{\text{cat2}}CaMCaMKII^* \cdot PP1}{K_{m2} + CaMCaMKII^*}, \quad (4.21)$$

where $PP1$ is a constant (the number of PP1 in the synapse).

The values of rate constants $(k_{1f}, k_{1b}, k_{6f}, k_{6b}, k_{7f}, k_{7b}, k_{8f}, k_{8b}, k_{9f}, k_{9b}, K_{\text{cat1}}, K_{\text{cat2}})$ and Michaelis constants (K_{m1}, K_{m2}) are obtained from the original model at 37°C [21] and converted into the unit of molecular numbers using protocol in Sec. 4.2.4.1 (the values of parameters are shown in Table 4.5). Other conditions, such as the total numbers of CaM and CaMKII, are modified to reflect the hippocampal synapses; these are discussed in Sec. 4.3.

4.2.4.1. *Unit conversion to molecular numbers*

Here, we consider a protein, x, involving in two types of reactions governing by two rate laws: mass action rate law (first-order reaction and second-order reaction to interact with another protein, y) and Michaelis–Menten rate law (a catalytic reaction by an enzyme, z).

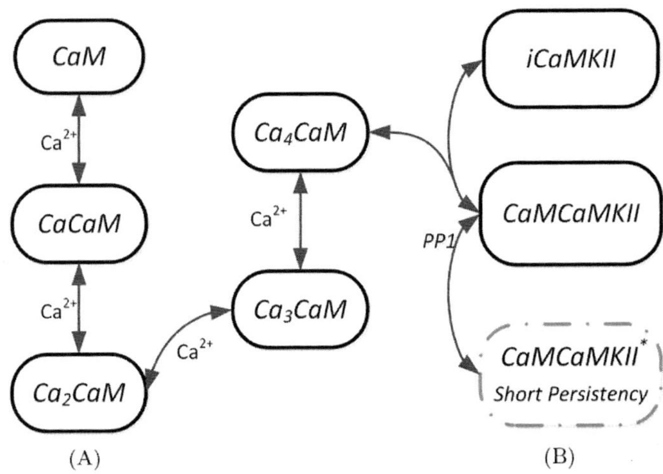

Fig. 4.2. Ca^{2+}–CaM interactions and the ISTs. (A) Ca^{2+} attaches to all four Ca^{2+} binding sites of CaM by four sequential steps [24]. (B) An inhibited CaMKII subunit (iCaMKII) binds to a Ca^{2+}/CaM complex (Ca$_4$CaM) to form a Ca$_4$CaM bound CaMKII subunit (CaMCaMKII). A CaMCaMKII is autophosphorylated by its active neighbour to form an autophosphorylated CaMKII subunit (CaMCaMKII*). Both CaMCaMKII and CaMCaMKII* are active CaMKII subunits, but CaMCaMKII* has a prolonged activity for up to 1 min after the disappearance of signal (Ca^{2+}) [26]. Dephosphorylating CaMCaMKII* by PP1 reverts CaMCaMKII* into CaMCaMKII. When Ca$_4$CaM dissociates CaMCaMKII, the CaMKII subunit becomes inactive (iCaMKII).

Concentration x, $[x]$, changes according to ODE in (4.22):

$$\frac{d[x]}{dt} = -k_1[x] - k_2[x][y] - \frac{K_{\text{cat}}[z][x]}{K_m + [x]}, \qquad (4.22)$$

where k_1 and k_2 are reaction rate constants for the first- and second-order reactions, respectively, K_{cat} and K_m are the turnover rate constant and the Michaelis constant, respectively, for the catalytic reaction. Here, we need to convert the units of these parameters, which is based on concentrations, into molecular numbers.

The relationship between the molecular number of x, $\#x$, and its concentration is stated as follows:

$$\#x = [x] \cdot N_A \cdot V, \qquad (4.23)$$

where N_A is Avogadro constant whose value is $6.02214179 \times 10^{23} \, \text{mol}^{-1}$. Equation (4.22) is rewritten in terms of molecular numbers:

$$\frac{d\left(\frac{\#x}{N_AV}\right)}{dt} = -k_1\left(\frac{\#x}{N_AV}\right) - k_2\left(\frac{\#x}{N_AV}\right)\left(\frac{\#y}{N_AV}\right) - \frac{K_{\text{cat}}\left(\frac{\#z}{N_AV}\right)\left(\frac{\#x}{N_AV}\right)}{K_m + \left(\frac{\#x}{N_AV}\right)}. \tag{4.24}$$

where $\#y$ and $\#z$ are molecular numbers of y and z, respectively. Multiplying both sides of (4.24) by N_AV, and with simplification, Eq. (4.24) becomes:

$$\frac{d\#x}{dt} = -k_1\#x - \frac{k_2}{N_AV}\#x\#y - \frac{K_{\text{cat}}\#z\#x}{K_mN_AV + \#x}. \tag{4.25}$$

Let $k_{2_{\text{new}}} = k_2/(N_AV)$, $K_{m_{\text{new}}} = K_mN_AV$, then the ODE of $\#x$ is given by

$$\frac{d\#x}{dt} = -k_1\#x - k_{2_{\text{new}}}\#x\#y - \frac{K_{\text{cat}}\#z\#x}{K_{m_{\text{new}}} + \#x}. \tag{4.26}$$

4.2.5. *A new model of HST of CaMKII*

In this section, we illustrate the development of the HSTs of MoHST (R_8–R_{19}, unit: $\#\,\text{s}^{-1}$). The schematic diagram of MoHST is shown in Fig. 4.3.

4.2.5.1. *Translocation of CaMKII into PSD*

The attachment of CaM to the CaM footprint controls the translocation of CaMKII into PSD, while the autophosphorylation delays the dissociation of CaMKII out of PSD [18]. But, the status of CaM attachment in a single CaMKII holoenzyme is difficult to model. In MoHST, we consider the active fraction of the CaMKII subunit ($\Gamma_2 + \Gamma_3$) as the control factor for the translocation of CaMKII into PSD (both active ISs have a CaM attached due to assumption (6)). We also consider the fraction of the autophosphorylated CaMKII

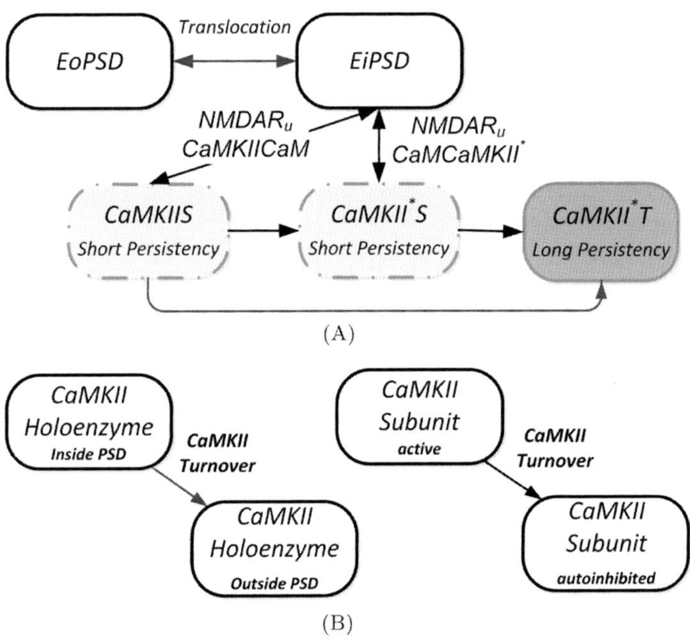

Fig. 4.3. The HSTs. (A) EoPSD is translocated into PSD and then, EiPSD binds to NMDAR through S site mediated T site binding [11], where a reversible binding is established first at the S site (CaMKIIS and CaMKII*S) and then is transferred to a long persisting binding at the T site (CaMKII*T). CaMKIIS can be phosphorylated to become CaMKII*S, which has a slower dissociation rate from NMDAR binding. (B) CaMKII turnover is modelled as a simple conversion process that CaMKII inside of PSD is converted into CaMKII outside of PSD. Similarly, active CaMKII subunits are converted into inhibited CaMKII subunits.

subunit (Γ_3) as the control factor for the dissociation of CaMKII out of PSD. The controls are formulated based on Hill functions and the reaction rates of the translocation (V_1) and the dissociation (V_2) are given, respectively, by

$$V_1 = V_{\max 1} \frac{\Gamma_2 + \Gamma_3}{K_{m3} + \Gamma_2 + \Gamma_3}, \quad \text{and} \tag{4.27}$$

$$V_2 = V_{\max 2} \frac{K_{m4}}{K_{m4} + \Gamma_3}, \tag{4.28}$$

where $V_{\max 1}$ is the max translocation rate and $V_{\max 2}$ is the max dissociation rate. According to mass action law, the max translocation rate is $k_{2f} \cdot EoPSD$ (if K_{m3} is very small), where k_{2f} is the translocation rate constant, and the max translocation rate occurs when the subunits of EoPSD are all active. Similarly, the max dissociation rate is $k_{2b} \cdot EiPSD$ (if K_{m4} is very small), where k_{2b} is the dissociation rate constant, and the max dissociation occurs when none of the subunits of EiPSD are autophosphorylated. Hence, the reaction rate (R_8) of the CaMKII translocation is given by

$$R_8 = k_{2f}EoPSD\frac{\Gamma_2+\Gamma_3}{K_{m3}+\Gamma_2+\Gamma_3} - k_{2b}EiPSD\frac{K_{m4}}{K_{m4}+\Gamma_3}, \quad (4.29)$$

where K_{m3} is the dissociation constant indicating the value of $\Gamma_2+\Gamma_3$ at which the translocation rate is half of its maximum and K_{m4} is the dissociation constant indicating the value of Γ_3 at which the dissociation rate is half of its maximum.

4.2.5.2. *Formation of CaMKII-NMDAR complex*

First, EiPSD attaches the S site of a subunit to NMDAR forming CaMKIIS or CaMKII*S if the subunit is a CaM-pBS or a T286-pBS, respectively (Fig. 4.3A). As discussed in Sec. 4.1.3.3, the average rate constants are $k_{3f} \cdot \Omega_1$ and $k_{3f} \cdot \Omega_2$ for the formation of CaMKIIS and CaMKIIS*S, respectively. The corresponding reaction rates for the formation of CaMKIIS (R_9) or CaMKII*S (R_{10}) are given by

$$R_9 = k_{3f}\Omega_1 EiPSD \cdot NMDAR_u - k_{3ba}CaMKIIS, \quad (4.30)$$

$$R_{10} = k_{3f}\Omega_2 EiPSD \cdot NMDAR_u - k_{3bb}CaMKII^*S, \quad (4.31)$$

where k_{3f} is the basal rate constant based on an elementary CaMKII (we assume the basal rate constant is the same between CaMKIIS and CaMKII*S), $k_{3b}s$ are the dissociation rate constants of the binding between CaMKII and NMDAR (k_{3ba} for CaMKIIS, k_{3bb} for CaMKII*S), and NMDAR$_u$ is the number of unbound NMDAR.

Next, the S site binding is transferred to the T site (Fig. 4.3A). We model the transfer as first-order irreversible reactions since the T site binding lasts throughout our simulation, i.e. 30 min. The rates of the transfer from CaMKIIS (R_{11}) and CaMKII*S (R_{12}) are given by

$$R_{11} = k_4 CaMKIIS, \quad \text{and} \tag{4.32}$$

$$R_{12} = k_4 CaMKII^*S, \tag{4.33}$$

where k_4 is the rate constant of the transfer. Again, we assume that the transfer rate constant is the same between CaMKIIS and CaMKII*S.

The transition from CaMKIIS to CaMKII*S occurs through autophosphorylation, the same process as in IST (Eq. (4.21)). Based on the same format, the rate of autophosphorylating CaMKIIS into CaMKII*S (R_{13}) is

$$R_{13} = \frac{K_{\text{cat1}} CaMKIIS \cdot P \cdot ATP}{K_{m1} + ATP}. \tag{4.34}$$

4.2.5.3. *CaMKII turnover*

Turnover replaces the old CaMKII with newly synthesised CaMKII [27]. Two concurrent processes underlie the turnover: the synthesis of new CaMKII and the degradation of old CaMKII. The newly synthesised CaMKII must be in the inhibited states. Hence, we consider turnover as the conversion of active CaMKII into inhibited states since the total number is conserved due to assumption (3) (Fig. 4.3B). The active HSs are EiPSD, CaMKIIS, CaMKII*S and CaMKII*T which are converted into EoPSD. The active ISs are CaMCaMKII and CaMCaMKII*, are converted into iCaMKII. The turnovers are modelled as a set of first-order reactions with rates (R_{14})–(R_{19}) at a constant rate constant, k_5, given by

$$R_{14} = k_5 CaMCaMKII, \tag{4.35}$$

$$R_{15} = k_5 CaMCaMKII^*, \tag{4.36}$$

$$R_{16} = k_5 EiPSD, \tag{4.37}$$

$$R_{17} = k_5 CaMKIIS, \tag{4.38}$$

$$R_{18} = k_5 CaMKII^*S, \quad \text{and} \tag{4.39}$$

$$R_{19} = k_5 CaMKII^*T. \tag{4.40}$$

4.2.6. *Complete model and constraints*

ODEs governing the transitions among states of CaMKII (Eqs. (4.33)–(4.46)) are formulated based on rates of ISTs and HSTs as given by $R_1 - R_{19}$:

(1) Ca^{2+}/CaM complex formation:

$$\frac{dCaCaM}{dt} = R_1 - R_2, \tag{4.41}$$

$$\frac{dCa_2CaM}{dt} = R_2 - R_3, \tag{4.42}$$

$$\frac{dCa_3CaM}{dt} = R_3 - R_4, \quad \text{and} \tag{4.43}$$

$$\frac{dCa_4CaM}{dt} = R_4 - R_5. \tag{4.44}$$

(2) IS changes:

$$\frac{dCaMCaMKII}{dt} = R_5 - R_6 + R_7 - R_9 - R_{14}, \quad \text{and} \tag{4.45}$$

$$\frac{dCaMCaMKII^*}{dt} = R_6 - R_7 - R_{10} - R_{15}. \tag{4.46}$$

(3) HS changes:

$$\frac{dEiPSD}{dt} = R_8 - R_9 - R_{10} - R_{16}, \tag{4.47}$$

$$\frac{dCaMKIIS}{dt} = R_9 - R_{11} - R_{13} - R_{17}, \tag{4.48}$$

$$\frac{dCaMKII^*S}{dt} = R_{10} - R_{12} + R_{13} - R_{18}, \quad \text{and} \tag{4.49}$$

$$\frac{dCaMKII^*T}{dt} = R_{11} + R_{12} - R_{19}. \tag{4.50}$$

(4) Constraints due to assumption (3):

$$iCaMKII = CaMKII_{\text{Total}} - CaMCaMKII$$
$$- CaMCaMKII^* - CaMKIIS$$
$$- CaMKII^*S - CaMKII^*T, \tag{4.51}$$
$$CaM = CaM_{\text{Total}} - CaCaM - Ca_2CaM$$
$$- Ca_3CaM - Ca_4CaM$$
$$- (CaMKII_{\text{Total}} - iCaMKII), \tag{4.52}$$
$$NMDAR_u = NMDAR_{\text{Total}} - CaMKIIS - CaMKII^*S$$
$$- CaMKII^*T, \quad \text{and} \tag{4.53}$$
$$EoPSD = CaMKII_{\text{Total}}/12 - EiPSD -$$
$$\times (NMDAR_{\text{Total}} - NMDAR). \tag{4.54}$$

The biological meanings of parameters and constants are summarised in Table 4.4. The values of parameters and the constants used in MoHST are discussed in Sec. 4.3.2.

4.3. Simulations with MoHST

In this section, we discuss the procedure of simulating MoHST, including the generation of Ca^{2+} signals at different frequencies, the parameter estimation according to experimental data, and conducting meaningful computational experiments.

4.3.1. *Input: Generation of Ca^{2+}*

We test MoHST with three Ca^{2+} patterns: (1) steady Ca^{2+} levels, (2) Ca^{2+} waves following prolonged tetanus at a low frequency, and (3) Ca^{2+} waves following a brief tetanus at high-frequency.

    The steady Ca^{2+} levels investigate the steady-state behaviour of the system at rest level $(0.07\,\mu\text{M})$ or following a period of gentle stimulation. Prolonged low frequency stimulation $(1\,\text{Hz})$ is a common protocol for LTD induction [28, 29] and we use it as the contrast with the brief high-frequency stimulation $(100\,\text{Hz in}$

Table 4.4. Biological meanings of parameters and constants.

Parameters	Biological meaning
k_{1f}	Binding rate constant of Ca_4CaM and iCaMKII
k_{1b}	Dissociation rate constant of CaMCaMKII into Ca_4CaM and iCaMKII
k_{2f}	Translocation rate constant of EoPSD into PSD
k_{2b}	Dissociation rate constant of EiPSD out of PSD
k_{3f}	Binding rate constant of elementary CaMKII to NMDAR at the S site
k_{3ba}	Dissociation rate constant of CaMKIIS into EiPSD and NMDAR
k_{3bb}	Dissociation rate constant of CaMKII*S into EiPSD and NMDAR
k_4	Transfer rate constant of NMDAR binding from S site to T site
k_5	Turnover rate constant of CaMKII
k_{6f}	Binding rate constant of Ca^{2+} and CaM to form CaCaM
k_{6b}	Dissociation rate constant of CaCaM into Ca^{2+} and CaM
k_{7f}	Binding rate constant of Ca^{2+} and CaCaM to form Ca_2CaM
k_{7b}	Dissociation rate constant of Ca_2CaM into Ca^{2+} and CaCaM
k_{8f}	Binding rate constant of Ca^{2+} and CaCaM to form Ca_3CaM
k_{8b}	Dissociation rate constant of Ca_3CaM into Ca^{2+} and Ca_2CaM
k_{9f}	Binding rate constant of Ca^{2+} and CaCaM to form Ca_4CaM
k_{9b}	Dissociation rate constant of Ca_4CaM into Ca^{2+} and Ca_3CaM
K_{cat1}	Autophosphorylation rate of CaMCaMKII into CaMCaMKII*
K_{cat2}	Dephosphorylation rate of CaMCaMKII* by PP1
K_{m1}	Michaelis constant of the autophosphorylation
K_{m2}	Michaelis constant of the dephosphorylation by PP1
K_{m3}	Fraction of active CaMKII subunits at half maximal translocation rate
K_{m4}	Fraction of autophosphorylated CaMKII subunits at half maximal dissociation rate

Constants	Biological meaning
$CaMKII_{Total}$	Total number of CaMKII subunits
CaM_{Total}	Total number of CaM molecule
$NMDAR_{Total}$	Total number of NMDAR (or NR2B)
ATP	Total number of ATP
PP1	Total number of PP1

Table 4.5. Ranges of the parameter for MCMC.

Parameter	Low	High
k_{2f}	$0.00625\,\mathrm{s}^{-1}$	$0.075\,\mathrm{s}^{-1}$
k_{2b}	$0.01\,\mathrm{s}^{-1}$	$0.3\,\mathrm{s}^{-1}$
K_{m3}	0	0.5
K_{m4}	0	0.1
k_{3f}	$0.001\,\#^{-1}\mathrm{s}^{-1}$	$0.02\,\#^{-1}\mathrm{s}^{-1}$
k_{3ba}	$0.001\,\mathrm{s}^{-1}$	$0.5\,\mathrm{s}^{-1}$
k_{3bb}	$0.001\,\mathrm{s}^{-1}$	$0.1\,\mathrm{s}^{-1}$
k_4	$0.001\,\mathrm{s}^{-1}$	$0.02\,\mathrm{s}^{-1}$

this study), which induces LTP. The critical system behaviours can be revealed through the comparison. The Ca^{2+} input is generated by a framework developed by Zhabotinsky [30] (see Chapter 3 for details).

4.3.2. *Estimation of parameters*

We use the estimated values of parameters that are available in the literature. But, the new HST model contains eight new parameters (k_{2f}, k_{2b}, K_{m3}, K_{m4}, k_{3f}, k_4, k_{3ba}, k_{3bb}) which have to be estimated. We estimate these parameters using Markov Chain Monte Carlo (MCMC) method [31, 32] by minimising the mean square error (MSE) between the model output and the experimentally established data [18]; both are normalised. The normalisation of is based on

$$\hat{x}_t = (x_t - x_{\min})/(x_{\max} - x_{\min}), \tag{4.55}$$

where x_t is the model output or experimental data at time, t, $x_{\min}$ and $x_{\max}$ are the corresponding minimum and maximum, respectively, over the period. MSE is calculated under the normalisation since the experimental data is presented as the normalisation of fluorescence intensity which lacks quantitative information in molecular numbers. In this case, the data suggests the trend of the

CaMKII translocation and we adapt to the normalisation to compare the trends of the prediction and real system.

The parameters are estimated with the ranges given from available information in the literature (Table 4.5). In addition, we assume the ratio between alpha and beta isoforms is 1:1. The ranges are summarised as follows: (1) k_{2f}: it takes approximately 20 s (alpha isoform) and 80 s (alpha/beta isoforms 1:1), respectively, for half of the CaMKII to translocate into PSD [18], corresponding to $0.00625\,\mathrm{s}^{-1}$ and $0.075\,\mathrm{s}^{-1}$ (with 50% uncertainty) for k_{2f}; (2) k_{2b}: it takes approximately 5 s (T286A mutant) and 50 s (wild type), respectively for half of the CaMKII to dissociate out of PSD [18, 33], corresponding to $0.01\,\mathrm{s}^{-1}$ and $0.3\,\mathrm{s}^{-1}$ (with 50% uncertainty) for k_{2b}; (3) at least three subunits of a holoenzyme have a CaM attached according to binomial distribution (probability is 0.98) when the probability of CaM attaching to a single unit is 0.5. Assuming the binding priority is given to the beta isoform so that a half of the beta subunits are active. A half of the CaMKII holoenzymes are free from the F-actin binding allowing them to translocate. The range of K_{m3} is set between 0 and 0.5; (4) we assume the dissociation is delayed when at least one subunit is autophosphorylated on average, given a peak of 0.1 (slightly higher than 1/12) for K_{m4}; (5) k_{3ba} and k_{3bb} are given ranges between $0.001\,\mathrm{s}^{-1}$ to $0.5\,\mathrm{s}^{-1}$ and $0.001\,\mathrm{s}^{-1}$ to $0.1\,\mathrm{s}^{-1}$, respectively [11]; and (6) the dissociation constant (K_d) of the binding between the CaMKII and NMDAR is approximately 1# based on estimated K_d of $138 \pm 60\,\mathrm{nM}$ [34] and PSD volume of 0.01 fL (assumption 1). K_d of the elementary CaMKII (calculated by k_{3b}/k_{3f}) may be much higher given the multiple pBS enhancement of the binding. Therefore, the range of k_{3f} is set at $0.001\,\#^{-1}\,\mathrm{s}^{-1}$ to $0.02\,\#^{-1}\,\mathrm{s}^{-1}$, given the aforementioned range of k_{3b}.

The CaMKII holoenzyme has 100 molecules, corresponding to 1200 subunits altogether [35]. NMDAR has 20 molecules, corresponding to 20 NR2B subunits (assumption 9) [35, 36]. The concentration of CaM in the hippocampus is $17\,\mu\mathrm{M}$ on average [37], corresponding

Table 4.6. Parameters and constants: their biological meaning, values, and sources.

Parameter	Value	Source
k_{1f}	$0.035\,\#^{-1}\,\mathrm{s}^{-1}$	[21]
k_{1b}	$0.14\,\mathrm{s}^{-1}$	[21]
k_{2f}	$0.0088\,\mathrm{s}^{-1}$	MCMC estimated
k_{2b}	$0.247\,\mathrm{s}^{-1}$	MCMC estimated
k_{3f}	$0.008\,\#^{-1}\,\mathrm{s}^{-1}$	MCMC estimated
k_{3ba}	$0.38\,\mathrm{s}^{-1}$	MCMC estimated
k_{3bb}	$0.024\,\mathrm{s}^{-1}$	MCMC estimated
k_4	$0.01\,\mathrm{s}^{-1}$	MCMC estimated
k_5	$1/108000\,\mathrm{s}^{-1}$	[27]
k_{6f}	$0.0415\,\#^{-1}\,\mathrm{s}^{-1}$	[21]
k_{6b}	$50\,\mathrm{s}^{-1}$	[21]
k_{7f}	$1.45\,\#^{-1}\,\mathrm{s}^{-1}$	[21]
k_{7b}	$50\,\mathrm{s}^{-1}$	[21]
k_{8f}	$0.2\,\#^{-1}\,\mathrm{s}^{-1}$	[21]
k_{8b}	$1250\,\mathrm{s}^{-1}$	[21]
k_{9f}	$4.15\,\#^{-1}\,\mathrm{s}^{-1}$	[21]
k_{9b}	$1250\,\mathrm{s}^{-1}$	[21]
K_{cat1}	$0.9\,\mathrm{s}^{-1}$	[21]
K_{cat2}	$1.72\,\mathrm{s}^{-1}$	[21]
K_{m1}	$1150\,\#$	[21]
K_{m2}	$660\,\#$	[21]
K_{m3}	0.39	MCMC estimated
K_{m4}	0.019	MCMC estimated

Constant	Value ($\#$)	Source
$\mathrm{CaMKII}_{\mathrm{Total}}$	1200	[35]
$\mathrm{CaM}_{\mathrm{Total}}$	500	MCMC estimated
$\mathrm{NMDAR}_{\mathrm{Total}}$	20	[35, 36]
ATP	240000	[38]
PP1	145	MCMC estimated

to 1000 molecules by Eq. (4.23). However, CaM binds not only to CaMKII. 500 molecules of CaM are assumed to dedicate to CaMKII. The molecular number of PP1 is estimated by MCMC with a range 0 to 200. Table 4.6 summarised the values of parameters used in MoHST.

4.3.3. *Computational implementation and experiments with MoHST*

MoHST is solved using Matlab ode solver: ode15 s. We select two model outputs and set up computational experiments based on them to obtain meaningful insights into the behaviour of synaptic plasticity.

4.3.3.1. *Time courses of CaMKII translocation*

First, the steady levels of all variables at the rest Ca^{2+} level are obtained by running MoHST for a sufficiently long time. The steady levels are used as initial values of the variables for the rest computations. Then, Ca^{2+} is increased to a higher level for 600 s. As a result, a pulse of CaMKII translocated is induced to drive CaMKII into PSD. The selection of the higher Ca^{2+} level should induce the peak CaMKII translocation since we assume experimental data were measured at the peak CaMKII translocation. Our analysis shows that for Ca^{2+} level greater than 2.5 μM (especially greater than 5 μM), the translocation of CaMKII reaches the peak. We use 20 μM to match up with the peak Ca^{2+} level generated by high-frequency stimulation using the protocol in Sec. 4.2.3.1. At the end, Ca^{2+} is set back to the rest level to induce the dissociation of CaMKII out of PSD. We record the number of CaMKII in PSD, which is the sum of EiPSD, CaMKIIS, CaMKII* S, and CaMKII* T, during both translocation into and dissociation out of PSD, and normalise the numbers by Eq. (4.55). The results are shown in Fig. 4.4, which are consistent with experimental data [18].

4.3.3.2. *Formation of CaMKII-NMDAR complex*

Similar, initial values of parameters are obtained by running MoHST for a sufficiently long time. Then, a Ca^{2+} signal following a tetanus stimulation is applied and the time of the application is marked as $t = 0$. The duration of the signal is dependent on the number of pulses carried by tetanus and the frequency of which the pulses is applied. For example, 100 pulses at 1 Hz, 10 Hz and 100 Hz take 100 s,

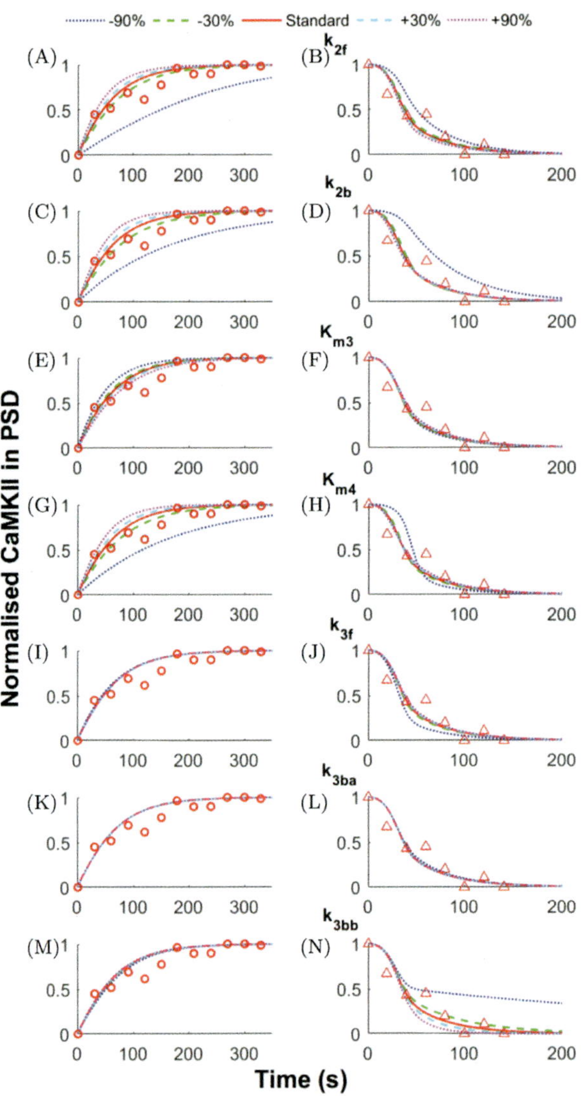

Fig. 4.4. Parameter sensitivity related to CaMKII translocation. The figures on the left-hand side (A, C, E, G, I, K, and M) show the changes in the time courses of CaMKII translocation into PSD and figures in the right-hand side (B, D, F, H, J, L, and N) show the changes in the time course of CaMKII dissociation from PSD. The solid lines show the time courses on standard values of the parameters while the dashed and dotted lines show those with 30% and 90% perturbations of the standard parameters, respectively. Circles and triangles are the corresponding experimental data retrieved from [18].

10 s, and 1 s, respectively. The simulation stops at 30 minutes after the disappearing of the Ca^{2+} signal, and the number of CaMKII-NMDAR complex (CaMKII*T) is recorded.

4.4. Parameter Perturbation

We want to understand the relationship between the new parameters related to HSTs (k_{2f}, k_{2b}, K_{m3}, K_{m4}, k_{3f}, k_4, k_{3ba}, and k_{3bb}) and the two model outputs. This relationship may help us to understand the significant factors of the CaMKII-NMDAR formation to gain insights into the induction of E-LTP. We achieve this objective by perturbing these new parameters through both local sensitivity analysis (LSA) and partial rank correlation coefficient (PRCC) [39]. Theoretically, a parameter, whose perturbation causes a much larger variation of the model outputs, has a prominent role in these model outputs. We perform only LSA on the CaMKII translocation and both LSA and PRCC on the formation of CaMKII-NMDAR complex.

4.4.1. *Methods of parameter perturbation*

4.4.1.1. *Local sensitivity analysis*

In LSA, one parameter is perturbed at a time, from 10% to 190% of its standard value listed in Table 4.6, and the other parameters are unchanged. The results are shown in Figs. 4.4 and 4.5 for the CaMKII translocation and the formation of CaMKII-NMDAR complex, respectively.

We quantify LSA on the formation of CaMKII-NMDAR complex through the perturbation of the jth parameter ($j = 1, 2, \ldots, 8$, each number corresponds to one of the eight new parameters), V_j, by (see [40])

$$V_j = \left| \frac{\left(C_j^{130\%} - C^{\text{Standard}}\right)/C^{\text{Standard}}}{\left(P_j^{130\%} - P_j^{\text{Standard}}\right)/P_j^{\text{Standard}}} \right|, \tag{4.56}$$

where P_j^{Standard} and $P_j^{130\%}$ are values of the jth parameter at the standard value or 30% increase from its standard value, respectively.

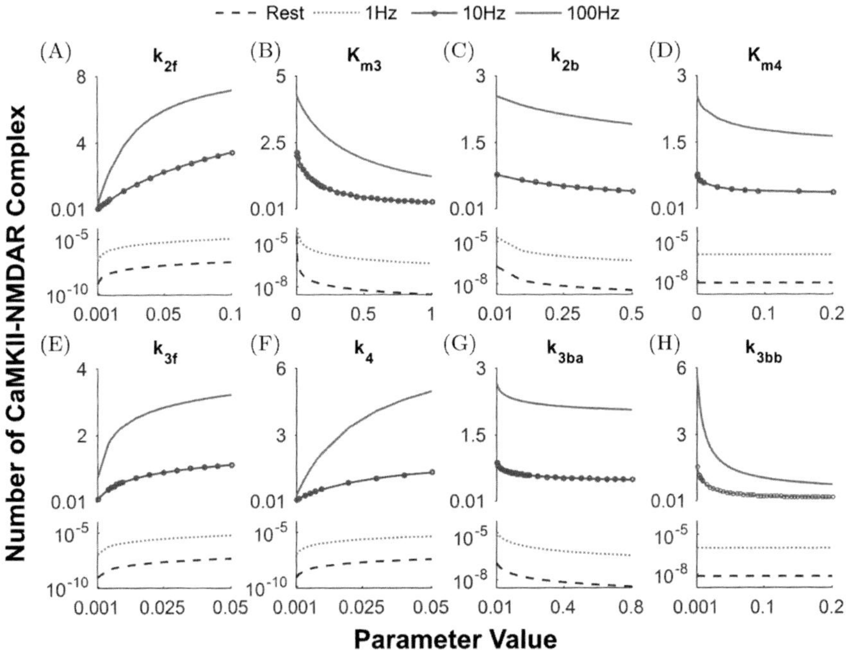

Fig. 4.5. Parameter sensitivity related to CaMKII-NMDAR complex formation in the wild-type CaMKII. Four stimulation patterns (rest, 1 Hz, 10 Hz, and 100 Hz) are tested for each parameter selected. The parameters are perturbed from 10% to 190% of its standard value and the results are the number of CaMKII-NMDAR complex at the end of the simulation (30 minutes after the end of the Ca^{2+} elevation). The lower panels are on the log scale and the upper panels are in the decimal scale. The parameters selected are: (A) k_{2f}, (B) K_{m3}, (C) k_{2b}, (D) K_{m4}, (E) k_{3f}, (F) k_4, (G) k_{3ba}, and (H) k_{3bb}.

C^{Standard} and $C_j^{130\%}$ are the numbers of CaMKII-NMDAR complex formed taking the jth parameter at P_j^{Standard} or $P_j^{130\%}$, respectively. A ranking of the eight parameters according to their V_j, from the highest to the lowest, is shown in Fig. 4.6.

4.4.1.2. *Partial rank correlation coefficient*

In PRCC, we simultaneously vary the new parameters and investigate the global effects caused by the simultaneous variation onto the model outputs. A total of 1000 sets of parameters, each set contains eight values corresponding to the eight new parameters,

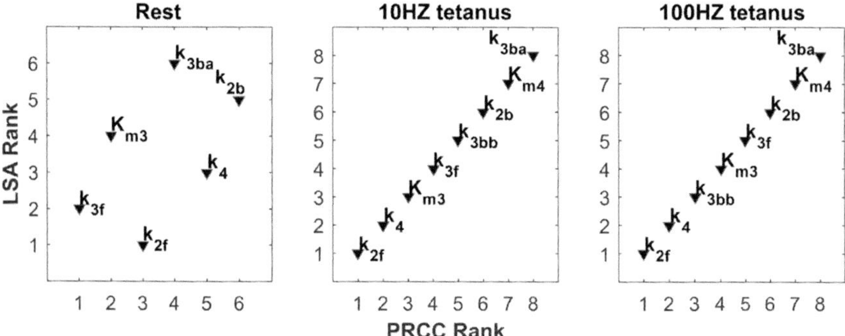

Fig. 4.6.   Parameter sensitivity rank. We rank the sensitivity of eight parameters $(k_{2f},\ k_{2b},\ K_{m3},\ K_{m4},\ k_{3f},\ k_4,\ k_{3ba},\ k_{3bb})$ to the formation of CaMKII-NMDAR complex by LSA and PRCC as discussed in Sec. 4.4.1 (lower rank denotes higher sensitivity). The insensitive parameters are removed based on the p-value from PRCC (0.05 significant level).

are generated using Latin hypercube sampling [41] from a predefined range 50–150% of their standard values. LHS gives simultaneously varied, evenly distributed and minimally correlated sets of parameters. MoHST is simulated using the 1000 sets as inputs, one set at a time, to replace the values of the eight new parameters while keeping other parameters at their standard values. For each simulation, the number of CaMKII-NMDAR complex formed is recorded; 1000 numbers are obtained corresponding to 1000 sets of parameters. Finally, we calculate PRCC of the number of CaMKII-NMDAR complex formed against the *jth* parameter ($j = 1$ to 8) and use the corresponding p-value to reject the null hypothesis where no relationship exists between the investigated output and the *jth* parameter. To investigate these parameters to see whether they have a significant relationship to the investigated output, we rank them based on PRCC$_j$, the absolute value of the PRCC, from the highest to the lowest (Fig. 4.6).

4.4.2. *Factors related to the formation of CaMKII-NMDAR complex*

Several studies suggest that LTP induction is strongly correlated to the stimulation frequency [29, 42] so that we first test the relationship between the stimulation frequency and the formation of CaMKII-NMDAR complex. Following experiments by Giese *et al.* [3], we simulate MoHST with four patterns of Ca^{2+} inputs, rest level at a steady 70 nM and tetanus with 100 pulses at 1 Hz (peak Ca^{2+} level at $0.4\,\mu M$), 10 Hz (peak at $2.5\,\mu M$) and 100 Hz (peak at $20\,\mu M$). The Ca^{2+} peaks are given by simulation using the protocol to generate Ca^{2+} input. Using the methods of parameter perturbation, we try to identify parameters, as well as the related processes, with a prominent role in triggering the formation of CaMKII-NMDAR complex.

The formation of CaMKII-NMDAR complex shows a relationship to the stimulation frequency (Fig. 4.5) and the relationship is consistent with the experimental observations [3]. No sign of the formation reveals at the rest and 1 Hz tetanus across the parameter regions we have analysed and the significant formation is shown at 10 Hz and 100 Hz tetanus within certain parameter regions in each case of the parameters. More importantly, the pattern of the formation displays significant differences between 10 Hz and 100 Hz tetanus.

Then, we rank the parameters related to the HSTs (Fig. 4.6) by both LSA and PRCC. The most sensitive parameters are k_{2f}, K_{m3}, k_{3f} and k_4 across the three patterns of Ca^{2+} inputs shown in the figure (1 Hz is not shown) and importantly, the two methods give consistent ranking at the high-frequency stimulations. Here, these sensitive parameters are related to the CaMKII translocation (k_{2f} and K_{m3}) and the CaMKII binding (k_{3f} and k_4) indicate that the two processes have the prominent roles in triggering the formation. Moreover, k_{3bb} and K_{m4}, which are related to the autophosphorylation, switch from insensitive at rest to very sensitive at high-frequency stimulations suggesting a significant contribution of the autophosphorylation to the formation following transient signals.

4.4.3. *Discussion and summary*

In this study, a theoretical model of HSTs of CaMKII (MoHST) is developed and carries several novel contributions: (1) the composition of CaMKII is expressed according to multinomial distribution based on the ratios of three possible conformations of CaMKII subunits over the total number of CaMKII subunits; (2) the first attempt to model the binding between CaMKII and NMDAR is based on a probability framework which is dynamically changing according to the composition of CaMKII; and (3) MoHST gives good agreement with experimental data in the time courses of the CaMKII translocation and the frequency dependence of the formation of CaMKII-NMDAR complex.

MoHST suggests two prominent factors governing the formation of CaMKII-NMDAR complex: the affinity of the binding between CaMKII and NMDAR and the rate of the CaMKII translocation into PSD (Fig. 4.6). Given the different translocation rates for CaMKII isoforms, where the translocation rate of CaMKIIα subunit is significantly faster than that of CaMKIIβ subunit [18, 43], MoHST suggests a potential mechanism to regulate the formation of CaMKII-NMDAR complex by controlling the ratios of the CaMKII isoforms which constitute the CaMKII holoenzyme. Moreover, there is evidence for the active regulation of the ratio between α and β isoforms of CaMKII in hippocampal neurons [43, 44].

The parameter perturbation suggests that the autophosphorylation may be critical to the formation of CaMKII-NMDAR complex, but only during transient signals. One possible explanation is that the translocation of CaMKII into PSD takes a relatively long time (Fig. 4.4), therefore, transient signals, which induce LTP, would only give a limited amount of CaMKII translocation. As a result, maintaining the translocated CaMKII in PSD increases the likelihood to form CaMKII-NMDAR complex. Given that the autophosphorylation delays the CaMKII dissociation out of PSD [18, 43], the autophosphorylation maintains CaMKII in PSD and prolongs the interaction between CaMKII and NMDAR during the transient signals. However, the level of the autophosphorylation is low at rest (data not shown), therefore more specific details of the autophosphorylation induced dynamics of CaMKII is needed and is discussed in the next chapter.

MoHST has a number of limitations: (1) 305/306 of CaMKII subunit can be autophosphorylated and is often called inhibitory autophosphorylation by blocking the activation of CaMKII. The inhibitory that autophosphorylation that accelerates that the CaMKII dissociation out of PSD [33, 45, 46] may regulate the formation of CaMKII-NMDAR complex, but the inhibitory autophosphorylation is not included in MoHST; (2) mechanisms other than the S site mediated T site binding, which is not specific to the autophosphorylation, are not tested by MoHST. Given that one of two CaMKII binding sites of NR2B depends on the autophosphorylation [17], there may be another autophosphorylation-dependent mechanism of the CaMKII-NMDAR binding; (3) other CaMKII binding partners in PSD [47], which may cause impact on the CaMKII translocation into PSD as well as the staying of CaMKII in PSD, are not considered in MoHST; (4) CaMKIIN, an inhibitor specific for CaMKII-NMDAR binding [8, 48], is not included in MoHST. But, CaMKIIN would restrict the formation of CaMKII-NMDAR complex and tighten the threshold for LTP induction; and (5) although the switching between CaMKII autoinhibited states is ignored in assumption (2), the switching needs to be considered especially when we study the CaMKII isoforms dependence of the formation of CaMKII-NMDAR complex.

4.5. Implication of the Autophosphorylation in LTP

There are disparate experimental evidence regarding the nature of the autophosphorylation in the formation of CaMKII-NMDAR complex as well as the linkage with the induction of E-LTP. On the one hand, the formation of CaMKII-NMDAR complex is not blocked by T286A mutants lacking the autophosphorylation [11, 17] or repressors of the autophosphorylation [49]. On the other hand, the strength of CaMKII-NMDAR binding decreases without the autophosphorylation [11] and the autophosphorylation is a critical component of LTP and spatial memory formation [3, 5], especially during E-LTP [2]. Furthermore, the autophosphorylation is irreversible in PSD [20] and the irreversibility may be essential for the induction of E-LTP. Overall, it is unclear for the nature of relationship among the autophosphorylation, the formation of CaMKII-NMDAR complex and the induction of E-LTP.

A better understanding of the autophosphorylation can be obtained by comparing the dynamic behaviours of the system between the wild-type and T286A mutant studies [3, 11, 26]. T286A mutant CaMKII subunit has the alanine residue at position 286, which replaces the threonine residue (T286 site) in normal CaMKII subunit [50]. As a result, the T286 site is no longer allowing the attaching of the phosphate group and thus the subunit lacks the autophosphorylation. We analyse MoHST regarding the relationship among the autophosphorylation, the formation of CaMKII-NMDAR complex and the induction of E-LTP by investigating the differences in CaMKII dynamics, specifically the CaMKII translocation time course and the formation of CaMKII-NMDAR complex, between wide-type and T286A mutant CaMKII.

4.5.1. *Computational experiments*

We set up two sets of computational experiments as described in Sec. 4.3.3 and in each set, we simulated twice, one by the wild-type CaMKII and one by T286A mutant, with the same pattern of Ca^{2+} input. The computational study of T286A mutant is performed by setting ATP level to be 0, which makes the autophosphorylation rate to be 0 according to Eq. (4.20). We compare the CaMKII dynamics between wild-type and T286A mutant in terms of (1) the CaMKII translocation, and (2) the formation of CaMKII-NMDAR complex. We then comment on the difference to understand its biological meaning.

4.5.2. *Role of the autophosphorylation related to CaMKII translocation*

The simulated time courses of CaMKII translocation and dissociation are consistent with experimental data. A good agreement in the time course of the CaMKII translocation is shown in Fig. 4.7A between the simulation and experimental data for wild type, but no T286A mutant data was found; reasonable agreements in the time courses of the CaMKII dissociation are shown in Fig. 4.7C for both wild-type and T286A mutant [18, 33]. In addition, the complete CaM dissociation from CaMKII subunit takes approximately 1 min in wild

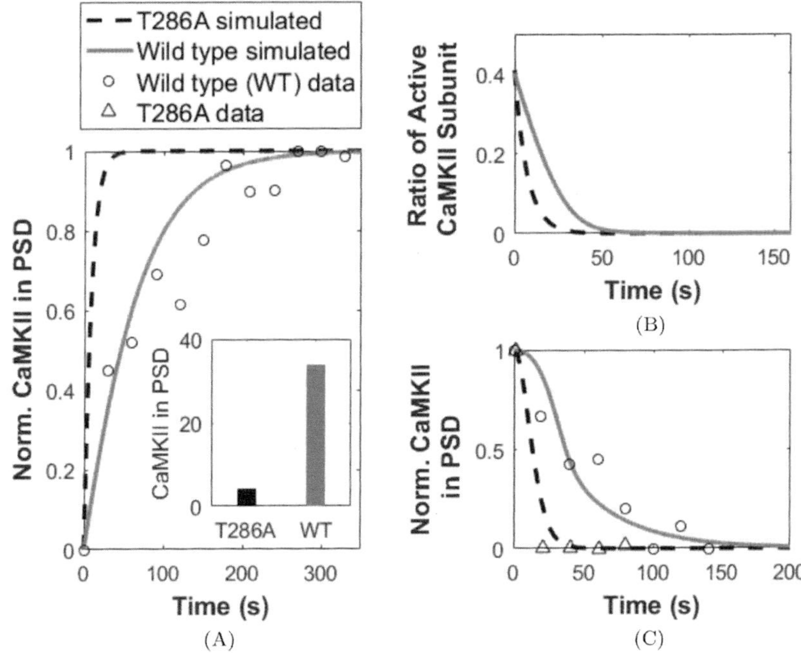

Fig. 4.7. The implication of the autophosphorylation in the CaMKII translocation into PSD. (A) The time course of CaMKII translocation into PSD. Inset: peak
number of CaMKII in PSD. (B) Time course of the CaM dissociation from the
CaMKII subunit. The level of CaM-bound (active) subunit is shown as a ratio
of the total number of CaMKII subunit. (C) The time course of the CaMKII
dissociation from PSD. Circles and triangles are the corresponding experimental
data for the time courses of CaMKII [18].

type (Fig. 4.7B) and which is consistent with the time duration
reported by the experimental study [26].

The autophosphorylation delays the CaMKII inactivation and
the CaMKII dissociation as observed in experiments [18, 26, 33].
As a result, a much smaller peak translocation is shown in T286A
mutant (Fig. 4.7A inset) and a less time is taken for the complete
translocation of T286A mutant into PSD (Fig. 4.7A). Moreover, the
T286A mutant dissociates much faster from PSD (Fig. 4.7C) that
may be a consequence from the faster CaM dissociation from CaMKII
subunits (Fig. 4.7B).

4.5.3. *Role of the autophosphorylation related to the formation of CaMKII-NMDAR complex*

4.5.3.1. *Frequency dependence*

The relationship between CaMKII-NMDAR complex formation and the stimulation frequency, as shown in Sec. 4.4.2, has impaired by T286A mutant. The difference in the number of CaMKII-NMDAR complex formed between 10 Hz and 100 Hz tetanus is removed by T286A mutant (Fig. 4.8). Now, 1–10% of the wild-type differences remain in T286A mutant across the 1000 sets of parameters generated (Fig. 4.8I) suggesting a significant inability of T286A

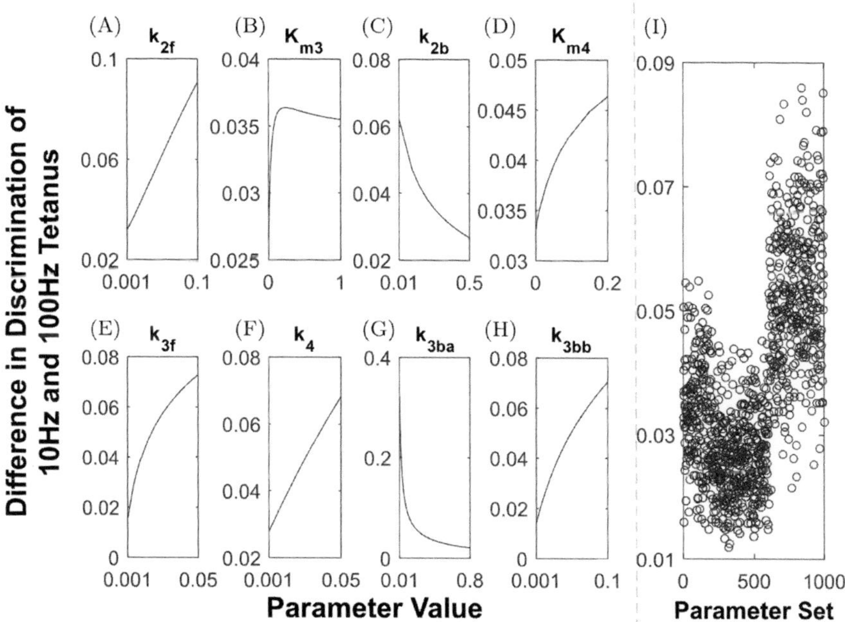

Fig. 4.8. Comparison of the capabilities in discrimination of 10 Hz and 100 Hz tetanus between T286A mutant and the wild-type CaMKII. The differences between the formation of CaMKII-NMDAR complex in response to 10 Hz and 100 Hz tetanus are calculated for both T286A mutant and the wild-type CaMKII. The results are presented as a quotient of dividing the difference for the T286A mutant by the difference for the wild type. Parameters are perturbed to search for the "entire" possible space of the quotient based on (1) LSA (A–H), and (2) LHS (I).

mutant in discriminating the stimulation frequency. Moreover, the deficit predicted is consistent with the impairment in LTP observed from T286 mutant mice (see [3, Fig. 2C]).

4.5.3.2. *The autophosphorylation in response to a single tetanus*

Single tetanus triggers similar peak numbers of active CaMKII subunits between wild-type and T286A mutant, but the CaMKII inactivation is much slower in wild-type due to the aforementioned function of the autophosphorylation (Fig. 4.9A). This small difference of active CaMKII subunits causes a significant difference in the downstream CaMKII-NMDAR complex formation between wild-type and T286A mutant (Fig. 4.9B). First, there is a phase shift between the CaMKII dynamics in wild-type and T286A mutant (Fig. 4.9B inset), which is very similar to the experimental observation (see, [11, Fig. 2B]). Secondly, the peak number of

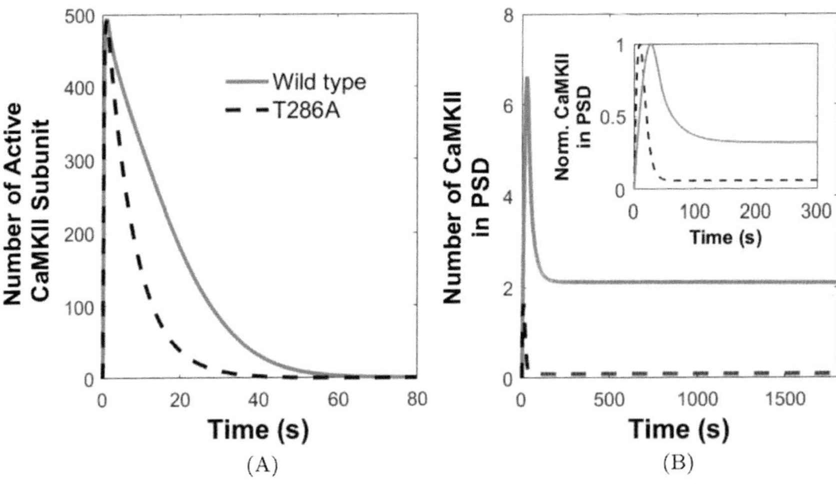

Fig. 4.9. The implication of the autophosphorylation in response to single tetanus. A single tetanus of 100 pulses at 100 Hz is applied. MoHST is simulated for 30 min for both T286A mutant and the wild type CaMKII. (A) The number of active CaMKII subunits over the first 100 s. (B) The CaMKII translocation and the formation of CaMKII-NMDAR complex. Inset: normalised CaMKII in PSD by the range over the first 300 s of the simulation.

CaMKII translocated into PSD is smaller (seven in wild type vs. four in T286A mutant) and less time is taken in T286A mutant to reach the peak translocation. Thirdly, the dissociation of CaMKII out of PSD is much faster in T286A mutant. And lastly, the number of CaMKII-NMDAR complex formed decreases significantly in T286A mutant. In conclusion, the postsynaptic response indicated here by the formation of CaMKII-NMDAR complex is amplified by the autophosphorylation in response to single tetanus.

4.5.3.3. *The autophosphorylation in response to multiple trains of tetanus*

In wild type, multiple trains of tetanus maintain active CaMKII subunits at a high level (Fig. 4.10A), enhance significantly the CaMKII translocation into PSD (Fig. 4.10B) and increase the formation of CaMKII-NMDAR complex (Fig. 4.10C). The limited number of NMDAR in PSD sets an upper limit of the increase of the CaMKII-NMDAR complex formation due to multiple trains. In T286A mutant, significant reductions in the CaMKII translocation (28 over 20 trains in wild type vs. 5 in T286A mutant) and the CaMKII-NMDAR complex formation (18 over 20 trains in wild type vs. 2.5 in T286A mutant) are shown (Figs. 4.10B and 4.10C).

There is a range of the inter-train interval in which successive tetanus receives a greater formation than that of completely separated tetanus (Fig. 4.10D). The sustained CaMKII-NMDAR complex formation requires different intervals between wild type (66 s, from 22 s to 88 s) and T286A mutant (32 s, from 10 s to 42 s). For intervals smaller than the given range, a significant reduction in the formation is shown since the trains arrive before the previous train is completely expressed. Comparing wild-type and T286A mutant, the response-coupling among multiple trains is improved in wild type since a wider range of intervals is allowed for the response-coupling. But no experimental evidence exists for us to verify these intervals, but E-LTP is generally induced with 10–20 s inter-train interval [51].

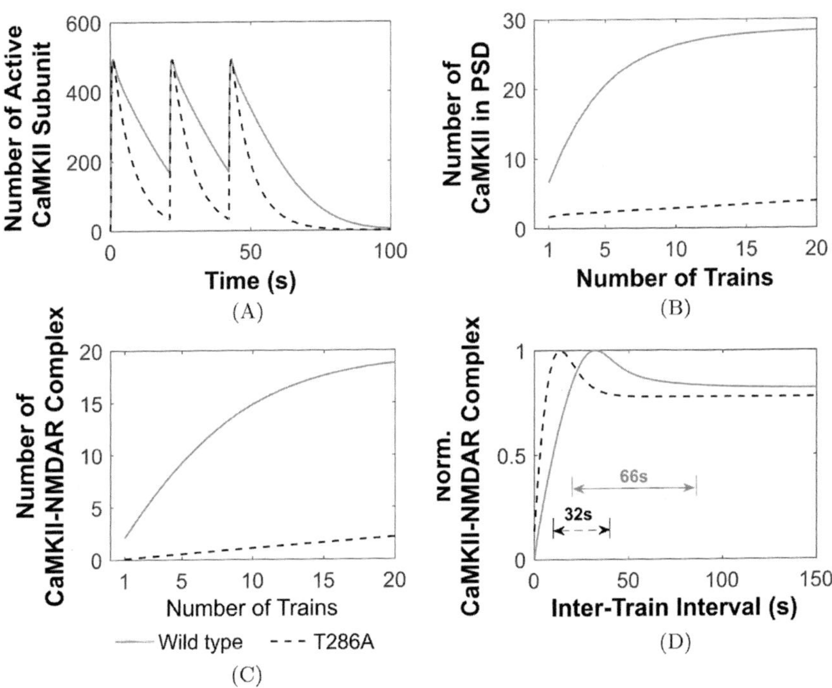

Fig. 4.10. The implication of the autophosphorylation in response to multiple trains of tetanus. (A) The number of active CaMKII subunits over the simulation of 100 s in response to three trains of tetanus with an inter-train interval of 20 s. Each tetanus lasts for 1 s and contains 100 pulses at 100 Hz. (B) The peak translocation of CaMKII in response to different numbers of tetanus with 20 s inter-train interval. (C) The formation of CaMKII-NMDAR complex in response to different numbers of tetanus with 20 s inter-train interval. The number of CaMKII-NMDAR complex presented is measured at the end of the simulation (simulation continues for 30 min after the signal is disappeared). (D) The normalised number of CaMKII-NMDAR complex in response to three trains of tetanus with different inter-train intervals. The number of CaMKII-NMDAR complex presented is recorded at the end of the simulation (simulation continues for 30 min after the signal is disappeared) and normalised by the maximum CaMKII-NMDAR complex formed among the different inter-train intervals. Arrows show the range of inter-train intervals in which the formation of CaMKII-NMDAR complex is greater than that of three completely separated trains (CaMKII-NMDAR complex formed in response to three completely separated trains is recorded at an inter-train interval of 150 s. A 1% up-adjustment is added to the record to balance the CaMKII turnover for the additional signal duration).

4.5.4. *Discussion and summary*

We analyse MoHST to gain insights into the relationship among the autophosphorylation, the formation of CaMKII-NMDAR complex and the induction of E-LTP by comparing behavioural differences between wild-type and T286A mutant on two of model outputs selected: the CaMKII translocation and the CaMKII-NMDAR complex formation.

MoHST shows good agreement of the time courses of CaMKII dissociation mutant against the experimental data (Fig. 4.7). In addition, MoHST predicts a deficit for T286A mutant to distinguish stimulation frequency, which yields consistency with observations in experiments (see [3, Figs. 4.8 and 2C]). The impairment may be a consequence of the faster dissociation of T286A mutant (Fig. 4.7C) so that a much shorter duration of CaMKII staying in PSD is insufficient for the attachment to NMDAR to form the CaMKII-NMDAR complex during the transient 10 Hz and 100 Hz tetanus.

MoHST shows some potential functions of the autophosphorylation: (1) amplification of the postsynaptic response, i.e. the formation of CaMKII-NMDAR complex, in reaction to high-frequency stimulation; (2) decoding the stimulation frequency which has been reported by several modelling studies [52, 53]; and (3) coupling largely separated trains of tetanus. Although our study shows an inter-train interval of up to 1 min would benefit from the coupling, the irreversibility of the autophosphorylation in PSD, which we do not consider in MoHST, allows an even wider range of the inter-train interval of trains to benefit from the coupling [20]. While repetitive learning has a relatively long separation between the successive repeats, the autophosphorylation may potentially link the learnings to trigger memory formation. As shown in experiments, T286A mutant mice have difficulties in forming the spatial memory to search the hidden platform (see [3, Fig. 4]).

References

[1] Barria A. *et al.* (1997). Regulatory phosphorylation of AMPA-type glutamate receptors by CaMKII during long-term potentiation. *Science*, 276, pp. 2042–2045.

[2] Buard I. *et al.* (2010). CaMKII "autonomy" is required for initiating but not for maintaining neuronal long-term information storage. *J Neurosci*, 30, pp. 8214–8220.

[3] Giese K.P., Fedorov N.B., Filipkowski R.K. and Silva A.J. (1998). Autophosphorylation at Thr286 of the α calcium-calmodulin kinase II in LTP and learning. *Science*, 279, pp. 870–873.

[4] Halt A.R. *et al.* (2012). CaMKII binding to GluN2B is critical during memory consolidation. *Embo J*, 31, pp. 1203–1216.

[5] Irvine E.E., Von Hertzen L.S.J., Plattner F. and Giese K.P. (2006). αCaMKII autophosphorylation: a fast track to memory. *Trends Neurosci*, 29, pp. 459–465.

[6] Otmakhov N. *et al.* (2004). Persistent accumulation of calcium/calmodulin-dependent protein kinase II in dendritic spines after induction of NMDA receptor-dependent chemical long-term potentiation. *J Neurosci*, 24, pp. 9324–9331.

[7] Sanhueza M. *et al.* (2011). Role of the CaMKII/NMDA receptor complex in the maintenance of synaptic strength. *J Neurosci*, 31, pp. 9170–9178.

[8] Vest R.S. *et al.* (2007). Dual mechanism of a natural CaMKII inhibitor. *Mol Biol Cell*, 18, pp. 5024–5033.

[9] Zhang Y.-P, Holbro N. and Oertner T.G. (2008). Optical induction of plasticity at single synapses reveals input-specific accumulation of αCaMKII. *Proc Natl Acad Sci USA*, 105, pp. 12039–12044.

[10] Zhou Y. *et al.* (2007). Interactions between the NR2B receptor and CaMKII modulate synaptic plasticity and spatial learning. *J Neurosci*, 27, pp. 13843–13853.

[11] Bayer K.U. *et al.* (2006). Transition from reversible to persistent binding of CaMKII to postsynaptic sites and NR2B. *J Neurosci*, 26, pp. 1164–1174.

[12] Appleby V.J. *et al.* (2011). LTP in hippocampal neurons is associated with a CaMKII-mediated increase in GluA1 surface expression. *J Neurochem*, 116, pp. 530–543.

[13] Strack S., McNeill R.B. and Colbran R.J. (2000). Mechanism and regulation of calcium/calmodulin-dependent protein kinase II targeting to the NR2B subunit of the N-methyl-D-aspartate receptor. *J Biol Chem*, 275, pp. 23798–23806.

[14] He Y., Kulasiri D. and Samarasinghe S. (2015). Modelling the dynamics of CaMKII-NMDAR complex related to memory formation in synapses: The possible roles of threonine 286 autophosphorylation of CaMKII in long term potentiation. *J Theor Biol*, 365, pp. 403–419.

[15] Augustine G.J., Santamaria F. and Tanaka K. (2003). Local calcium signaling in neurons. *Neuron*, 40, pp. 331–346.

[16] Hudmon A. and Schulman H. (2002). Structure–function of the multifunctional Ca^{2+}/calmodulin-dependent protein kinase II. *Biochem J*, 364, pp. 593–611.

[17] Bayer K.U., Paul De Koninck A., Hell J.W. and Schulman H. (2001). Interaction with the NMDA receptor locks CaMKII in an active conformation. *Nature*, 411, pp. 801–805.

[18] Shen K. and Meyer T. (1999). Dynamic control of CaMKII translocation and localization in hippocampal neurons by NMDA receptor stimulation. *Science*, 284, pp. 162–167.

[19] Shinohara Y. *et al.* (2008). Left-right asymmetry of the hippocampal synapses with differential subunit allocation of glutamate receptors. *Proc Natl Acad Sci USA*, 105, pp. 19498–19503.

[20] Mullasseril P., Dosemeci A., Lisman J.E. and Griffith L.C. (2007). A structural mechanism for maintaining the "on-state"of the CaMKII memory switch in the post-synaptic density. *J Neurochem*, 103, pp. 357–364.

[21] Chiba H., Schneider N.S., Matsuoka S. and Noma A. (2008). A simulation study on the activation of cardiac CaMKIIδ-Isoform and its regulation by phosphatases. *Biophys J*, 95, pp. 2139–2149.

[22] Benjamin A.T. and Quinn J.J. (2003). *Proofs that Really Count: The Art of Combinatorial Proof* (MAA).

[23] Goldberg S. (1986). *Probability: An Introduction* (Dover Publications).

[24] Holmes W.R. (2000). Models of calmodulin trapping and CaM kinase II activation in a dendritic spine. *J Comput Neurosci*, 8, pp. 65–86.

[25] Peersen O.B., Madsen T.S. and Falke J.J. (1997). Intermolecular tuning of calmodulin by target peptides and proteins: differential effects on Ca^{2+} binding and implications for kinase activation. *Protein Sci*, 6, pp. 794–807.

[26] Lee S.J., Escobedo-Lozoya Y., Szatmari E.M. and Yasuda R. (2009). Activation of CaMKII in single dendritic spines during long-term potentiation. *Nature*, 458, pp. 299–304.

[27] Ehlers M.D. (2003). Activity level controls postsynaptic composition and signaling via the ubiquitin-proteasome system. *Nat Neurosci*, 6, pp. 231–242.

[28] Bliss T.V.P. and Collingridge G.L. (1993). A synaptic model of memory: long-term potentiation in the hippocampus. *Nature*, 361, pp. 31–39.

[29] Mayford M., Siegelbaum S.A. and Kandel E.R. (2012). Synapses and memory storage. *Cold Spring Harb Perspect Biol*, 4, p. a005751.

[30] Zhabotinsky A.M. (2000). Bistability in the Ca^{2+}/calmodulin-dependent protein kinase-phosphatase system. *Biophys J*, 79, pp. 2211–2221.

[31] Haario H., Laine M., Mira A. and Saksman E. (2006). DRAM: efficient adaptive MCMC. *Stat Comput*, 16, pp. 339–354.

[32] Haario H., Saksman E. and Tamminen J. (2001). An adaptive metropolis algorithm. Bernoulli, pp. 223–242.

[33] Shen K. *et al.* (2000). Molecular memory by reversible translocation of calcium/calmodulin-dependent protein kinase II. *Nat Neurosci*, 3, pp. 881–886.

[34] Strack S. and Colbran R.J. (1998). Autophosphorylation-dependent targeting of calcium/calmodulin-dependent protein kinase II by the NR2B subunit of the *N*-methyl-*D*-aspartate receptor. *J Biol Chem*, 273, pp. 20689–20692.

[35] Ribrault C., Sekimoto K. and Triller A. (2011). From the stochasticity of molecular processes to the variability of synaptic transmission. *Nat Rev Neurosci*, 12, pp. 375–387.

[36] Sheng M. and Hoogenraad C.C. (2007). The postsynaptic architecture of excitatory synapses: a more quantitative view. *Annu Rev Biochem*, 76, pp. 823–847.

[37] Kakiuchi S. *et al.* (1982). Quantitative determinations of calmodulin in the supernatant and particulate fractions of mammalian tissues. *J Biochem*, 92, pp. 1041–1048.

[38] Coultrap S.J., Barcomb K. and Bayer K.U. (2012). A significant but rather mild contribution of T286 autophosphorylation to Ca^{2+}/CaM-stimulated CaMKII activity. *PLoS One*, 7, pp. e37176.

[39] Marino S., Hogue I.B., Ray C.J. and Kirschner D.E. (2008). A methodology for performing global uncertainty and sensitivity analysis in systems biology. *J Theor Biol*, 254, pp. 178–196.

[40] Ling H., Kulasiri D. and Samarasinghe S. (2010). Robustness of G1/S checkpoint pathways in cell cycle regulation based on probability of DNA-damaged cells passing as healthy cells. *Biosystems*, 101, pp. 213–221.

[41] McKay M.D., Beckman R.J. and Conover W.J. (1979). A comparison of three methods for selecting values of input variables in the analysis of output from a computer code. *Technometrics*, 21, pp. 239–245.

[42] Kandel E.R. (2009). The biology of memory: a forty-year perspective. *J Neurosci*, 29, pp. 12748–12756.

[43] Shen K., Teruel M.N., Subramanian K. and Meyer T. (1998). CaMKIIβ functions as an F-actin targeting module that localizes CaMKIIα/β heterooligomers to dendritic spines. *Neuron*, 21, pp. 593–606.

[44] Thiagarajan T.C., Piedras-Renteria E.S. and Tsien R.W. (2002). α-and βCaMKII: inverse regulation by neuronal activity and opposing effects on synaptic strength. *Neuron*, 36, pp. 1103–1114.

[45] Elgersma Y. *et al.* (2002). Inhibitory autophosphorylation of CaMKII controls PSD association, plasticity, and learning. *Neuron*, 36, pp. 493–505.

[46] Goh J.J. and Manahan-Vaughan D. (2015). Role of inhibitory autophosphorylation of calcium/calmodulin-dependent kinase II (αCAMKII) in persistent (>24 h) hippocampal LTP and in LTD facilitated by novel object-place learning and recognition in mice. *Behav Brain Res*, 285, pp. 79–88.

[47] Robison A.J. *et al.* (2005). Multivalent interactions of calcium/calmodulin-dependent protein kinase II with the postsynaptic density proteins NR2B, densin-180, and α-actinin-2. *J Biol Chem*, 280, pp. 35329–35336.

[48] Scozzari R. *et al.* (2012). On the mechanism of synaptic depression induced by CaMKIIN, an endogenous inhibitor of CaMKII. *PLoS One*, 7, p. e49293.

[49] Barcomb K., Coultrap S.J. and Bayer K.U. (2013). Enzymatic activity of CaMKII is not required for its interaction with the glutamate receptor subunit GluN2B. *Mol Pharmacol*, 84, pp. 834–843.

[50] Fong Y.L., Taylor W.L., Means A.R. and Soderling T.R. (1989). Studies of the regulatory mechanism of Ca2+/calmodulin-dependent protein kinase II. Mutation of threonine 286 to alanine and aspartate. *J Biol Chem*, 264, pp. 16759–16763.

[51] Bhalla U.S. (2013). *20 Years of Computational Neuroscience.* ed. Bower J.M. Chapter 9 "Still looking for the memories: molecules and synaptic plasticity" (Springer Science+Business Media, New York), pp. 187–205.

[52] Dosemeci A. and Albers R.W. (1996). A mechanism for synaptic frequency detection through autophosphorylation of CaM kinase II. *Biophys J*, 70, pp. 2493–2501.

[53] Kubota Y. and Bower J.M. (2001). Transient versus asymptotic dynamics of CaM kinase II: possible roles of phosphatase. *J Comput Neurosci*, 11, pp. 263–279.

Chapter 5

Bidirectionality of Synaptic Pathways Related to LTP and LTD

5.1. Introduction

Age-related changes in brain structure and function cause impairments in basic cognitive functions, including attention, working and long-term memory, and perception, and may lead to deficits in higher level cognitive functions, including speech and language, decision making and executive control [1, 2]. Early works have established connections between brain regions and cognitive functions, but complementary insights into the biomolecular pathways related to the cognitive functions may help better understand the age-related changes in brain and the associated functional impairments. Over the last two decades, compelling experimental evidence indicates that the synapse may accumulate memory as synaptic strength, which reflects the degree of association between the environmental stimulus and the magnitude of the induced postsynaptic response by the synaptic transmission [3, 4]. The synaptic strength is modulated by synaptic plasticity [5–12]. The major forms of long-term synaptic plasticity, long-term potentiation (LTP) and long-term depression (LTD) have been shown *in vivo* to have direct roles in memory formation and removal, respectively [13]. Furthermore, emerging evidence has suggested the strong connection between brain diseases and abnormality in synaptic plasticity, where memory loss is accompanied by the increase in LTD induction [14, 15] and impairment in LTP [15–17]. In addition, normal aging is shown to cause the increased susceptibility to LTD induction [18–20]. Hence, it is clear that the memory system is at least partially related to the emergence of LTP and LTD and aging as well as brain diseases may alter the modulators which coordinate the LTP/LTD inductions. We can hypothesise that the major modulators coordinating LTP/LTD expressions may be the main targets in aging. To understand the major factors, we need to investigate the potential pathways that give rise to LTP and LTD and their behaviours during aging.

The mechanisms of NMDAR-dependent synaptic plasticity are discussed in detail in Chapter 3. Here we summarise the mechanisms: (1) the glutamate released from presynaptic neuron depolarises the postsynaptic membrane and activates NMDAR [21, 22]; (2) a transient Ca^{2+} influx through the activated NMDAR attaches to calmodulin (CaM) to form Ca^{2+}/CaM complex which triggers the activation of a postsynaptic cascade [23]; (3) the postsynaptic cascade causes specific postsynaptic changes, i.e. alteration of the property of AMPAR, resulting in the modulation of the magnitude of the postsynaptic response in reaction to a synaptic transmission [24–26]; and (4) the bidirectional selection of the LTP or LTD expression depends on both the pattern and amplitude of the Ca^{2+} elevation following the Ca^{2+} influx, where a brief high Ca^{2+} elevation ends in LTP leading to an increase in postsynaptic response and a prolonged moderate Ca^{2+} elevation ends in LTD leading to a decrease in postsynaptic response [27–29]. The strong association between the bidirectional expression of LTP or LTD and the morphology of the Ca^{2+} signal, both in the pattern and amplitude, is a crucial characteristic of synaptic plasticity. The complex mixture of protein–protein interaction networks giving rise to the complex behaviours of synaptic plasticity is difficult to understand using barely *in vitro* and *in vivo* experiments [30]; a mathematical model can be very useful to obtain insights at this level.

Many models have been published to investigate the emergent properties of synaptic plasticity [31–37]. Models of complete pathways as we have introduced in Chapter 3 (see [31–34]) try to capture the detailed responses of the postsynaptic terminal following the signals for the induction of synaptic plasticity. These approaches are very useful to obtain insights into the complex behaviour of synaptic plasticity at the system level. What are the major modulators coordinating LTP/LTD expressions, or bidirectionality? The potential candidates are the most sensitivity parameters, as well as the processes they belong to, related to the emergence of both LTP and LTD. In this case, a sensitivity analysis is most appropriate to access the sensitivity of parameters related to the bidirectionality. But the models of complete pathways are comprehensive, where Hayer and Bhalla's model [32] and Kim *et al.*'s model [34] have approximately 256 variables with 173 parameters and 69 variables with 136 parameters, respectively. The computational times of global sensitivity analysis (GSA) using these comprehensive models are not affordable, for example, Hayer and Bhalla's model takes 3 months of simulation time to compute over 50,000 sets of parameters on a multi-core high-performance computer (E5 1650 V3).

A new mathematical model is developed in this chapter to capture the dynamics of the essential pathways of the Ca^{2+}-dependent NMDAR-mediated synaptic plasticity and we call the model MoNP. MoNP consists of five sub-models, which are shown in Fig. 5.1 as boxes A to E, demarcating the interacting pathways for the activation of CaM, protein kinase A (PKA), calcineurin (PP2B), protein phosphatase 1 (PP1), and Ca^{2+}/CaM-dependent protein kinase II (CaMKII), respectively. We develop MoNP based on the up-to-date experimental established protein–protein interactions and kinetic properties of proteins. We ensure the minimised selection of states included in each sub-models through the following procedures: (1) in each case of the sub-models, select Ca^{2+} driven processes, which display essential functions in the emergence of LTP or LTD, from the literature; (2) develop sub-models based on the selected processes, validate the sub-models against the observed phenomena of synaptic plasticity, and understand the potential emergent property of the interacting network constituting the sub-model to support the emergence of synaptic plasticity; and (3) integrate the sub-models through the interactions among the active states of the modulators as shown in Fig. 5.2.

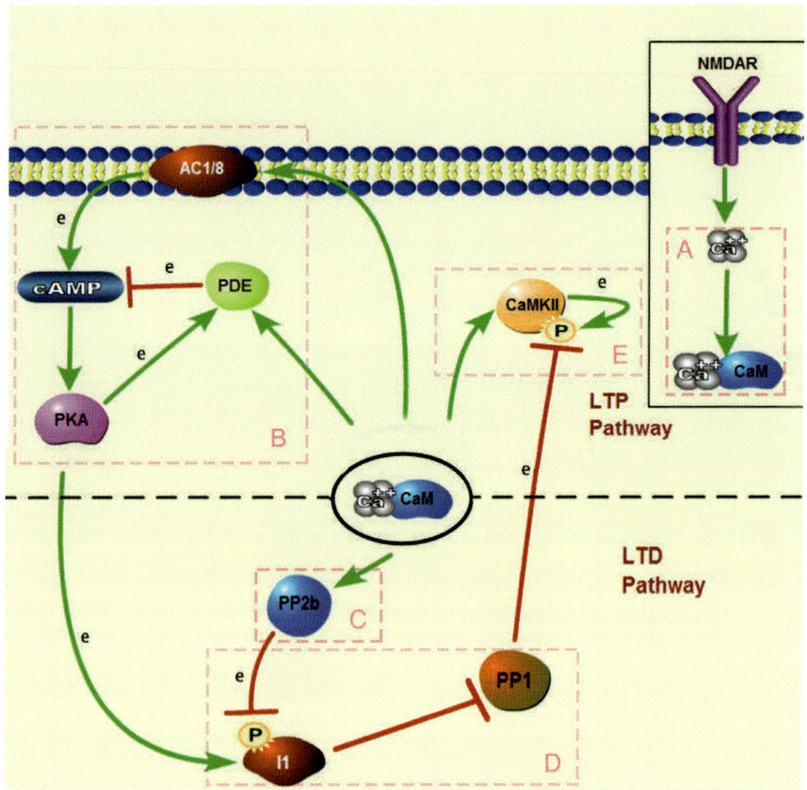

Fig. 5.1. NMDAR-mediated pathway of synaptic plasticity. Inset: The activated NMDAR triggers transient Ca^{2+} influx to free calmodulin and form Ca^{2+}/CaM complex. Main: Ca^{2+}/CaM complex triggers activation of LTP and LTD pathways (e denotes an enzymatic reaction). (A) Ca^{2+} and CaM interaction submodel: Ca^{2+} binds to CaM to form Ca^{2+}/CaM complex; (B) PKA activation sub-model: cyclic adenosine monophosphate (cAMP), is activated through a signalling cascade: Ca^{2+}/CaM complex activates adenylyl cyclase (AC), and AC further activates cAMP. The activity of cAMP is inhibited by phosphodiesterase (PDE), whose activity is promoted by both PKA and Ca^{2+}/CaM complex; (C) Calcineurin (PP2B) activation sub-model: PP2B is activated at a low level by binding to Ca^{2+} ions and then is transferred to full activation by binding to Ca^{2+}/CaM complex; (D) Protein phosphatase 1 (PP1) activation sub-model: PP1 is inhibited by the phosphorylated inhibitor 1 (I1), while the phosphorylation level of I1 is regulated by PKA and PP2B — PKA phosphorylates I1 and PP2B dephosphorylates I1; and (E) Ca^{2+}/CaM-dependent protein kinase II (CaMKII) activation and autophosphorylation: Ca^{2+}/CaM activates CaMKII and the activated CaMKII can undergo the autophosphorylation to maintain the activation. The autophosphorylated CaMKII is dephosphorylated by PP1.

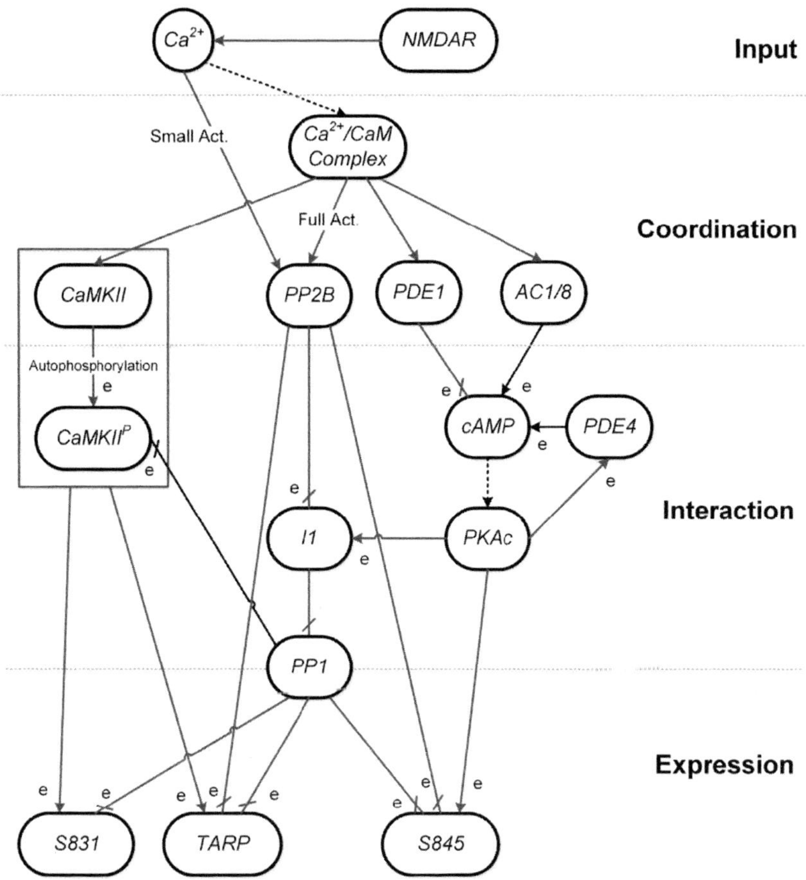

Fig. 5.2. The process modelled by MoNP. (1) Input: activation of NMDAR induces a Ca^{2+} influx to increase the intracellular Ca^{2+} concentration. (2) Coordination: Ca^{2+} ions bind to CaM to form Ca^{2+}/CaM complex which then binds to CaM-regulated proteins, including AC isoforms 1/8 (AC1/8), PDE isoform 1 (PDE1), PP2B and CaMKII. (3) Interaction: CaM-regulated proteins interact through downstream signalling pathways, including PKA activation pathway involving AC1/8, PDE1/4, cAMP, and catalytic subunits of PKA (PKAc); PP1 activation pathway involving PP1, PP2B, and PKAc; and CaMKII autophosphorylation pathway involving CaMKII, the autophosphory-lated CaMKII (CaMKII^P), and PP1. (4) Expression: the phosphorylation level of the target sites is dependent on the activation strength of the proteins, including PKA, CaMKII, PP2B, and PP1. The critical sites include but not limited to S831 of AMPAR regulating the single channel conductance of AMPAR, S845 of AMPAR regulating surface expression of AMPAR, and transmembrane AMPAR regulatory protein (TARP) regulating postsynaptic density (PSD) anchoring of surface AMPAR. (Here e denotes an enzymatic reaction.)

As a result, MoNP expresses biologically meaningful LTP and LTD behaviours with only 20 variables and 70 parameters. MoNP takes approximately 5 days to compute GSA over 50,000 sets of parameters on the same platform that is 20 times faster than Hayer and Bhalla's model. MoNP provides us with an affordable mathematical framework to conduct GSA as well as other required computational analyses on limited computational resources.

Interpretation of the computational results relies on two measures, ΔKBI and ΔI1$^\text{P}$, which are developed to capture the relative activation levels of modulators in contrary roles for the emergence of LTP and LTD, respectively. The results suggest the relative activity strength of the modulators underlying synaptic plasticity, where LTD is induced by prolonged moderate Ca^{2+} elevation having the relative activity strength shifting towards the phosphatase and LTP is induced by brief high Ca^{2+} elevation having the strength shifting towards the kinase. By investigating the effective time scales of modulators aligned with the changing in relative activity strength using the aforementioned measures, two conclusions are made: (1) PP2B, a phosphatase, has a high Ca^{2+} sensitivity which allows it to be activated over the prolonged moderate Ca^{2+} elevation to facilitate the induction of LTD; and (2) the abundance of kinases and their slower deactivation rates allow them to maintain a significant activation after the brief high Ca^{2+} elevation to facilitate the induction of LTP. From the most sensitive parameters suggested by GSA, three factors are marked significant for the bidirectionality: the cooperativity of binding between PP2B and Ca^{2+}, the rate of phosphorylation on I1 and the CaM pool size. Visualising the changes of the measures aligned with the variation of these factors, we propose that the CaM pool size is a single most important factor for the bidirectionality. Moreover, experimental evidence shows age-related reduction in the activity level of CaM [38–40]. Thus, using MoNP to conduct computational experiments, we can gain insights which are both novel and biologically meaningful in the literature.

5.2. Background of the Key Modulators

The detailed pathway of NMDAR-dependent synaptic plasticity is explained in Chapter 3. Here, we give a brief description of the biochemical properties of the key modulators to be considered by MoNP.

5.2.1. *Proteins related to LTP*

PKA is inactive when its two catalytic subunits (C_2 for the dimer or PKAc for each subunit) are attached to its regulatory dimer (PKAr) [41]. Two PKAc are released when cyclic adenosine monophosphate (cAMP) binds to both of the two tandem cAMP-binding domains in the regulatory dimer, each domain has two cAMP-binding sites. The released PKAc phosphorylates S845 of AMPAR [42–44] and threonine 67 of inhibitor 1 (I1) [45, 46], which are critical in the AMPAR surface expression and the PP1 activation, respectively. As the release of C_2 is mediated by cAMP, a tight regulation on cAMP underlies the regulation of PKA. cAMP is produced by conversion, catalysed by adenylyl cyclase (AC), from adenosine triphosphate (ATP) [47, 48] and inhibited by conversion, catalysed by phosphodiesterase (PDE), into adenosine monophosphate (AMP) [49]. Most importantly, the activity of PDE is promoted by both Ca^{2+}/CaM complex and PKA [50, 51], forming a dual inhibition loop on cAMP, with a direct inhibition by Ca^{2+}/CaM complex and an inhibitory feedback loop by PKA.

CaMKII subunit has a kinase domain, a regulatory domain, and a hub domain with a linker linking the regulatory and hub domains [52]. An inactive subunit has the kinase domain attached to the regulatory domain. Within the domains, there are four critical sites: (1) substrate binding site (S site); (2) T site; (3) threonine 286 site (T286); and (4) pseudosubstrate segment contains CaM footprint and threonine 305/306 sites (T305/T306).

Activation of the subunit occurs when Ca^{2+}/CaM complex attaches to CaM footprint resulting in a conformational change forcing the regulatory domain to come off the kinase domain [52].

The active CaMKII subunit can phosphorylate a number of target sites, including S831 of AMPAR [45, 53, 54] and transmembrane AMPAR regulatory protein (TARP) [55, 56] both of which are critical for LTP. With the detachment of mains, T286 of the CaMKII subunit can be autophosphorylated to maintain the detachment and the activation for up to one minute [57].

5.2.2. *Proteins related to LTD*

PP2B has calcineurin A (CaNA), a CaM-regulated subunit, attached to calcineurin B (CaNB), a Ca^{2+} regulated subunit [58]. Binding of Ca^{2+}/CaM complex to CaNA is prevented until CaNB is fully bound by four Ca^{2+} ions [59]. A conformational change occurs when Ca^{2+}/CaM complex attaches to CaNA to allow PP2B to dephosphorylate target substrates, including TARP [56, 60] and I1 [61]. Moreover, PP2B may remove PKA to prevent the rephosphorylation [62, 63].

Dephosphorylation of I1 by PP2B breaks the inhibition of PP1 by I1 binding. The released PP1 dephosphorylates many targets, including S831/S845 of AMPAR [60, 64–66], TARP [56, 60], and phosphorylations on CaMKII subunit [67]. It should be emphasised that PKA phosphorylates I1, as previously mentioned, to tighten the binding between I1 and PP1 to secure the inhibition of PP1.

5.3. The Overview of MoNP

In this section, we provide an overview of MoNP, including the general assumptions of MoNP, the schematic diagram of MoNP to give a comprehensive picture to support the MoNP development in the next section.

5.3.1. *General assumptions*

We have four general assumptions applied to all sub-models and each sub-model has its own assumptions.

(1) A single compartment is considered as the whole postsynaptic terminal (a cylinder with volume $0.1\,\text{fL} = 10^{-16}\,\text{L} = 10^{-19}\,\text{m}^3$). Biochemical species are well mixed across the compartment.

(2) We consider processes which are completely insulated from Ca^{2+} concentration change, to remain at equilibrium or near equilibrium since we investigate on how do Ca^{2+} signals change the dynamics of postsynaptic pathways which lead to synaptic plasticity. But the treatments of these processes are different in each case: for example, Ca^{2+}-independent AC is held at the equilibrium level while PP2A is held at a constant retrieved from [32]. Differential equations are expressed for processes relating to Ca^{2+} concentration, either directly or indirectly, except small active PP2B, which forms much faster than the formation of Ca^{2+}/CaM complex. Therefore, the concentration of small active PP2B is expressed by a function in terms of Ca^{2+} concentration.

(3) Protein turnovers are ignored since the time scale of the turnover exceeds the length of our simulation, which is 30 min at maximum [68].

(4) A CaM-regulated protein is activated only when its four Ca^{2+} sites are all bound by Ca^{2+} ions. The fully bound CaM is called Ca^{2+}/CaM complex here.

(5) Catalytic reactions, i.e. phosphorylation and dephosphorylation, obey the rules of Michaelis–Menten kinetics and all other reactions obey the rules of mass action kinetics.

Biochemical reactions, their parameters and reaction rates, which form the biological bases of the developed mathematical equations, are searched from the literature and databases, including Brenda (http://www.brenda-enzymes.info) and DOQCS (http://doqcs.ncbs.res.in).

5.3.2. *Schematic diagram of MoNP*

NMDAR-dependent synaptic plasticity can be summarised to four levels of operations in connecting Ca^{2+} input, Ca^{2+}/CaM complex coordination, modulators interaction, and target sites expression.

Input level considers the Ca^{2+} elevation caused by Ca^{2+} influx through NMDAR in response to different patterns of stimulations.

The mathematical equations of this level are formulated based on Zhabotinsky's study [69].

Coordination level considers the transduction of the Ca^{2+} elevation to the activation of modulators. The formation of Ca^{2+}/CaM complex following the Ca^{2+} elevation is essential in the transduction while most of the modulators are CaM-regulated, including CaNA of PP2B, PDE isoform1, AC isoforms 1/8, and CaMKII. The limited CaM concentration in the neuron sets competition among the modulators for the activation.

Interaction level considers the interactions among the activated modulators for regulatory purposes. For example, CaMKII autophosphorylation is reversed by PP1 and PP1 activation can be reversed by phosphorylation I1 by PKA.

Expression level considers resulting alteration on target sites related to synaptic plasticity. However, it may be unnecessary to consider the exact dynamics of this level since we are uncertain about the synaptic functions and mechanisms for the many expression sites except the three that are listed in Fig. 5.2. We can assume that the modulators included in the interaction level are the only ones dependent on Ca^{2+}, at least for the early phase of synaptic plasticity where our study is focused on. Hence, it is more prudent to consider the relative activity strength between kinases and phosphatases which reflects the phosphorylation level on the target sites. As shown in Fig. 5.2, CaMKII and PKA, two key kinases, phosphorylate different target sites that need to be considered separately. CaMKII, at a similar enzymatic activity (turnover rate (K_{cat}):$1.2-10\,\mathrm{s}^{-1}$ [70]) to that of phosphatases, is more abundant than phosphatases. Hence we can predict the phosphorylation level on the target sites of CaMKII by looking at the activation strength between CaMKII and phosphatases. Furthermore, the activation strength can be replaced by the ratio of Ca^{2+}/CaM complex competitively bound by CaMKII, since CaMKII is the only kinase considered in this study to directly bind to Ca^{2+}/CaM complex. On the other hand, PKA is both lower level and lower enzymatic activity than those of phosphatases, so that a target of PKA should be considered to reflect the activation strength between PKA and phosphatases. I1 is chosen because it is involved in the interaction level.

5.4. Development of MoNP

In this section, we discuss the development of sub-models, the equations of MoNP, and validation of MoNP.

5.4.1. *Sub-model A — Ca^{2+}/CaM complex formation*

According to Holmes's approach [71], Ca^{2+}/CaM complex is the final product of four sequential reactions where each reaction attaches one Ca^{2+} ion to a CaM molecule. The reaction schema is shown as follows [71]:

$$iCaM + Ca^{2+} \underset{k_{1b}}{\overset{k_{1f}}{\rightleftharpoons}} CaCaM, \tag{5.1}$$

$$CaCaM + Ca^{2+} \underset{k_{2b}}{\overset{k_{2f}}{\rightleftharpoons}} Ca_2CaM, \tag{5.2}$$

$$Ca_2CaM + Ca^{2+} \underset{k_{3b}}{\overset{k_{3f}}{\rightleftharpoons}} Ca_3CaM, \quad \text{and} \tag{5.3}$$

$$Ca_3CaM + Ca^{2+} \underset{k_{4b}}{\overset{k_{4f}}{\rightleftharpoons}} Ca_4CaM, \tag{5.4}$$

where all reactions are elementary reactions. The free CaM is denoted by iCaM and the Ca^{2+} bound CaM is denoted by Ca_nCaM, where n is the number of Ca^{2+} ions attached. Thus, Ca^{2+}/CaM complex is denoted by Ca_4CaM in this notation. Here k_{if} and k_{ib} $(i = 1, 2, 3, 4)$ are association and dissociation rate constants, respectively, of attaching the ith Ca^{2+} ion to CaM.

The reaction rates are formulated based on mass action rate law with unit micromolar per second (μM/s), which is applied for all reaction rates in this chapter. These reaction rates are notated R_1–R_4 corresponding to attaching 1–4 Ca^{2+} ions to CaM, respectively, and their equations are given by Eqs. (5.5)–(5.8):

$$R_1 = k_{1f}[Ca^{2+}][iCaM] - k_{1b}[CaCaM], \tag{5.5}$$

$$R_2 = k_{2f}[Ca^{2+}][CaCaM] - k_{2b}[Ca_2CaM], \tag{5.6}$$

$$R_3 = k_{3f}[Ca^{2+}][Ca_2CaM] - k_{3b}[Ca_3CaM], \quad \text{and} \tag{5.7}$$

$$R_4 = k_{4f}[Ca^{2+}][Ca_3CaM] - k_{4b}[Ca_4CaM]. \tag{5.8}$$

5.4.2. *Sub-model B — PKA activation pathway*

The activation of PKA involves several processes as discussed previously, including the regulated production and inhibition of cAMP by ACs and PDE, respectively, and the regulated activation of PKA by cAMP [72, 73]. Due to the complexity of the pathway, we divide the development into three parts: (1) the production of cAMP; (2) the inhibition of cAMP; and (3) the activation of PKA.

5.4.2.1. *Production of cAMP*

AC is stimulated by many sources to produce cAMP. We differentiate between two sources in MoNP: Ca^{2+}/CaM complex and Ca^{2+} independent.

From previous studies [32, 33], the reaction schema is summarised as follows:

$$iAC^* \underset{k_{5b}}{\overset{k_{5j}}{\rightleftarrows}} AC^*, \tag{5.9}$$

$$Ca_4CaM + iAC1 \underset{k_{c1b}}{\overset{k_{c1f}}{\rightleftarrows}} CaMAC1, \tag{5.10}$$

$$Ca_4CaM + iAC8 \underset{k_{c2b}}{\overset{k_{c2f}}{\rightleftarrows}} CaMAC8, \tag{5.11}$$

$$ATP + CaMAC1 \underset{k_{6b}}{\overset{k_{6f}}{\rightleftarrows}} CaMAC1ATP$$

$$\xrightarrow{K_{cat1}} cAMP + CaMAC1, \tag{5.12}$$

$$ATP + CaMAC8 \underset{k_{7b}}{\overset{k_{7f}}{\rightleftharpoons}} CaMAC1ATP$$

$$CaMAC1ATP \xrightarrow{K_{cat2}} cAMP + CaMAC8 \qquad , \quad \text{and} \quad (5.13)$$

$$ATP + AC^* \underset{k_{8b}}{\overset{k_{8f}}{\rightleftharpoons}} AC^*ATP \xrightarrow{K_{cat3}} cAMP + AC^*. \qquad (5.14)$$

Reaction (5.9) is the transition between inactive (iAC) and active (AC^*) forms of Ca^{2+}-independent ACs. Since Ca^{2+}-independent sources do not respond to Ca^{2+} elevation, the only input of MoNP, an equilibrium level underlies Ca^{2+}-independent ACs according to the assumption 2. Hence, the production of cAMP by AC^* is assumed to remain at an equilibrium rate during our simulation. The equilibrium level of active $AC^*, [AC^*]$, is given by

$$[AC^*] = \frac{[AC_T^*]}{1 + K}, \qquad (5.15)$$

where $K = k_{5b}/k_{5f}$ and $[AC_T^*]$, the total concentration of Ca^{2+}-independent ACs, is conserved according to mass conservation law: $[AC_T^*] = [AC^*] + [iAC^*]$.

Reactions (5.10) and (5.11) are the bindings between Ca^{2+}/CaM and AC isoform 1/8, two Ca^{2+}-dependent ACs, where iAC1/8 and CaMAC1/8 denote the inactive and active forms of AC isoforms 1 and 8, respectively. Hence, the activation rates of AC1/8, notated R_5 and R_6, are given, respectively, by

$$R_5 = k_{c1f}[Ca_4CaM]([AC1_T] - [CaMAC1])$$

$$-k_{c1b}[CaMAC1], \quad \text{and} \qquad (5.16)$$

$$R_6 = k_{c2f}[Ca_4CaM]([AC8_T] - [CaMAC8])$$

$$-k_{c2b}[CaMAC8], \qquad (5.17)$$

where $[AC1_T]$, the total concentration of AC1, is conserved: $[AC1_T] = [iAC1] + [CaMAC1], [AC8_T]$, the total concentration of AC8, is conserved: $[AC8_T] = [iAC8] + [CaMAC8], k_{c1f}/k_{c2f}$ and k_{c1b}/k_{c2b} are association rate constant and dissociation rate constant for the bindings, respectively. The letter c in the rate constant indicates the involvement of the reactions in the competitiveness of the binding to Ca^{2+}/CaM complex.

Reactions (5.12)–(5.14) are three channels of production of cAMP through conversion from ATP catalysed by AC1, AC8 and AC^*, respectively. Their rates are denoted by R_7-R_9, respectively, and are formulated based on Michaelis and Menten rate law given, respectively, by

$$R_7 = \frac{K_{\text{cat1}}[CaMAC1][ATP]}{K_{m1} + [ATP]}, \tag{5.18}$$

$$R_8 = \frac{K_{\text{cat2}}[CaMAC8][ATP]}{K_{m2} + [ATP]}, \quad \text{and} \tag{5.19}$$

$$R_9 = \frac{K_{\text{cat3}}[AC^*][ATP]}{K_{m3} + [ATP]}, \tag{5.20}$$

where $K_{m1} = (k_{6b}+K_{\text{cat1}})/k_{6f}, K_{m2} = (k_{7b}+K_{\text{cat2}})/k_{7f}$, and $K_{m3} = (k_{8b} + K_{\text{cat3}})/k_{8f}$. We made a further simplification that,

$$\frac{[ATP]}{K_m + [ATP]} \approx 1, \tag{5.21}$$

by taking the concentration of ATP as a very large constant ($[ATP] \gg K_{mi}, i = 1, 2, 3$) since ATP is extremely rich in all cells:

$$R_7 = K_{\text{cat1}}[CaMAC1], \tag{5.22}$$

$$R_8 = K_{\text{cat2}}[CaMAC8], \quad \text{and} \tag{5.23}$$

$$R_9 = K_{\text{cat3}}\frac{[AC_T^*]}{1 + K}. \tag{5.24}$$

5.4.2.2. *Inhibition of cAMP*

cAMP is inhibited through the conversion into AMP with catalysis by PDEs. There are several PDE isoforms having different conversion rates and the rates can be promoted by different proteins. We consider two major PDEs in the brain, PDE isoform 1 (PDE1) and isoform 4 (PDE4), which are promoted by Ca^{2+}/CaM complex and PKA, respectively [49, 74, 75]. PDE1 has a higher catalytic activity than PDE4. The reaction schema is summarised as follows [32]:

$$iPDE1 + Ca_4CaM \underset{k_{c3b}}{\overset{k_{c3f}}{\rightleftharpoons}} CaMPDE1, \tag{5.25}$$

$$PDE4B + PKAc \underset{k_{9b}}{\overset{k_{9f}}{\rightleftharpoons}} PDE4BPKAc$$

$$\xrightarrow{K_{\text{cat3}}} PDE4B^P + PKAc,$$

$$PDE4B^P \xrightarrow{k_{10}} PDE4B, \tag{5.26}$$

$$PDE4D + PKAc \underset{k_{9b}}{\overset{k_{9f}}{\rightleftharpoons}} PDE4DPKAc$$

$$\xrightarrow{K_{\text{cat4}}} PDE4D^P + PKAc,$$

$$PDE4D^P \xrightarrow{k_{10}} PDE4D, \tag{5.27}$$

$$cAMP + iPDE1 \underset{k_{11b}}{\overset{k_{11f}}{\rightleftharpoons}} cAMPiPDE1$$

$$\xrightarrow{K_{\text{cat5}}} AMP + iPDE1, \tag{5.28}$$

$$cAMP + CaMPDE1 \underset{k_{12b}}{\overset{k_{12f}}{\rightleftharpoons}} cAMPCaMPDE1$$

$$\xrightarrow{K_{\text{cat6}}} AMP + CaMPDE1, \tag{5.29}$$

$$cAMP + PDE4B \underset{k_{13b}}{\overset{k_{13f}}{\rightleftharpoons}} cAMPPDE4B$$

$$\xrightarrow{K_{\text{cat7}}} AMP + PDE4B, \tag{5.30}$$

$$cAMP + PDE4B^P \underset{k_{14b}}{\overset{k_{14f}}{\rightleftharpoons}} cAMPPDE4B^P$$

$$\xrightarrow{K_{\text{cat8}}} AMP + PDE4B^P, \tag{5.31}$$

$$cAMP + PDE4D \underset{k_{15b}}{\overset{k_{15f}}{\rightleftarrows}} cAMPPDE4D$$

$$\xrightarrow{K_{\text{cat9}}} AMP + PDE4D, \quad \text{and} \tag{5.32}$$

$$cAMP + PDE4D^P \underset{k_{16b}}{\overset{k_{16f}}{\rightleftarrows}} cAMPPDE4D^P$$

$$\xrightarrow{K_{\text{cat10}}} AMP + PDE4D^P \tag{5.33}$$

Reaction (5.25) is the binding between PDE1 and Ca^{2+}/CaM complex ending in the bound PDE1 (CaMPDE1) which has a higher catalytic activity. The reaction rate of the binding (R_{10}) is formulated based on mass action rate law as given by

$$R_{10} = k_{c3f}[Ca_4CaM]([PDE1_T] - [CaMPDE1])$$
$$- k_{c3b}[CaMPDE1], \tag{5.34}$$

where $[PDE1_T]$, the total concentration of PDE1, is conserved: $[PDE1_T] = [iPDE1] + [CaMPDE1]$, and k_{c3f} and k_{c3b} are the association and dissociation rate constants for the binding, respectively.

Reactions (5.26) and (5.27) are the promotion of the activity of PDE4 by PKA. PDE4 is given by two isoforms, PDE4B and PDE4D in [34], hence we have two reactions account for the promotion of PDE4. We made a further simplification to remove $PDE4BPKAc$ and $PDE4DPKAc$ because their peak concentrations at the equilibrium are much lower than other states. For example, the peak of $[PDE4BPKAc]$ is around $0.096\,\mu M$, obtained when reaction (5.26) is in equilibrium assuming all PKA subunits are active:

$$(k_{9b} + K_{\text{cat4}})[PDE4BPKAc] = \frac{k_{9b} + K_{\text{cat4}}}{K_{m4}}[PDE4B_T][PKAc_T], \tag{5.35}$$

where $k_b \approx 4K_{\text{cat}}$ [31], and other parameters are obtained from [34] and given in Table 5.8. However, the real peak is much lower than $0.096\,\mu M$ given the concentration of PKAc is much lower than the total due to the strong regulation of PKAc by PKAr [41]. The final simplified reaction rates of the promotions on PDE4B (R_{11}) and PDE4D (R_{12}) by PKA are given, respectively, by

$$R_{11} = \frac{K_{\text{cat4}}[PKAc]([PDE4B_T] - [PDE4B^P])}{K_{m4} + ([PDE4B_T] - [PDE4B^P])}$$

$$-k_{10}[PDE4B^P], \quad \text{and} \tag{5.36}$$

$$R_{12} = \frac{K_{\text{cat4}}[PKAc]([PDE4D_T] - [PDE4D^P])}{K_{m4} + ([PDE4D_T] - [PDE4D^P])}$$

$$-k_{10}[PDE4D^P], \tag{5.37}$$

where $[PDE4B_T]$, the total concentration of PDE4B, is conserved: $[PDE4B_T] = [PDE4B] + [PDE4B^P], [PDE4D_T]$, the total concentration of PDE4D, is conserved: $[PDE4D_T] = [PDE4D] + [PDE4D^P]$, and $K_{m4} = (k_{9b} + K_{\text{cat4}})/k_{9f}$.

Reactions (5.28)–(5.33) are the conversion of cAMP into AMP catalysed by different isoforms as well as different activity status of PDEs. All these reactions are modelled according to Michaelis and Menten rate law with the reaction rates (V_1)–(V_6) for catalysis by iPDE1, CaMPDE1, PDE4B, PDE4B^P, PDE4D, and PDE4D^P, respectively, given, respectively, by

$$V_1 = \frac{K_{\text{cat5}}[iPDE1][cAMP]}{K_{m5} + [cAMP]}$$

$$= ([PDE1_T] - [CaMPDE1])\frac{K_{\text{cat5}}[cAMP]}{K_{m5} + [cAMP]}, \tag{5.38}$$

$$V_2 = \frac{K_{\text{cat6}}[CaMPDE1][cAMP]}{K_{m6} + [cAMP]}, \tag{5.39}$$

$$V_3 = \frac{K_{\text{cat7}}[PDE4B][cAMP]}{K_{m7} + [cAMP]}$$

$$= ([PDE4B_T] - [PDE4B^P])\frac{K_{\text{cat7}}[cAMP]}{K_{m7} + [cAMP]}, \tag{5.40}$$

$$V_4 = \frac{K_{\text{cat8}}[PDE4B^P][cAMP]}{K_{m8} + [cAMP]}, \tag{5.41}$$

$$V_5 = \frac{K_{\text{cat9}}[PDE4D][cAMP]}{K_{m9} + [cAMP]}$$

$$= ([PDE4D_T] - [PDE4D^P])\frac{K_{\text{cat9}}[cAMP]}{K_{m9} + [cAMP]}, \quad \text{and} \quad (5.42)$$

$$V_6 = \frac{K_{\text{cat10}}[PDE4D^P][cAMP]}{K_{m10} + [cAMP]}, \tag{5.43}$$

where $K_{m5} = (k_{11b} + K_{\text{cat5}})/k_{11f}, K_{m6} = (k_{12b} + K_{\text{cat6}})/k_{12f}, K_{m7} = (k_{13b} + K_{\text{cat7}})/k_{13f}, K_{m8} = (k_{14b} + K_{\text{cat8}})/k_{14f}, K_{m9} = (k_{15b} + K_{\text{cat9}})/k_{15f}$, and $K_{m10} = (k_{16b} + K_{\text{cat10}})/k_{16f}$.

Literature suggests the same K_m for PDE1 after the promotion ($K_{m5} = K_{m6}$), but $V_{\max}$ is increased significantly ($K_{\text{cat6}} \gg K_{\text{cat5}}$) [49]. The similar trends for PDE4 are included in the previous computational studies [32, 34]. Hence, we made a simplification by assuming K_m stays the same before and after the promotion for both PDE1 and PDE4. As a result, reaction rates before and after the promotion for both types of PDEs have the same denominators, which allows the two rates to be combined as one. The combined reactions rates $(R_{13})-(R_{14})$ for the conversions of cAMP into AMP with catalysis by PDE1, PDE4B, and PDE4D, respectively, are given, respectively, by

$$R_{13} = V_1 + V_2 = \frac{K_{\text{cat5}}[PDE1_T][cAMP]}{K_{m5} + [cAMP]}$$

$$\times \left(1 + \left(\frac{K_{\text{cat6}}}{K_{\text{cat5}}} - 1\right)\frac{[CaMPDE1]}{[PDE1_T]}\right), \tag{5.44}$$

$$R_{14} = V_3 + V_4 = \frac{K_{\text{cat7}}[PDE4B_T][cAMP]}{K_{m7} + [cAMP]}$$

$$\times \left(1 + \left(\frac{K_{\text{cat8}}}{K_{\text{cat7}}} - 1\right)\frac{[PDE4B^P]}{[PDE4B_T]}\right), \quad \text{and} \quad (5.45)$$

$$R_{15} = V_5 + V_6 = \frac{K_{\text{cat9}}[PDE4D_T][cAMP]}{K_{m9} + [cAMP]}$$

$$\times \left(1 + \left(\frac{K_{\text{cat10}}}{K_{\text{cat9}}} - 1\right)\frac{[PDE4D^P]}{[PDE4D_T]}\right). \tag{5.46}$$

5.4.2.3. *Activation of PKA*

The attachment between the regulatory dimer and the catalytic subunits is broken by binding four cAMP to the two tandem cAMP-binding domains (A and B domains) in the regulatory dimer. It is clear that binding to domain A is required for the release and activation in RIIβ, the dominant isoform of PKA expressed in the brain [41, 76], but it is unclear whether an order needs to be followed for the bindings to A and B domains. Moreover, domain B seems to inhibit the activation through a structural feature resulting in a switch-like behaviour causing the cooperative activation of *PKA* [41, 76]. A switch-like behaviour may be very important for LTP which is induced with restrictions in the frequency of stimulation. With these uncertainties in mind, we adapt an established framework of binding two cAMP to the regulatory dimer with assumptions as follows [33]: (1) cAMP binds to the domains through a two-step mechanism where domain B is first bound by two cAMP followed by domain A binds the other two cAMP. Here, we bring in the binding to domain B as the first step to reproduce the cooperative activation observed in the experiment [41]. This structure does not consider the inhibitory structural feature of the domain B, but the reaction rate constants should reflect some of the differences between the domains; (2) once the domains are fully bound by cAMP, the two catalytic subunits are released simultaneously; and (3) before the cAMP fully bound, the regulatory dimer and the catalytic subunits are always attached together. The reaction schema is summarised as follows [33]:

$$R_2C_2 + 2cAMP \underset{k_{17b}}{\overset{k_{17f}}{\rightleftharpoons}} R_2C_2cAMP_2, \tag{5.47}$$

$$R_2C_2cAMP_2 + 2cAMP \underset{k_{18b}}{\overset{k_{18f}}{\rightleftharpoons}} R_2C_2cAMP_4, \quad \text{and} \tag{5.48}$$

$$R_2C_2cAMP_4 \underset{k_{19b}}{\overset{k_{19f}}{\rightleftharpoons}} 2PKAc + R_2cAMP_4. \tag{5.49}$$

Reactions (5.47) and (5.48) are the binding of cAMP to domain B and domain A, respectively, where R_2C_2 is the unbound PKA,

$R_2C_2cAMP_2$ is the PKA with two cAMP bound to domain B, and $R_2C_2cAMP_4$ are the fully bound PKA at both domains. In reaction (5.49), we have two catalytic subunits (denote by PKAc variable) released leaving behind the regulatory dimer fully bound by cAMP (R_2cAMP_4). The reaction rates of the domain B binding (R_{16}), the full binding (R_{17}), and the release (R_{18}) are formulated according to mass action law as follows:

$$R_{16} = k_{17f}[R_2C_2][cAMP]^2 - k_{17b}[R_2C_2cAMP_2], \tag{5.50}$$

$$R_{17} = k_{18f}[R_2C_2cAMP_2][cAMP]^2 - k_{18b}[R_2C_2cAMP_4], \quad \text{and} \tag{5.51}$$

$$R_{18} = k_{19b}[R_2C_2cAMP_4] - k_{19f}[PKAc]^3/2, \tag{5.52}$$

where $[R_2C_{2T}]$, the total concentration of PKA, is conserved, expressed as $[R_2C_{2T}] = [R_2C_2] + [R_2C_2cAMP_2] + [R_2C_2cAMP_4] + [R_2cAMP_4]$ for regulatory dimers or $[R_2C_{2T}] = [R_2C_2] + [R_2C_2 cAMP_2] + [R_2C_2cAMP_4] + 2[PKAc]$ for catalytic subunits. The simplification of the expressions gives $[R_2cAMP_4] = [PKAc]/2$.

5.4.3. *Sub-model C — PP2B activation pathway*

During the activation, Ca^{2+} ions first bind to four Ca^{2+} sites in CaNB of PP2B, where two sites have a high Ca^{2+} affinity (around $0.07\,\mu M$) and the other two have a low Ca^{2+} affinity ($0.5-1\,\mu M$) [58]. CaNB carries a basal activity for PP2B, which is regulated by the filled low Ca^{2+} affinity sites with a Hill coefficient of 1.8 [59]. When CaNB is fully bound by Ca^{2+} ions, CaNA of PP2B is allowed to bind to Ca^{2+}/CaM complex that leads to a conformational change to expose the catalytic domain of CaNA [59]. The catalytic domain, once exposed, increases the basal activity of PP2B by 20 folds to the full activation, while K_m remains the same [58].

We consider three states for PP2B: (1) iPP2B for inactivation; (2) Ca_4CaNB for basal activation; and (3) CaMCaNA for full activation. We assume that the high Ca^{2+} affinity sites have Ca^{2+} ions fully bound when any of the low Ca^{2+} affinity sites have a Ca^{2+} ion attached. Hence, the basal activation and the exposure of the catalytic domain in CaNA depend on the filling of two low Ca^{2+} affinity sites in CaNB. Moreover, a faster binding rate governs the binding between CaNB and Ca^{2+} ions according to the rate constants of the binding in the previous computational study [32]. Hence, the binding can be thought of instantly occurring following the changes of Ca^{2+} concentration that Ca_4CaNB can be approximately by a Hill function with Ca^{2+} concentration, the control parameter as given by

$$[Ca_4CaNB] = \frac{([PP2B_T] - [CaMCaNA])[Ca^{2+}]^{n1}}{K_{d1}^{n1} + [Ca^{2+}]^{n1}}, \qquad (5.53)$$

where $[PP2B_T]$, the total concentration of PP2B, is conserved: $[PP2B_T] = [iPP2B] + [Ca_4CaNB] + [CaMCaNA], n1$ is the Hill coefficient and K_{d1} is the dissociation constant. Note that the expression $[PP2B_T] - [CaMCaNA] = [iPP2B] + [Ca_4CaNB]$ gives the concentration of free CaNA.

Following the basal activation, Ca^{2+}/CaM complex binds to CaNA to trigger the full activation, given by reaction schema as follows:

$$Ca_4CaM + Ca_4CaNB \xrightarrow{K_{c4f}} CaMCaNA, \qquad (5.54)$$

with the reaction rates (R_{19}) given by the following equation, according to mass action rate law:

$$R_{19} = k_{c4f}[Ca_4CaM][Ca_4CaNB]$$

$$= k_{c4f}[Ca_4CaM]\frac{([PP2B_T] - [CaMCaNA])[Ca^{2+}]^{n1}}{K_{d1}^{n1} + [Ca^{2+}]^{n1}}.$$

$$(5.55)$$

CaM may dissociate from CaNA through two pathways; one pathway has direct dissociation of Ca^{2+}/CaM complex from CaNA and the other pathway has the dissociation of Ca^{2+} from CaM followed by dissociation of CaM from CaNA [77]. The two pathways are given by reaction schema as follows:

$$\text{Pathway 1: } CaMCaNA \xrightarrow{K_{c4b1}} Ca_4CaM + Ca_4CaNB, \quad \text{and}$$
(5.56)

$$\text{Pathway 2: } CaMCaNA \xrightarrow{K_{c4b2}} iCaM + Ca_4CaNB + 4Ca^{2+}.$$
(5.57)

The second pathway may occur only when Ca^{2+} concentration is low [77]. Therefore, a Hill function in terms of Ca^{2+} concentration is formulated to approximate the probability of dissociation by the first pathway, which may be dominant during Ca^{2+} elevation:

$$\frac{[Ca^{2+}]^{n2}}{K_{d2}^{n2} + [Ca^{2+}]^{n2}},$$
(5.58)

and the probability of dissociation by the second pathway is approximated by

$$1 - \frac{[Ca^{2+}]^{n2}}{K_{d2}^{n2} + [Ca^{2+}]^{n2}},$$
(5.59)

where $n2$ is the Hill coefficient and K_{d2} is the dissociation constant. The rates of the dissociation, R_{20} and R_{21}, through pathways one and two, respectively, are given, respectively, by

$$R_{20} = k_{c4b1}[CaMCaNA]\frac{[Ca^{2+}]^{n2}}{K_{d2}^{n2} + [Ca^{2+}]^{n2}},$$
(5.60)

$$R_{21} = k_{c4b2}[CaMCaNA]\left(1 - \frac{[Ca^{2+}]^{n2}}{K_{d2}^{n2} + [Ca^{2+}]^{n2}}\right).$$
(5.61)

5.4.4. *Sub-model D — PP1 activation pathway*

PP1 is activated as the result of decrease I1 phosphorylation to free PP1 from the I1 binding [61]. Hence, the activation pathway involves three major regulators: I1, which binds and inhibits PP1; PKA, which phosphorylates I1; and PP2B, which dephosphorylates I1 [32].

The reaction schema is summarised as follows [32, 34]:

$$I1 + PKAc \underset{k_{20b}}{\overset{k_{20f}}{\rightleftharpoons}} I1PKAc \xrightarrow{K_{\text{cat}11}} I1^P PP1 + PKAc,$$

(5.62)

$$I1^P PP1 + PP2A \underset{k_{21b}}{\overset{k_{21f}}{\rightleftharpoons}} I1^P PP1PP2A$$

$$\xrightarrow{K_{\text{cat}12}} I1 + PP1 + PP2A,$$

(5.63)

$$I1^P PP1 + Ca_4CaNB \underset{k_{22b}}{\overset{k_{22f}}{\rightleftharpoons}} I1^P PP1Ca_4CaNB$$

$$I1^P PP1Ca_4CaNB \underset{k_{20}}{\overset{k_{\text{cat}13}}{\rightleftharpoons}} I1 + PP1 + Ca_4CaNB, \quad \text{and}$$

(5.64)

$$I1^P PP1 + CaMCaNA \underset{k_{22b}}{\overset{k_{22f}}{\rightleftharpoons}} I1^P PP1CaMCaNA$$

$$I1^P PP1CaMCaNA \xrightarrow{K_{\text{cat}13}} I1 + PP1 + CaMCaNA.$$

(5.65)

Here, PP1 has two states, active (PP1) and inactive ($I1^P$PP1), by including an assumption that phosphorylated I1 binds to PP1 immediately and unphosphorylated I1 releases PP1 immediately. This assumption is made on the basis of an extremely large change of binding affinity between PP1 and I1 before and after the phosphorylation as indicated by previous computational studies, for instance, the affinity is changed from extremely high (less than $1\,\text{nM}$) for phosphorylated I1 to extremely low (no binding) for unphosphorylated I1 [32, 34]. As a result, for the possible intermediate states, phosphorylated unbound and unphosphorylated bound have negligible concentrations both at equilibrium and during transitional changes following the phosphorylation or dephosphorylation when the total concentration of PP1, $[PP1_T]$, is greater than or equal to the total centration of I1, $[I1_T]([PP1_T] > [I1_T])$. If $[I1_T] \gg [PP1_T]$, this framework may lose accuracy because PP1 is insufficient to bind to I1 when the concentration of phosphorylated

I1 is greater than that of PP1. But, the situation is expected to be rare given the relatively low concentration and catalytic activity of PKA.

Since CaMCaNA and Ca_4CaNB states of PP2B have same K_m of their catalytic activity as discussed in the development of PP2B activation, three reaction rates (R_{22})–(R_{24}) are formulated for the phosphorylation of I1 by PKA, the dephosphorylation by PP2A, and the dephosphorylation by PP2B, respectively, according to Michaelis–Menten rate law. PP2A is modelled as a Ca^{2+}-independent phosphatase so that its level is a constant. The reaction rates are given, respectively, by

$$R_{22} = \frac{K_{\text{cat11}}[PKAc]([I1_T] - [I1^P PP1])}{K_{m11} + ([I1_T] - [I1^P PP1])}, \tag{5.66}$$

$$R_{23} = \frac{K_{\text{cat12}}[PP2A][I1^P PP1]}{K_{m12} + [I1^P PP1]}, \quad \text{and} \tag{5.67}$$

$$R_{24} = \frac{K_{\text{cat13}}([CaMCaNA] + 1/20[Ca_4CaNB])[I1^P PP1]}{K_{m13} + [I1^P PP1]}, \tag{5.68}$$

where total concentration of I1 and PP1 are conserved: $[I1_T] = [I1] + [I1^P PP1], [PP1_T] = [PP1] + [I1^P PP1], K_{m11} = (k_{20b} + K_{\text{cat11}})/k_{20f}, K_{m12} = (k_{21b} + K_{\text{cat12}})/k_{21f}$ and $K_{m13} = (k_{22b} + K_{\text{cat13}})/k_{22f}$.

5.4.5. *Sub-model E — CaMKII activation and autophosphorylation*

Chapter 4 gives details for the development of models for CaMKII state transitions. MoNP includes the state transitions of CaMKII subunits since the translocation between compartments is not considered. We modify CaMKII subunit model in Chapter 4 to include the pathway of dephosphorylation by PP2A [78]. The CaMKII subunits include state transitions among four states of the CaMKII subunit: inhibited CaMKII (iCaMKII); Ca^{2+}/CaM bound CaMKII (CaMCaMKII); Ca^{2+}/CaM bound autophosphorylated CaMKII (CaMCaMKII^P); and Ca^{2+}/CaM dissociated autophosphorylated CaMKII ($CaMKII^P$). The reaction rates of the state transitions are given as follows:

(1) Activation of the CaMKII subunit with Ca^{2+}/CaM complex binding (R_{25}):

$$R_{25} = k_{c5f}[iCaMKII][Ca_4CaM] - k_{c5b1}[CaMCaMKII],$$

(5.69)

where $[CaMKII^T]$, the total concentration of CaMKII subunit, is conserved: $[iCaMKII] = [CaMKII^T] - [CaMCaMKII] - [CaMCaMKII^P] - [CaMKII^P]$.

(2) Re-association of $CaMKII^P$ and Ca^{2+}/CaM complex (R_{26}):

$$R_{26} = k_{c5f}[CaMKII^P][Ca_4CaM] - k_{c5b2}[CaMCaMKII^P].$$

(5.70)

(3) Autophosphorylation (R_{27}):

$$R_{27} = \frac{K_{\mathrm{cat}14}P[CaMCaMKII][ATP]}{K_{m14} + [ATP]}, \quad \text{and} \quad (5.71)$$

$$P = \left(1 - \left(\frac{[iCaMKII]}{[CaMKII_T]}\right)^2\right).$$

(5.72)

So that, following the assumption of $[ATP]$ in Eq. (5.21), we have

$$R_{27} = K_{\mathrm{cat}14}P[CaMCaMKII].$$

(5.73)

(4) Dephosphorylation of $CaMCaMKII^P$ by PP1 (R_{28}) and PP2A (R_{29}):

$$R_{28} = \frac{K_{\mathrm{cat}15}[PP1][CaMCaMKII^P]}{K_{m15} + [CaMCaMKII^P]}, \quad \text{and} \quad (5.74)$$

$$R_{29} = \frac{K_{\mathrm{cat}16}[PP2A][CaMCaMKII^P]}{K_{m16} + [CaMCaMKII^P]}.$$

(5.75)

(5) Dephosphorylation of $CaMKII^P$ by PP1 (R_{30}) and PP2A (R_{31}):

$$R_{30} = \frac{K_{\mathrm{cat}15}[PP1][CaMKII^P]}{K_{m15} + [CaMKII^P]}, \quad \text{and} \tag{5.76}$$

$$R_{31} = \frac{K_{\mathrm{cat}16}[PP2A][CaMKII^P]}{K_{m16} + [CaMKII^P]}. \tag{5.77}$$

5.4.6. *The complete model of MoNP*

5.4.6.1. *Reaction rates*

This section gives the summary of reaction rates, including their mathematical equations, variables, constants and their biological meaning. The mathematical equations are given in Eqs. (5.78)–(5.109). The biological meaning of reaction rates are summarised in Table 5.1. The biological meaning and mathematical approaches for variables are summarised in Table 5.2. The biological meaning of constants are given in Table 5.3. Moreover,

$$R_1 = k_{1f}[Ca^{2+}][iCaM] - k_{1b}[CaCaM], \tag{5.78}$$

$$R_2 = k_{2f}[Ca^{2+}][CaCaM] - k_{2b}[Ca_2CaM], \tag{5.79}$$

$$R_3 = k_{3f}[Ca^{2+}][Ca_2CaM] - k_{3b}[Ca_3CaM], \tag{5.80}$$

$$R_4 = k_{4f}[Ca^{2+}][Ca_3CaM] - k_{4b}[Ca_4CaM], \tag{5.81}$$

$$R_5 = k_{c1f}[Ca_4CaM]([AC1_T] - [CaMAC1])$$

$$-k_{c1b}[CaMAC1], \tag{5.82}$$

$$R_6 = k_{c2f}[Ca_4CaM]([AC8_T] - [CaMAC8])$$

$$-k_{c2b}[CaMAC8], \tag{5.83}$$

$$R_7 = K_{\mathrm{cat}1}[CaMAC1], \tag{5.84}$$

$$R_8 = K_{\mathrm{cat}2}[CaMAC8], \tag{5.85}$$

Table 5.1. Variables of MoNP, their biological meaning, approach, and initial values.

Name	Biological meaning	Approach	Initial (μM)
Ca^{2+}	intracellular calcium ion	APRX.	
CaCaM	one Ca^{2+} ion bound to CaM	ODE	0
Ca_2CaM	two Ca^{2+} ions bound to CaM	ODE	0
Ca_3CaM	three Ca^{2+} ions bound to CaM	ODE	0
Ca_4CaM	Ca^{2+}/CaM complex	ODE	0
iCaM	CaM without Ca^{2+} bound	ODE	[CaMT]
CaMAC1	Ca_4CaM bound to AC isoform type 1	ODE	0
CaMAC8	Ca_4CaM bound to AC isoform type 8	ODE	0
CaMCaMKII	Ca_4CaM bound to CaMKII	ODE	0
$CaMCaMKII^P$	autophosphorylated CaMKII with Ca_4CaM	ODE	0
$CaMKII^P$	autophosphorylated CaMKII without Ca_4CaM	ODE	0
Ca_4CaNB	basal activated PP2B	APRX.	
CaMCaNA	full activated PP2B	ODE	0
cAMP	active cAMP	ODE	0
CaMPDE1	Ca_4CaM bound to PDE isoform type 1	ODE	0
iAC1	inactive AC isoform type 1	M.C.	$[AC1_T]$
iAC8	inactive AC isoform type 8	M.C.	$[AC8_T]$
iCaMKII	inactive CaMKII	M.C.	$[CaMKII_T]$
iPDE1	inactive PDE isoform type 1	M.C.	$[PDE1_T]$
iPP2B	inactive PP2B	M.C.	$\approx[PP2B_T]$
PDE4B	unphosphorylated PDE isoform type 4B	M.C.	$[PDE4B_T]$
PDE4D	unphosphorylated PDE isoform type 4D	M.C.	$[PDE4D_T]$
$PDE4B^P$	phosphorylated PDE isoform type 4B	ODE	0

(*Continued*)

Table 5.1.   (*Continued*)

Name	Biological meaning	Approach	Initial (μM)
PDE4D^P	phosphorylated PDE isoform type 4D	ODE	0
PKAc	released PKA catalytic subunit	ODE	0
PP1	active PP1	M.C.	[PP1$_T$]
R$_2$C$_2$	unreleased PKA complex	ODE	[R$_2$C$_{2T}$]
R$_2$C$_2$cAMP$_2$	unreleased PKA complex bound to 2 cAMP	ODE	0
R$_2$C$_2$cAMP$_4$	unreleased PKA complex bound to 4 cAMP	ODE	0
I1^{P}PP1	phosphorylated I1 bound to PP1	ODE	0
AC*	Ca^{2+} independent AC	APRX.	0

APRX.: approximation;
M.C.: Mass conservation.

$$R_9 = \frac{K_{\text{cat3}}[AC_T^*]}{1 + K}, \tag{5.86}$$

$$R_{10} = k_{c3f}[Ca_4CaM]([PDE1_T] - [CaMPDE1])$$

$$- k_{c3b}[CaMPDE1], \tag{5.87}$$

$$R_{11} = \frac{K_{\text{cat4}}[PKAc]([PDE4B_T] - [PDE4B^P])}{K_{m4} + ([PDE4B_T] - [PDE4B^P])}$$

$$- k_{10}[PDE4B^P], \tag{5.88}$$

$$R_{12} = \frac{K_{\text{cat4}}[PKAc]([PDE4D_T] - [PDE4D^P])}{K_{m4} + ([PDE4D_T] - [PDE4D^P])}$$

$$- k_{10}[PDE4D^P], \tag{5.89}$$

$$R_{13} = \frac{K_{\text{cat5}}[PDE1_T][cAMP]}{K_{m5} + [cAMP]}$$

$$\times \left(1 + \left(\frac{K_{\text{cat6}}}{K_{\text{cat5}}}\right)\frac{[CaMPDE1]}{[PDE1_T]}\right), \tag{5.90}$$

Table 5.2.   Biological meaning of reaction rates.

R_i	Biological meaning
1	Formation of CaCaM
2	Formation of Ca_2CaM
3	Formation of Ca_3CaM
4	Formation of Ca^{2+}/CaM complex
5	Activation of AC1
6	Activation of AC8
7	Converting ATP into cAMP by AC1
8	Converting ATP into cAMP by AC8
9	Converting ATP into cAMP by AC^*
10	Activation of PDE1
11	Phosphorylation of PDE4B by PKA
12	Phosphorylation of PDE4D by PKA
13	Converting cAMP into AMP by PDE1
14	Converting cAMP into AMP by PDE4B
15	Converting cAMP into AMP by PDE4D
16	Association of cAMP and PKA complex (1)
17	Association of cAMP and PKA complex (2)
18	Releasing of PKA catalytic subunit
19	Full activation of PP2B
20	Dissociation of Ca_4CaM from PP2B (high Ca^{2+})
21	Dissociation of Ca_4CaM from PP2B (low Ca^{2+})
22	Phosphorylation of I1by PKA
23	Dephosphorylation of $I1^P$ by PP2A
24	Dephosphorylation of $I1^P$ by PP2B
25	Activation of CaMKII
26	Association of Ca_4CaM and $CaMKII^P$
27	Autophosphorylation of CaMCaMKII
28	Dephosphorylation of $CaMCaMKII^P$ by PP1
29	Dephosphorylation of $CaMCaMKII^P$ by PP2A
30	Dephosphorylation of $CaMKII^P$ by PP1
31	Dephosphorylation of $CaMKII^P$ by PP2A

$$R_{14} = \frac{K_{\text{cat7}}[PDE4B_T][cAMP]}{K_{m7} + [cAMP]}$$

$$\times \left(1 + \left(\frac{K_{\text{cat8}}}{K_{\text{cat7}}} - 1\right)\frac{[PDE4B^P]}{[PDE4B_T]}\right), \tag{5.91}$$

$$R_{15} = \frac{K_{\text{cat9}}[PDE4D_T][cAMP]}{K_{m9} + [cAMP]}$$

Table 5.3. Biological meaning of constants.

Constants	Biological meaning
$[AC1_T]$	Total concentration of AC1
$[AC8_T]$	Total concentration of AC8
$[AC_T^*]$	Total concentration of Ca^{2+} independent AC
$[PDE1_T]$	Total concentration of PDE1
$[PDE4B_T]$	Total concentration of PDE4B
$[PDE4D_T]$	Total concentration of PDE4D
$[R_2C_{2T}]$	Total concentration of PKA complex
$[PP2B_T]$	Total concentration of PP2B
$[PP2A]$	Concentration of PP2A
$[PP1_T]$	Total concentration of PP1
$[I1_T]$	Total concentration of I1
$[CaMKII_T]$	Total concentration of CaMKII
$[CaM_T]$	Total concentration of CaM/CaM pool size

$$\times \left(1 + \left(\frac{K_{\text{cat}10}}{K_{\text{cat}9}} - 1 \right) \frac{[PDE4D^P]}{[PDE4D_T]} \right), \tag{5.92}$$

$$R_{16} = k_{17f}[R_2C_2][cAMP]^2 - k_{17b}[R_2C_2cAMP_2], \tag{5.93}$$

$$R_{17} = k_{18f}[R_2C_2cAMP_2][cAMP]^2$$

$$-k_{18b}[R_2C_2cAMP_4], \tag{5.94}$$

$$R_{18} = k_{19b}[R_2C_2cAMP_4] - k_{19f}\frac{[PKAc]^3}{2}, \tag{5.95}$$

$$[Ca_4CaNB] = \frac{([PP2B_T] - [CaMCaNA])[Ca^{2+}]^{n1}}{K_{d1}^{n1} + [Ca^{2+}]^{n1}}, \tag{5.96}$$

$$R_{19} = k_{c4f}[Ca_4CaM][Ca_4CaNB], \tag{5.97}$$

$$R_{20} = k_{c4b1}[CaMCaNA]\frac{[Ca^{2+}]^{n2}}{K_{d2}^{n2} + [Ca^{2+}]^{n2}}, \tag{5.98}$$

$$R_{21} = k_{c4b2}[CaMCaNA]\left(1 - \frac{[Ca^{2+}]^{n2}}{K_{d2}^{n2} + [Ca^{2+}]^{n2}} \right), \tag{5.99}$$

$$R_{22} = \frac{K_{\text{cat11}}[PKAc]([I1_T] - [I1^P PP1])}{K_{m11} + ([I1_T] - [I1^P PP1])}, \tag{5.100}$$

$$R_{23} = \frac{K_{\text{cat12}}[PP2A][I1^P PP1]}{K_{m12} + [I1^P PP1]}, \tag{5.101}$$

$$R_{24} = \frac{K_{\text{cat13}}([CaMCaNA] + 1/20[Ca_4CaNB])[I1^P PP1]}{K_{m13} + [I1^P PP1]}, \tag{5.102}$$

$$[iCaMKII] = [CaMKII_T] - [CaMCaMKII]$$

$$-[CaMCaMKII^P] - [CaMKII^P],$$

$$R_{25} = k_{c5f}[iCaMKII][Ca_4CaM]$$

$$-k_{c5b1}[CaMCaMKII], \tag{5.103}$$

$$R_{26} = k_{c5f}[CaMKII^P][Ca_4CaM]$$

$$-k_{c5b2}[CaMCaMKII^P], \tag{5.104}$$

$$R_{27} = K_{\text{cat14}}P[CaMCaMKII], \text{ where}$$

$$P = \left(1 - \frac{[iCaMKII]^2}{[CaMKII_T]^2}\right), \tag{5.105}$$

$$R_{28} = \frac{K_{\text{cat15}}[PP1][CaMCaMKII^P]}{K_{m15} + [CaMCaMKII^P]}, \tag{5.106}$$

$$R_{29} = \frac{K_{\text{cat16}}[PP2A][CaMCaMKII^P]}{K_{m16} + [CaMCaMKII^P]}, \tag{5.107}$$

$$R_{30} = \frac{K_{\text{cat15}}[PP1][CaMKII^P]}{K_{m15} + [CaMKII^P]}, \quad \text{and} \tag{5.108}$$

$$R_{31} = \frac{K_{\text{cat16}}[PP2A][CaMKII^P]}{K_{m16} + [CaMKII^P]}. \tag{5.109}$$

5.4.6.2. *ODEs*

This section gives ODEs for corresponding variables indicated "ODE approach" in Table 5.1:

$$\frac{d[iCaM]}{dt} = -R_1 + R_{21}, \tag{5.110}$$

$$\frac{d[CaCaM]}{dt} = R_1 - R_2, \tag{5.111}$$

$$\frac{d[Ca_2CaM]}{dt} = R_2 - R_3, \tag{5.112}$$

$$\frac{d[Ca_3CaM]}{dt} = R_3 - R_4, \tag{5.113}$$

$$\frac{d[Ca_4CaM]}{dt} = R_4 - R_5 - R_6 - R_7 - R_{19}$$

$$+ R_{20} - R_{25} - R_{26}, \tag{5.114}$$

$$\frac{d[CaMAC1]}{dt} = R_5, \tag{5.115}$$

$$\frac{d[CaMAC8]}{dt} = R_6, \tag{5.116}$$

$$\frac{d[CaMPDE1]}{dt} = R_{10}, \tag{5.117}$$

$$\frac{d[PDE4B^P]}{dt} = R_{11}, \tag{5.118}$$

$$\frac{d[PDE4D^P]}{dt} = R_{12}, \tag{5.119}$$

$$\frac{d[R_2C_2]}{dt} = -R_{16}, \tag{5.120}$$

$$\frac{d[R_2C_2cAMP_2]}{dt} = R_{16} - R_{17}, \tag{5.121}$$

$$\frac{d[R_2C_2cAMP_4]}{dt} = R_{17} - R_{18}, \tag{5.122}$$

$$\frac{d[cAMP]}{dt} = R_7 + R_8 + R_9 - R_{13} - R_{14}$$

$$-R_{15} - 2R_{16} - 2R_{17}, \tag{5.123}$$

$$\frac{d[PKAc]}{dt} = 2R_{18}, \tag{5.124}$$

$$\frac{d[CaMCaNA]}{dt} = R_{19} - R_{20} - R_{21}, \tag{5.125}$$

$$\frac{d[I1^P PP1]}{dt} = R_{22} - R_{23} - R_{24}, \tag{5.126}$$

$$\frac{d[CaMCaMKII]}{dt} = R_{25} - R_{27} + R_{28} + R_{29}, \tag{5.127}$$

$$\frac{d[CaMCaMKII^P]}{dt} = R_{26} + R_{27} - R_{28} - R_{29}, \quad \text{and} \tag{5.128}$$

$$\frac{d[CaMKII^P]}{dt} = -R_{26} - R_{30} - R_{31}. \tag{5.129}$$

5.5. Computational Procedures

5.5.1. *Simulating the sub-models*

A steady Ca^{2+} concentration, $[Ca^{2+}]$, is applied to drive sub-models (B, C, and D) to their steady states. For the steady $[Ca^{2+}]$, the initial concentration of Ca^{2+}/CaM complex, $[Ca_4CaM_0]$, is formulated as a function of $[Ca^{2+}]$ which is given by [79, 80]:

$$[Ca_4CaM_0] = \frac{[CaM_T][Ca^{2+}]^4}{K_{d1}K_{d2}K_{d3}K_{d4} + K_{d2}K_{d3}K_{d4}[Ca^{2+}]}{+ K_{d3}K_{d4}[Ca^{2+}]^2 + K_{d4}[Ca^{2+}]^3 + [Ca^{2+}]^4}, \tag{5.130}$$

where $[CaM_T]$ denotes the total concentration of CaM and K_{d1}, K_{d2}, K_{d3} and K_{d4} denote the dissociation constants of the bindings and have values, $20\,\mu\mathrm{M}$, $0.56\,\mu\mathrm{M}$, $100\,\mu\mathrm{M}$, $5\,\mu\mathrm{M}$, respectively [78]. The steady $[Ca^{2+}]$ and the corresponding $[Ca_4CaM_0]$ are used as the inputs for sub-models to allow us to compare the behaviours of sub-models under the identical stimulation environments since different modulators have different Ca^{2+} or CaM affinities, which may have a feedback interaction to the formation of Ca^{2+}/CaM complex. The length of the simulation should be sufficient to let sub-models to reach their steady states.

5.5.2. *Simulating MoNP*

Several computational experiments are designed to test MoNP under steady and transient signals. For computational experiments designed for steady and transient behaviours of MoNP, we call them steady activation level analysis (steady state in abbreviation) and transient activation levels analysis (transient in abbreviation), respectively. For steady state analysis, the input of MoNP is a steady $[Ca^{2+}]$ over a sufficient simulation time to let the modulators reach their steady states. The steady $[Ca^{2+}]$ is taken across a wide range starting from $0.07\,\mu$M, the rest concentration in hippocampal neurons. For transient analysis, the following procedure is followed: (1) $0.07\,\mu$M of $[Ca^{2+}]$ is applied over a sufficient simulation time to let the modulators reach their steady states. The steady states correspond to the rest state of a hippocampal neuron before the induction signals of synaptic plasticity reach the neuron, and (2) a special pattern of Ca^{2+} signal is applied to cause transient activations of the modulators. The Ca^{2+} patterns following the common protocols for the induction of synaptic plasticity are selected to simulate MoNP, where low-frequency stimulation (LFS) is for LTD induction and high-frequency stimulation (HFS) and theta burst stimulation (TBS) are for LTP induction [12, 81–83]. These Ca^{2+} patterns are generated using Zhabotinsky's approach [69] (see Chapter 3 for details).

5.5.3. *Development of indices for the effects*
of kinases and phosphatases

Kinase related Ca^{2+}/CaM complex binding index (KBI) is developed based on the ratio of Ca^{2+}/CaM complex competitively bound by CaMKII, which reflects the relative activation strength between CaMKII and phosphatases as discussed previously. KBI is given by

$$KBI = \frac{[Ca_4CaM_{LTP}]_B}{[Ca_4CaM_T]_B}. \tag{5.131}$$

$[Ca_4CaM_{LTP}]_B$ is the concentration of Ca^{2+}/CaM complex bound by proteins related to LTP, including AC isoforms 1/8 and two CaM attached states of CaMKII. $[Ca_4CaM_{LTP}]_B$ is expressed as follows: $[Ca_4CaM_{LTP}]_B = [CaMAC1] + [CaMAC8] + [CaMCaMKII] + [CaMCaMKII^P].[Ca_4CaM_T]_B$ is the concentration of Ca^{2+}/CaM complex bound by any proteins, expressed as follows: $[Ca_4CaM_T]_B = [CaMAC1] + [CaMAC8] + [CaMCaMKII] + [CaMCaMKII^P] + [CaMPDE1] + [CaMCaNA]$. It must be emphasised that KBI is defined in closely association with the structure with MoNP, hence it is only applied for the discussion of the behaviours of MoNP.

Net change of KBI is developed based on KBI to investigate the relative changes of the activation strength of CaMKII. Net change of KBI is the difference between recorded level of KBI and a base (KBI_{base}) which is set to the level of KBI at the rest Ca^{2+} concentration of $0.07\,\mu$M for steady state, or at the rest state of a hippocampal neuron before the induction signals are applied for transient. At the end, the normalised net change of $KBI, \Delta KBI$, is calculated as follows:

$$\Delta KBI = \frac{KBI - KBI_{\text{base}}}{KBI_{\text{base}}}. \tag{5.132}$$

Similarly, the normalised net change of concentration of the phosphorylated I1, $\Delta I1^P$, is used to reflect the relative activation strength between PKA and phosphatases following our discussion in Sec. 5.3.2. Now $\Delta I1^P$ is given by

$$\Delta I1^P = \frac{[I1^P PP1] - [I1^P PP1]_{\text{base}}}{[I1^P PP1]_{\text{base}}}, \tag{5.133}$$

where $I1^P PP1$ is the only state of phosphorylated I1 considered in MoNP.

The positive or negative $\Delta KBI/\Delta I1^P$, respectively, shows an increased or decreased relative activation strength of kinases against phosphatases, in relation to the basal relative activation strength.

In terms of LTP/LTD induction, positive or negative values of $\Delta KBI/\Delta \Pi^P$ indicate a higher probability of LTP or LTD induction, respectively, since the relative activation strength suggests the phosphorylation levels on target sites, where an increase may correlate to LTP induction and a decrease may correlate to LTD induction. The interpretation of $\Delta KBI/\Delta \Pi^P$ follows the use of the percentage change of EPSP from the basal level to indicate the existence of synaptic plasticity in experiments [12].

5.5.4. *Theoretical conditions for the bidirectionality*

An approximately 50% increase and an approximately 30% increase, respectively, of EPSP slope are observed following LTP induction by TBS or HFS and LTD induction by LFS [12, 83, 84]. We assume a strong correlation between $\Delta KBI/\Delta \Pi^P$ and EPSP slope so that we propose the following theoretical conditions which define the behaviours of MoNP to be LTP or LTD, based on the transient analysis. The theoretical conditions for LTP are: with subject to TBS or HFS, ΔKBI and $\Delta \Pi^P$ peak at greater than $\alpha(0 \leq \alpha \leq 1)$ and ΔKBI remains positive after the termination of the signal. The theoretical conditions for LTD are: with subject to LFS, $\Delta \Pi^P$ is smaller than $-\beta(0 \leq \beta \leq 1)$ at bottom and ΔKBI is negative during the signal. This is little direct evidence about the behaviour of ΔKBI during LTD, but we expect Ca^{2+}/CaM complex bound by phosphatases to increase since the activity of phosphatases is increased during LTD. In the computational experiments, both α and β are fixed at 0.4, which is within the range of EPSP slope change [12, 83, 84].

5.5.5. *GSA*

We use regionalised sensitivity analysis (RSA) [85–87] to access the sensitivity of the 70 parameters of MoNP (57 kinetic parameters and 13 constants) for the expression of LTP or LTD. We generate 50,000 sets of parameters and each set contains 70 randomly picked values where each value corresponds to a parameter of MoNP. The generation of the 70 values is using Latin Hypercube Sampling

(LHS) from the predefined ranges, 50% to 150% of the reference values, of parameters to ensure the uniform distribution and minimal correlation of samples [88]. MoNP is simulated by the 50,000 sets, one at a time. The acceptable set has MoNP to exhibit the corresponding behaviour investigated, LTP or LTD, by obeying the theoretical conditions as defined in Sec. 5.5.4. As a result, parameter sets are divided into two categories: "acceptable" and "unacceptable", for a behaviour investigated. A parameter is sensitive to the behaviour investigated if its cumulative frequency distributions (CFD) for the two categories are different, which means that some sub-regions of the parameter make the behaviour more or less acceptable compared to other regions [89]. We use Kolmogorov–Smirnov (KS) test to access the difference between the CFDs [87].

The results of RSA are used to understand the major factors coordinating LTP/LTD expressions based on the assumption that parameters related to the major factors must be sensitive to both LTP and LTD expressions. But, several precautions need to be considered carefully during the interpretation of the results of RSA:

(1) The "abnormal neurons" need to be excluded before we conduct further analysis of the results since "abnormal neurons" may not give useful information for the normal operation of a neuron. The parameter sets belonging to "normal neurons" should obey the following criteria:

(a) The rest state of *cAMP* is between 0.045 and 0.135 μM at 50% variation of the rest level (0.09 μM) from experiments [90].

(b) Ca^{2+}/CaM complex is expected to bind to LTP and LTD related proteins equally at rest state under the assumption of the balanced kinase and phosphatase activities at rest. As a result, basal *KBI* is between 0.4 and 0.6.

(c) Given 15% to 50% of S845 of AMPAR, (a target site of PKA) are phosphorylated at rest state [91], the acceptable ratio of phosphorylated I1 at rest is between 0.15 and 0.5.

It must be emphasised that the parameters are no longer uniformly distributed after the removal of "abnormal neurons".

Hence, the approach by CDFs takes advantage in this situation since it has no restriction in sampling distribution [89].

(2) The "normal neurons" should be sufficient and behaviours of LTP and LTD are independent. 6102 sets out of the 50,000 sets generated are "normal". 1164 and 4188 sets out of the 6102 "normal" sets give LTP and LTD behaviours, respectively. While 822 sets out of the 6102 "normal" sets give both LTP and LTD behaviours, we have

$$\frac{822}{6102} = P(LTP \cap LTD) \approx P(LTP) \times P(LTD)$$
$$= \frac{1164}{6102} \times \frac{4188}{6102}, \tag{5.134}$$

which suggests LTP and LTD behaviours are independent.

(3) We must not access the sensitivity of a parameter on the bidirectionality by comparing its CFD within sets which give both LTP and LTD behaviours and its CFD within sets which do not. The reason is that a parameter which is sensitive to either of LTP or LTD, but not both, also has its CFDs displaying a significant difference because some sub-regions of the parameter make either LTP or LTD behaviour more/less acceptable as well as the joint behaviour more/less acceptable because of the independence. Hence, the appropriate way is to find the intersection of separate sets of sensitive parameters for LTP and LTD, respectively.

Table 5.4 shows the 70 parameters of MoNP and their biological meaning for a better understanding of GSA results.

5.5.5.1. *Approximation of CFD*

It is difficult to calculate the exact CFDs for a parameter being accepted or rejected over a continuous range, but whether the CFDs are different can be obtained through approximations of the CFDs over a large number of samples drawn from the range. Given the m sets of accepted parameters and n sets of rejected parameters, the accepted space is an $m \times 70$ matrix, **AS**, and the rejected space is an $n \times 70$ matrix, **RS**. The rows of **AS** and **RS** denote sets of parameters and the 70 columns of **AS** and **RS** denote the corresponding samples of the 70 parameters which are accepted and rejected, respectively. For the *ith* parameter (or *ith* column of **AS**

Table 5.4.　Parameters, constants and their biological meanings.

(a) Name	Biological meaning
$[\mathrm{AC1_T}]$	Total concentration of AC1
$[\mathrm{AC8_T}]$	Total concentration of AC8
$[\mathrm{AC_T^*}]$	Total concentration of Ca^{2+} independent AC
$[\mathrm{PDE1_T}]$	Total concentration of PDE1
$[\mathrm{PDE4B_T}]$	Total concentration of PDE4B
$[\mathrm{PDE4D_T}]$	Total concentration of PDE4D
$[\mathrm{R2C2_T}]$	Total concentration of PKA complex
$[\mathrm{PP2B_T}]$	Total concentration of PP2B
$[\mathrm{PP2A_T}]$	Concentration of PP2A
$[\mathrm{CaM_T}]$	Total concentration of CaM/CaM pool size
$[\mathrm{PP1_T}]$	Total concentration of PP1
$[\mathrm{I1_T}]$	Total concentration of I1
$[\mathrm{CaMKII_T}]$	Total concentration of CaMKII
K	Active ratio of AC^*
K_{d1}	Dissociation constant for Ca^{2+} binding to CaNB
K_{d2}	Dissociation constant for Ca^{2+}/CaM complex binding to CaNA
$n1$	Hill coefficient for Ca^{2+} binding to CaNB
$n2$	Hill coefficient for Ca^{2+}/CaM complex binding to CaNA
k_{c1f}	Binding rate constant for Ca^{2+}/CaM complex and AC1
k_{c1b}	Dissociation rate constant of Ca^{2+}/CaM complex from AC1
k_{c2f}	Binding rate constant for Ca^{2+}/CaM complex and AC8
k_{c2b}	Dissociation rate constant of Ca^{2+}/CaM complex from AC8
k_{c3f}	Binding rate constant for Ca^{2+}/CaM complex and PDE1
k_{c3b}	Dissociation rate constant of Ca^{2+}/CaM complex from PDE1
k_{c4f}	Binding rate constant for Ca^{2+}/CaM complex and CaNA
k_{c4b1}	Dissociation rate constant of Ca^{2+}/CaM complex from CaNA at high Ca^{2+}
k_{c4b2}	Dissociation rate constant of Ca^{2+}/CaM complex from CaNA at low Ca^{2+}
k_{c5f}	Binding rate constant for Ca^{2+}/CaM complex and CaMKII
k_{c5b1}	Dissociation rate constant of Ca^{2+}/CaM complex from CaMKII subunit
k_{c5b2}	Dissociation rate constant of Ca^{2+}/CaM complex from autophosphorylated CaMKII subunit

(Continued)

Table 5.4.   (*Continued*)

(b) Parameter name	Biological meaning
K_{cat1}	Production rate of cAMP by active AC1
K_{cat2}	Production rate of cAMP by active AC8
K_{cat3}	Production rate of cAMP by AC*
K_{cat4}	Phosphorylation rate of PDE4s by PKAc
K_{cat5}	Converting rate of cAMP into AMP by inactive PDE1
K_{cat6}	Converting rate of cAMP into AMP by active PDE1
K_{cat7}	Converting rate of cAMP into AMP by unphosphorylated PDE4B
K_{cat8}	Converting rate of cAMP into AMP by phosphorylated PDE4B
K_{cat9}	Converting rate of cAMP into AMP by unphosphorylated PDE4D
K_{cat10}	Converting rate of cAMP into AMP by phosphorylated PDE4D
K_{cat11}	Phosphorylation rate of I1 by PKAc
K_{cat12}	Dephosphorylation rate of I1PP1 by PP2A
K_{cat13}	Dephosphorylation rate of I1PP1 by PP2B
K_{cat14}	Autophosphorylation rate of CaMKII
K_{cat15}	Dephosphorylation rate of autophosphorylated CaMKII by PP1
K_{cat16}	Dephosphorylation rate of autophosphorylated CaMKII by PP2A
K_{m4}	Michaelis constant for phosphorylating PDE4s by PKAc
K_{m5}	Michaelis constant for converting cAMP into AMP by PDE1
K_{m7}	Michaelis constant for converting cAMP into AMP by PDE4B
K_{m9}	Michaelis constant for converting cAMP into AMP by PDE4D
K_{m11}	Michaelis constant for phosphorylating I1 by PKAc
K_{m12}	Michaelis constant for dephosphorylating I1PP1 by PP2A
K_{m13}	Michaelis constant for dephosphorylating I1PP1 by PP2B
K_{m15}	Michaelis constant for dephosphorylating autophosphorylated CaMKII by PP1
K_{m16}	Michaelis constant for dephosphorylating autophosphorylated CaMKII by PP2A
k_{1f}	Binding rate constant of Ca^{2+} and CaM to form CaCaM
k_{1b}	Dissociation rate constant of CaCaM into Ca^{2+} and CaM

(*Continued*)

Table 5.4.   (*Continued*)

(b) Parameter name	Biological meaning
k_{2f}	Binding rate constant of Ca^{2+} and CaCaM to form Ca_2CaM
k_{2b}	Dissociation rate constant of Ca_2CaM into Ca^{2+} and CaCaM
k_{3f}	Binding rate constant of Ca^{2+} and CaCaM to form Ca_3CaM
k_{3b}	Dissociation rate constant of Ca_3CaM into Ca^{2+} and Ca_2CaM
k_{4f}	Binding rate constant of Ca^{2+} and CaCaM to form Ca_4CaM

(c) Parameter name	Biological meaning
k_{4b}	Dissociation rate constant of Ca_4CaM into Ca^{2+} and Ca_3CaM
k_5	Dephosphorylation rate of phosphorylated PDE4s
k_{6f}	Binding rate constant between cAMP and R_2C_2 to form R_2cAMP_2
k_{6b}	Dissociation rate constant of $R_2C_2cAMP_2$ into cAMP and R_2C_2
k_{7f}	Binding rate constant between cAMP and $R_2C_2cAMP_2$
k_{7b}	Dissociation rate constant of $R_2C_2cAMP_4$ into cAMP and $R_2C_2cAMP_2$
k_{8f}	Binding rate constant between PKAc and R2cAMP4
k_{8b}	Release rate constant of PKAc

and **RS**; $i = 1, 2, \ldots, 70$), it is accepted for m times and rejected for n times. Let us define AP_i as a vector containing the m values accepted, RP_i as a vector containing the n values rejected, AUP_i as a set containing the unique values in AP_i and RUP_i as a set containing the unique values in RP_i. Then we can calculate AF_i, the frequency of members of AUP_i appearing in AP_i, and RF_i, the frequency of members of RUP_i appearing in RP_i, as follows:

$$AF_i(x) = \begin{cases} \text{frequency of } x \text{ appeared in } APi, & x \in AUP_i, \\ 0, & x \notin AUP, \end{cases} \tag{5.135}$$

$$RF_i(x) = \begin{cases} \text{frequency of x appeared in } RPi, & x \in RUP_i, \\ 0, & x \notin RUP. \end{cases} \tag{5.136}$$

The cumulative frequencies, ACF_i and RCF_i, for being accepted and rejected, respectively, are given by

$$ACF_i(\beta) = \sum_{x=0}^{x \leq \beta} AF_i(x), \tag{5.137}$$

$$RCF_i(\beta) = \sum_{x=0}^{x \leq \beta} RF_i(x). \tag{5.138}$$

Here, β corresponds to parameter values and is drawn from a range as defined, but is unnecessary to be the same as samples drawn from LHS, to calculate corresponding ACF_i and RCF_i. The drawn βs are ordered from low to high and aligned with their corresponding ACF_is and RCF_is. Last, ACF_is are divided by m and RCF_is are divided by n. When m and n are large enough, ACF_is and RCF_is are approaching to the exact values of CFDs for being accepted and rejected, respectively.

5.5.5.2. *Examples of significant and insignificant parameters*

At the end of this section, we show some examples of the sensitivity of parameters to LTP and LTD in Figs. 5.3–5.6. The difference between the two CFDs tells the significance of a parameter regarding LTD and LTP behaviours.

5.6. Parameter Estimation and Validation of MoNP

5.6.1. *Kinetic characteristics of essential modulators*

In this section, the kinetic characteristics of the modulators are summarised. The kinetic characteristics include the affinities of bindings, mainly to Ca^{2+} or CaM; the enzymatic properties of kinases and phosphatases; the approximate concentrations in the neuron; and other relevant to the development of MoNP. Most kinetic characteristics of the modulators are estimated experimentally while

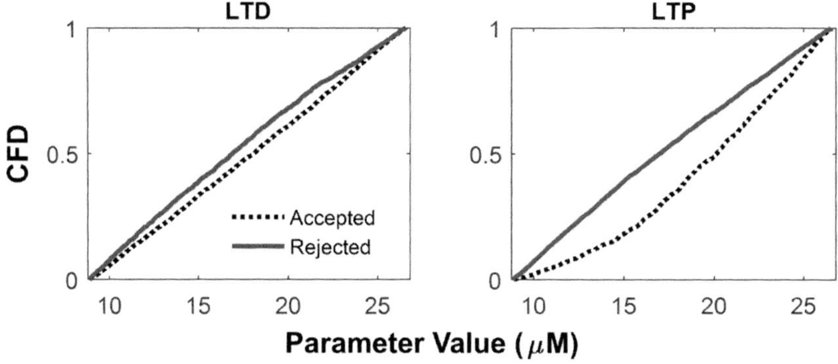

Fig. 5.3. $[CaM_T]$, sensitive to both LTP and LTD.

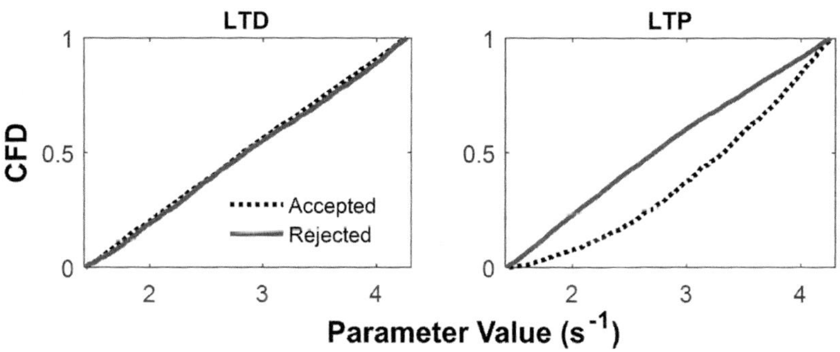

Fig. 5.4. K_{cat1}, sensitive to LTP, but not LTD.

only a few remain unknown, but are estimated by computational procedures [32, 34, 78]. When both experimental and computational estimations exist, we obtain the ranges of the experimentally estimated kinetic characteristics and compare them to the computational estimations. If they match (within 30% margin of difference), we use the computational estimations; otherwise, we use the mean of experimental estimations.

5.6.1.1. *Determining turnover rate of enzymatic reactions*

The enzymatic activity of an enzyme is usually estimated experimentally based on Michaelis–Menten kinetics and the estimation

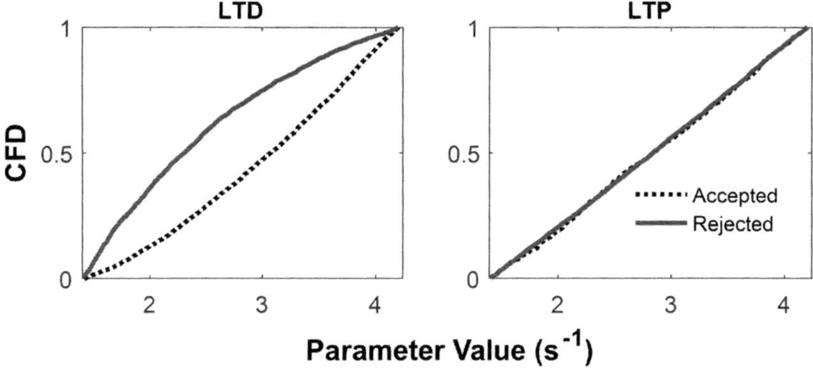

Fig. 5.5. $K_{\text{cat}13}$, sensitive to LTD, but not LTP.

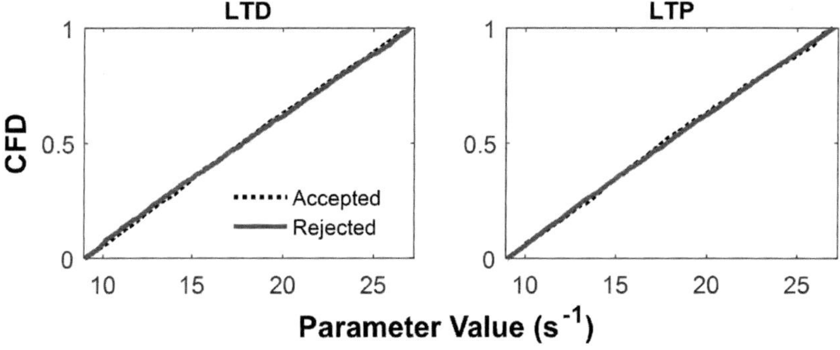

Fig. 5.6. $K_{\text{cat}4}$, insensitive to both LTP and LTD.

establishes two parameters: K_m and $V_{\max}$. $V_{\max}$ is the maximum
reaction rate of the enzymatic reaction to turn a substrate into a
product and K_m is the concentration of the substrate when the
reaction rate is half of $V_{\max}$. $V_{\max}$ can be approximated by a function
in terms of the concentration of the active enzyme, $[E_a]$, given as
$V_{\max} = K_{\text{cat}}[E_a]$, where K_{cat} is called the turnover rate of the
reaction equivalent to the rate constant of turning the substrate into
the product.

Since $[E_a]$ is dynamically changing due to the dynamic protein
interaction, determination of K_{cat} is essential for estimating the
reaction rate. The unified $V_{\max}$ unit is mol/(min.mg), expressed as
the number of moles of the product formed per milligram of the

enzyme over one minute. (mol/(min.mg) is convertible to the usual unit of the reaction rate, mol $L^{-1}s^{-1}$, given the volume and the mass of the enzyme). To calculate K_{cat} from V_{max}, the atomic mass of the enzyme, m_a (in Dalton), must be known. Hence, the K_{cat} (s^{-1}) can be calculated as follows:

$$K_{\text{cat}} = 1000V_{\text{max}}m_a/60. \qquad (5.139)$$

The K_m and K_{cat} of enzymatic reactions associated with the essential modulators are summarised in Table 5.5.

5.6.1.2. *Determining dissociation constant of bindings*

The dissociation constant, K_d, of a binding between two proteins can be determined by the quotient of their dissociation rate constant, k_{off}, divided by their association rate constant, k_{on}, given as $K_d = k_{\text{off}}/k_{\text{on}}$. K_d is a good indicator of the binding strength (or the affinity of the

Table 5.5.   Summary of the enzymatic characteristics of the modulators.

Substrate	Product	Enzyme	$K_{\text{cat}}(s^{-1})$	$K_m(\mu M)$	Source
I1	I1*	PKA	1.4	5	[46, 92]
I1*	I1	PP2A	2	16	[32]
		PP2B	2.8	3	[34, 58]
CaMKII	CaMKII*	CaMKII	1.2~10	19	[70, 78]
CaMKII*	CaMKII	PP2A	2	16	[32]
		PP1	1.72	11	[78]
ATP	cAMP	AC1/AC8	2.843	230	[32, 34]
		AC*	2	230	[32, 34]
cAMP	AMP	PDE1	1.7	10	[93]
		CaM-PDE1	10	10	[93]
		PDE4B	1.56	4.5	[94]
		PDE4D	5.4	1.5	[94]
		PDE4B*	3.12	4.5	[95]
		PDE4D*	10.8	1.5	[95]
PDE4	PDE4*	PKA	18	25	[96]

*Phosphorylated.

binding) between two proteins (the lower the stronger) and provides further information for the competitive dynamics among a mixture of proteins, which have the same binding partner. The modulators of synaptic plasticity usually require the binding to their activators to induce a conformational change to be activated and the CaM-regulated binding modulators may bind competitively to CaM for the activation.

The kinetic properties of the bindings are summarised in Table 5.6. As shown in the table, PP2B has the strongest affinity (28 pM) for the CaM binding among the CaM binding proteins listed; however, the autophosphorylation of CaMKII strengthens the binding affinity between CaMKII and CaM (52 pM) to the same level of PP2B. While PP2B and CaMKII are critical for the induction of LTD and LTP, respectively, the CaM competition may have a role in the bidirectional behaviour of synaptic plasticity. Furthermore, the binding affinity between PKA regulatory (PKAr) and catalytic (PKAc) subunits is strong, suggesting a low level of the PKA catalytic subunits existing alone.

5.6.1.3. *Concentrations of modulators in brain*

Most of the modulators have their brain concentration measured in the unit of μM. But the concentrations of PP2B, ACs and CaM-dependent PDEs are given in the units of μg per mg proteins [97],

Table 5.6. Summary of the binding affinities between the modulators.

Protein A	Protein B	k_{on} (μM^{-1}s^{-1})	k_{off} (s^{-1})	K_d	Source
	CaNA	46	$0.0012^{\mathrm{H}}/$ 0.2–2^{L}	$28\,pM/$ $\approx 5\,nM$	[77]
	PDE1	100	1	10 nM	[34]
Ca^{2+}/CaM	AC1	50	1	20 nM	[32]
Complex	AC8	20	1	50 nM	[47]
	CaMKII	21	$1.1/0.0011^{\mathrm{A}}$	51 nM/52 pM	[77]
PKAr	PKAc	0.17	0.0016	9 nM	[41]

^{L}Low $[Ca^{2+}]$; ^{H} high $[Ca^{2+}]$; ^{A} autophosphorylated.

the percentage of total membrane proteins [98], and percentage of total cellular proteins [74], respectively. However, the concentration of CaM is given in both units of μg per mg proteins and μM [99] that gives us an estimated concentration, $2.1\,\mu$M, of PP2B. For ACs and PDEs, we take the values used by a previous computational study [33], since the exact total amount of synaptic proteins is unclear. There are two types of ACs in hippocampus: Ca^{2+}-dependent ACs, including isoform type 1/8 (AC1 and AC8); and Ca^{2+}-independent ACs. AC8 accounts for less than 20% of Ca^{2+}-dependent ACs in the hippocampus [100]. The previous computational study uses $2.5\,\mu$M for the concentration of AC1, so that the concentration of AC8 is $0.625\,\mu$M (25% of $2.5\,\mu$M). The concentration of Ca^{2+}-independent ACs is estimated to be $2.5\,\mu$M to produce the "normal neuron" condition given in Section 5.4.5. We also assume equal concentrations between the PDE isoforms type 4B (PDE4B) and type 4D (PDE4D). The concentrations of the modulators are summarised in Table 5.7.

5.6.2. *Parameter estimation*

k_{17f}, k_{18f} and k_{19f} which are not given in the literature are estimated using Markov chain Monte Carlo (MCMC) [102–104]. Some studies

Table 5.7.　Summary of the concentrations of the modulators.

Protein	Concentration (μM)	Source
CaM	17	[99]
PKA	1.2	[101]
PP2B	2.1	[97]
PP2A	0.1111	[32]
PP1	3.5	[33]
I1	1.5	[33]
CaMKII	20	[32]
PDE1	4	[33][74]
PDE4	2	[33]
AC*	2.5	Estimated
AC1	2.5	[98]
AC8	0.625	[100]

give these parameters in $\mu\,\mathrm{M}^{-1}\mathrm{s}^{-1}$, but our framework requires values in $\mu\mathrm{M}^{-2}\mathrm{s}^{-1}$. Two sets of data are obtained for the normalised PKA activity against the concentration of cAMP [41, 76]. We perform computational experiments following the same procedure on which the data is obtained. Each experiment has two steps: (1) 10 nM PKA is incubated with cAMP for 2 minutes; and (2) the PKA activity, the concentration of PKA catalytic subunits, is measured at the end of the experiment. Many experiments are performed for different concentrations of cAMP from $0.01\,\mu\mathrm{M}$ to $100\,\mu\mathrm{M}$. At the end, the PKA activities among the experiments are normalised by PKA activity at the maximum cAMP concentration used among the experiments. One set of the data is used for the estimation and the other set is used for the validation. The result of the estimation is shown in Fig. 5.7. All values of parameters and constants are summarised in Tables 5.8 and 5.9, respectively.

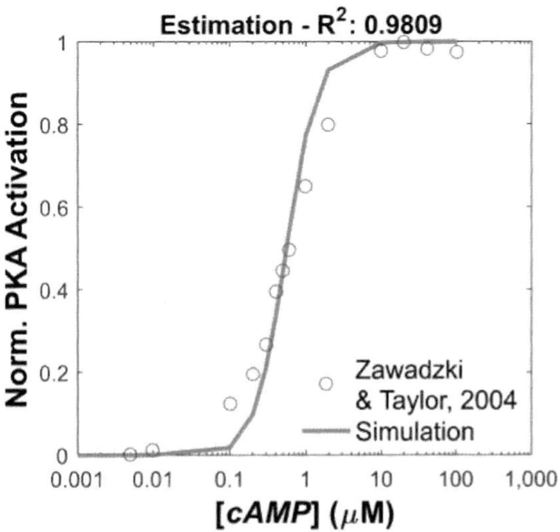

Fig. 5.7. Estimation of parameters for the activation of PKA. The estimation of the unknown parameters is based on the experimental data [41]. The R^2 of the estimation is shown in the title. The known parameters are $k_{17b}(0.02\,\mathrm{s}^{-1})$, $k_{18b}(0.2\,\mathrm{s}^{-1})$ and $k_{19b}(0.0016\,\mathrm{s}^{-1})$ [34, 41]. The estimated parameters are $k_{17f}(8\,\mu\mathrm{M}^{-2}\mathrm{s}^{-1})$, $k_{18f}(0.7\,\mu\mathrm{M}^{-2}\mathrm{s}^{-1})$, and $k_{19f}(0.25\,\mu\mathrm{M}^{-2}\mathrm{s}^{-1})$.

Table 5.8. Parameters of reaction rates, R_i, and their sources.

R_i	Kinetic parameters			Source
1	$k_{1f} : 2.5\,\mu\mathrm{M}^{-1}\mathrm{s}^{-1}$		$k_{1b} : 50\,\mathrm{s}^{-1}$	[78]
2	$k_{2f} : 88.25\,\mu\mathrm{M}^{-1}\mathrm{s}^{-1}$		$k_{2b} : 50\,\mathrm{s}^{-1}$	[78]
3	$k_{3f} : 12.5\,\mu\mathrm{M}^{-1}\mathrm{s}^{-1}$		$k_{3b} : 1250\,\mathrm{s}^{-1}$	[78]
3	$k_{4f} : 250\,\mu\mathrm{M}^{-1}\mathrm{s}^{-1}$		$k_{4b} : 1250\,\mathrm{s}^{-1}$	[78]
5	$k_{c1f} : 50\,\mu\mathrm{M}^{-1}\mathrm{s}^{-1}$		$k_{c1b} : 1\,\mathrm{s}^{-1}$	[32]
6	$k_{c2f} : 20\,\mu\mathrm{M}^{-1}\mathrm{s}^{-1}$		$k_{c2b} : 1\,\mathrm{s}^{-1}$	[47]
7	$K_{\mathrm{cat1}} : 2.843\,\mathrm{s}^{-1}$			[32, 34]
8	$K_{\mathrm{cat2}} : 2.843\,\mathrm{s}^{-1}$			[32, 34]
9	$K_{\mathrm{cat3}} : 2\,\mathrm{s}^{-1}$		$K : 20$	[32, 34]
10	$k_{c3f} : 100\,\mu\mathrm{M}^{-1}\mathrm{s}^{-1}$		$k_{c3b} : 1\,\mathrm{s}^{-1}$	[34]
11	$K_{\mathrm{cat4}} : 18\,\mathrm{s}^{-1}$	$K_{m4} : 25\,\mu\mathrm{M}$	$k_{10} : 0.25\,\mathrm{s}^{-1}$	[96]
12	$K_{\mathrm{cat4}} : 18\,\mathrm{s}^{-1}$	$K_{m4} : 25\,\mu\mathrm{M}$	$k_{10} : 0.25\,\mathrm{s}^{-1}$	[96]
13	$K_{\mathrm{cat5}} : 1.7\,\mathrm{s}^{-1}$	$K_{\mathrm{cat6}} : 10\,\mathrm{s}^{-1}$	$K_{m5} : 10\,\mu\mathrm{M}$	[93]
14	$K_{\mathrm{cat7}} : 1.56\,\mathrm{s}^{-1}$	$K_{\mathrm{cat8}} : 3.12\,\mathrm{s}^{-1}$	$K_{m7} : 4.5\,\mu\mathrm{M}$	[94]
15	$K_{\mathrm{cat9}} : 5.4\,\mathrm{s}^{-1}$	$K_{\mathrm{cat10}} : 10.8\,\mathrm{s}^{-1}$	$K_{m9} : 1.5\,\mu\mathrm{M}$	[94]
16	$k_{17f}^{*} : 8\,\mu\mathrm{M}^{-2}\mathrm{s}^{-1}$		$k_{17b} : 0.02\,\mathrm{s}^{-1}$	[33]
17	$k_{18f}^{*} : 0.7\,\mu\mathrm{M}^{-2}\mathrm{s}^{-1}$		$k_{18b} : 0.2\,\mathrm{s}^{-1}$	[33]
18	$k_{19f}^{*} : 0.25\,\mu\mathrm{M}^{-2}\mathrm{s}^{-1}$		$k_{19b} : 0.0016\,\mathrm{s}^{-1}$	[41]
19	$k_{c4f} : 46\,\mu\mathrm{M}^{-1}\mathrm{s}^{-1}$		$K_{d1} : 0.5\,\mu\mathrm{M}\,n1 : 1.8$	[59, 77]
20	$k_{c4b1} : 0.0012\,\mathrm{s}^{-1}$		$K_{d2} : 0.1\,\mu\mathrm{M}\,n2 : 3$	[59, 77]
21	$k_{c4b2} : 2\,\mathrm{s}^{-1}$		$K_{d2} : 0.1\,\mu\mathrm{M}\,n2 : 3$	[59, 77]
22	$K_{\mathrm{cat11}} : 1.4\,\mathrm{s}^{-1}$		$K_{m11} : 5\,\mu\mathrm{M}$	[46, 92]
23	$K_{\mathrm{cat12}} : 2\,\mathrm{s}^{-1}$		$K_{m12} : 16\,\mu\mathrm{M}$	[32]
24	$K_{\mathrm{cat13}} : 2.8\,\mathrm{s}^{-1}$		$K_{m13} : 3\,\mu\mathrm{M}$	[34, 58]
25	$k_{c5f} : 21\,\mu\mathrm{M}^{-1}\mathrm{s}^{-1}$		$k_{c5b1} : 1.1\,\mathrm{s}^{-1}$	[77, 78]
26	$k_{c5f} : 21\,\mu\mathrm{M}^{-1}\mathrm{s}^{-1}$		$k_{c5b2} : 0.0011\,\mathrm{s}^{-1}$	[105]
27	$K_{\mathrm{cat14}} : 1.2\,\mathrm{s}^{-1}$			[70, 78]
28	$K_{\mathrm{cat15}} : 1.72\,\mathrm{s}^{-1}$		$K_{m15} : 11\,\mu\mathrm{M}$	[78]
29	$K_{\mathrm{cat16}} : 2\,\mathrm{s}^{-1}$		$K_{m16} : 16\,\mu\mathrm{M}$	[32]
30	$K_{\mathrm{cat15}} : 1.72\,\mathrm{s}^{-1}$		$K_{m15} : 11\,\mu\mathrm{M}$	[78]
31	$K_{\mathrm{cat16}} : 2\,\mathrm{s}^{-1}$		$K_{m16} : 16\,\mu\mathrm{M}$	[32]

*:MCMC estimation. For example, row 18 gives values for two parameters: the left most one is estimated by MCMC and the other one is obtained from the source.

5.6.3. *Validating the sub-models*

In this section, the sub-models are validated against available experimental data. Since Ca^{2+}/CaM dynamics and CaMKII dynamics are extensively validated in Chiba *et al.*'s work [78], therefore,

Table 5.9.   Values of constants and their sources.

Constant	Value (μM)	Source
$[\text{AC1}_T]$	2.5	[98]
$[\text{AC8}_T]$	0.625	[100]
$[\text{AC}^*_T]$	2.5	Estimated
$[\text{PDE1}_T]$	4	[33, 74]
$[\text{PDE4B}_T]$	1	[33]
$[\text{PDE4D}_T]$	1	[33]
$[\text{R2C2}_T]$	1.2	[101]
$[\text{PP2B}_T]$	2.1	[97]
$[\text{PP2A}]$	0.11111	[32]
$[\text{PP1}_T]$	3.5	[33]
$[\text{I1}_T]$	1.5	[33]
$[\text{CaMKII}_T]$	20	[32, 106]
$[\text{CaM}_T]$	17.7	[99]

it is unnecessary to validate them again (sub-models A and E), since these sub-models are formulated based on their work. For sub-models lacking experimental data, we compare their dynamics against previous models in the next section.

5.6.3.1. *Validating PKA sub-model*

Following the same procedure as to estimate the parameters for PKA in Sec. 5.6.2, the PKA activation simulated is validated against experimental published in [76]. The result of the validation is shown in Fig. 5.8, which shows a good consistency between the simulation and data.

5.6.3.2. *Validating PP2B sub-model*

First, we compare PP2B activation against the experimental data published in [59]. The procedure, which is adapted from the experimental steps, is followed: (1) simulating MoNP with 30 nM of PP2B mixing with CaM at either of the two concentration, 300 nM or $20\,\mu$M, across different concentrations of Ca^{2+}. The simulation time should be sufficient to reach equilibrium; (2) at the equilibrium, the active PP2B concentration is measured for each of the Ca^{2+}

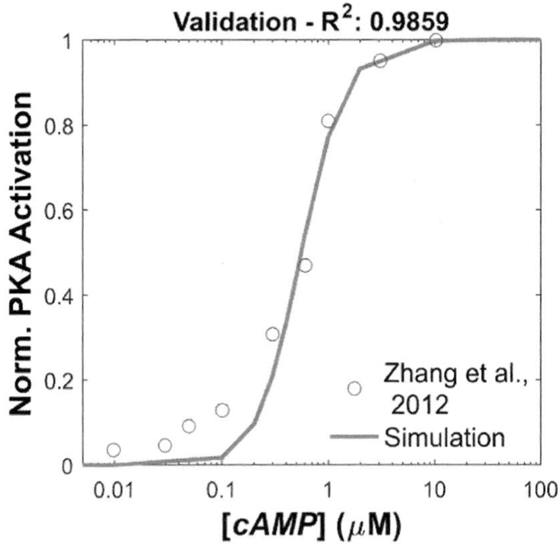

Fig. 5.8. Validation of PKA activation. The validation is based on the experimental data [76]. The R^2 of the validation is 0.9859.

concentrations used; and (3) in the PP2B concentrations measured, find the maximum and normalised all the measured concentration by the maximum. The validation result is shown in Fig. 5.9, which gives good agreement to the trends shown in experimental data. However, the curves show displacement from the data. Since the validation includes not only the PP2B activation by binding to Ca^{2+} and Ca^{2+}/CaM complex, but also the Ca^{2+} and CaM dynamics to form Ca^{2+}/CaM complex, one of the reasons for the displacement may be the simplified pathways for Ca^{2+}/CaM complex formation. To get rid of the Ca^{2+}/CaM complex formation pathway for the validation, the dissociation of CaM from PP2B is validated at next.

The dissociation of CaM from PP2B is validated under two condition: High Ca^{2+} level at $10\,\mu$M and low Ca^{2+} level at $5\,$nM. For high Ca^{2+} level, $0.1\,\mu$M of PP2B is first mixed with a small CaM pool of $0.06\,\mu$M until equilibrium. Then, a large CaM pool of $3\,\mu$M is added to force the small pool CaM to dissociate from PP2B. The concentration of PP2B bound by CaM from the small pool is

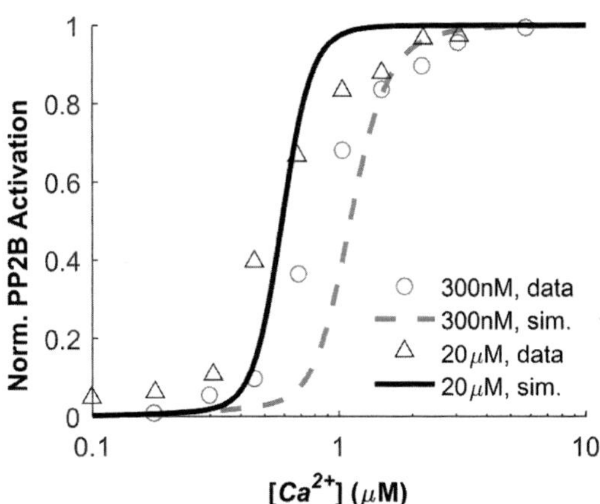

Fig. 5.9. Validation of PP2B activation. The validation is based on the experimental data [59].

recorded for 10,000s after the addition of the large CaM pool. For low Ca^{2+} level, $0.2\,\mu M$ of PP2B is mixed with $0.12\,\mu M$ of CaM under $10\,\mu M$ of Ca^{2+} until equilibrium. Then, Ca^{2+} concentration decreases to less than $5\,nM$ [77] (but, we use $5\,nM$ in the simulation) to force CaM to dissociate from PP2B. The concentration of PP2B bound by CaM is recorded for $20\,s$ after the Ca^{2+} concentration decrease. At the end, both records are normalised by the corresponding range of the maximum and minimum concentrations. The results are shown in Fig. 5.10 yielding excellent consistency against experiment data [77].

5.6.4. *Validation against previous models*

In this section, we compare the activation patterns of modulators, including PKA, PP2B, PP1 and CaMKII, simulated by MoNP against previous models [32, 33]. The quantitative change of the concentration is not an option for the comparison since these models use different protein concentration. Therefore, we focus on the trend of the change by the following procedure: (1) 100 sets of the 70 parameters are generated using LHS from a predefined range, 10% to 190% of their reference values; (2) MoNP is simulated with the 100 sets of parameters, one set at a time. The simulation

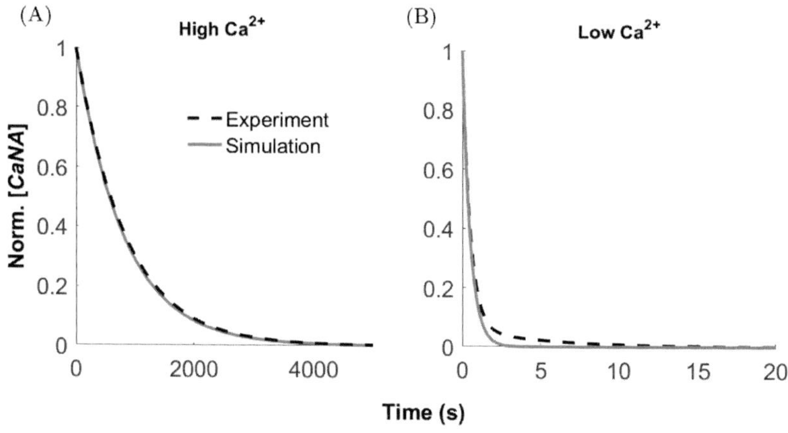

Fig. 5.10. Validation of CaM dissociation time course from PP2B. The validate is based on the experimental data [77]. The dissociation time courses of CaM from PP2B are shown following the procedures described in the text as (A) high Ca^{2+} level and (B) low Ca^{2+} level.

contains two pieces, one with LTP signal and the other with LTD signal. The dynamics of the four modulators during the two signals are recorded; and (3) each recorded dynamic is normalised by its corresponding range (maximum level–minimum level) into 0 to 1. The same procedure is also used to simulate previous models with their reference parameters only and Kim *et al.*'s model [33] is only simulated with LTP signal since they only investigate LTP behaviours. The results of the comparison are shown in Figs. 5.11 and 5.12. The dynamics which are far from previous models are removed for better visualisation. The trends of modulators in MoNP follow at least one of the previous models except for CaMKII, which exhibit a much greater oscillation in response to LTP signal. The greater oscillation may be caused by the structure of the CaMKII sub-models in these models. MoNP only considers CaMKII subunit state transitions in one compartment while Hayer and Bhalla [32] consider a separate PSD compartment as well as the translocation of CaMKII between PSD and postsynaptic cytoplasm. The slower translocation may offset the rapid activation of CaMKII subunits following the stimulations. Moreover, the time scales of these processes are different; the CaMKII autophosphorylation lasts for one

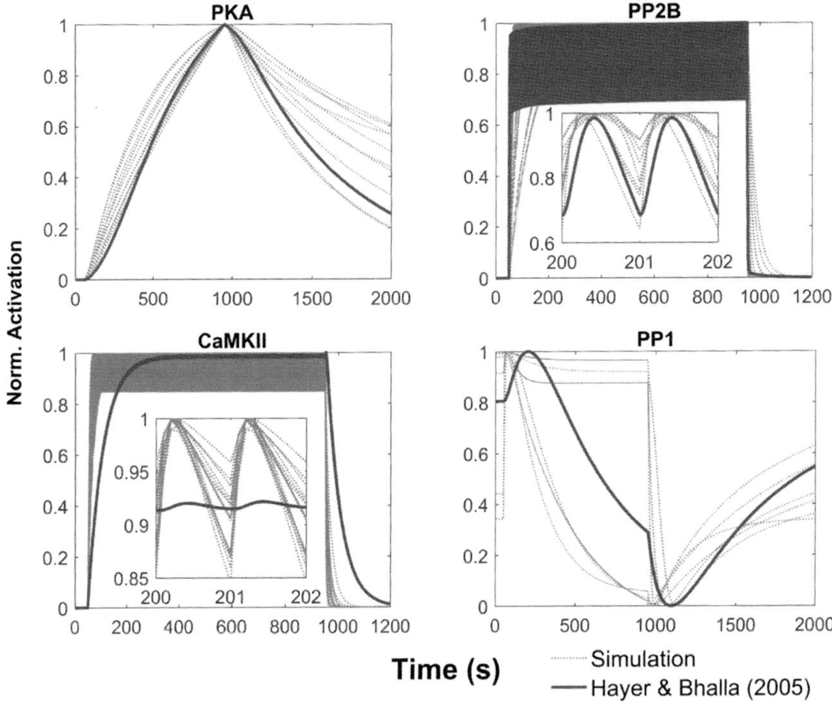

Fig. 5.11.   Comparison between MoNP and previous model during LTD signal.

minute [57] while the PSD components of CaMKII lasts for more than 30 minutes [107]. Since Chiba *et al.*'s model [78], which CaMKII sub-model of MoNP is developed based on, has excellent consistency with experimental data on CaMKII subunit state transitions, the different between MoNP and previous models in CaMKII dynamics is very likely to be caused by the different levels of the consideration on CaMKII processes.

5.7. Evaluating the Sub-models

In this section, we evaluate the sub-models and comment on aspects of the sub-models which are important for the bidirectionality. Sub-models A and D are extensively studied previously, please refer to [78] for their behaviours.

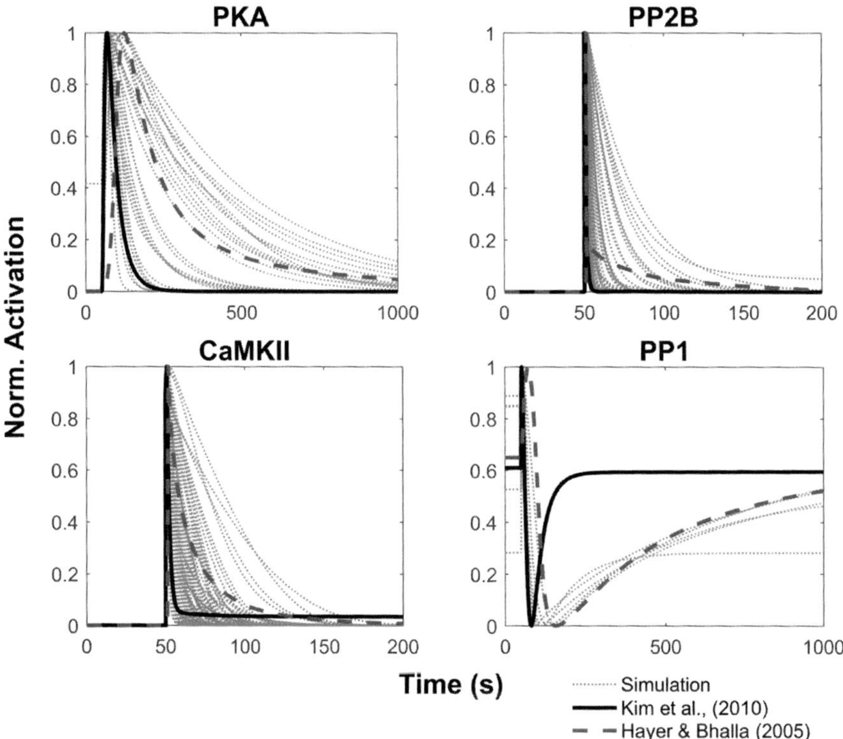

Fig. 5.12.   Comparison between MoNP and previous models during LTP signal.

5.7.1. *Sub-model B: Switch behaviour of PKA*

As shown in Fig. 5.13, PKA exhibits a switch-like activation having Ca^{2+} as the control parameter. PKA activation is prevented for Ca^{2+} concentration less than $2\,\mu M$ and rapidly elevated for Ca^{2+} concentration beyond $2\,\mu M$. The released PKA catalytic subunit reaches a peak at $0.25\,\mu M$, 1/10 of the total PKA concentration, given the strong binding affinity between the regulator dimer and catalytic subunits as discussed previously [41]. The PDEs are essential in the switch-like behaviour of PKA, where removing both PDEs, Ca^{2+} change has little effect on the PKA activation. With the removal of PDE1 or PDE4 alone, the peak of PKA activation is slightly increased approaching the peak at which both PDEs are removed, but the switch-like behaviour of PKA activation remains.

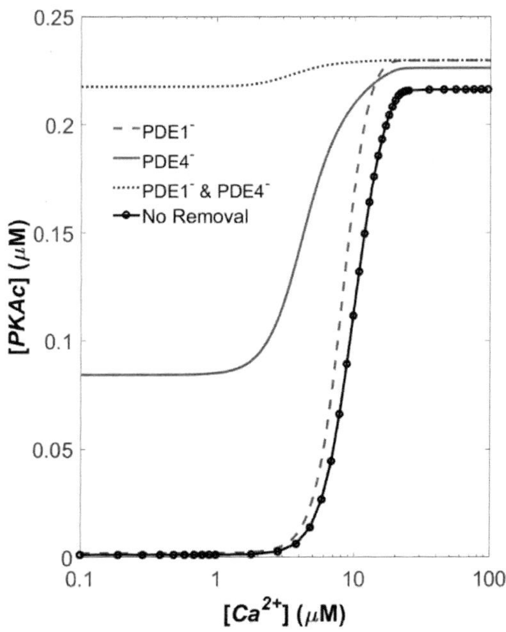

Fig. 5.13. Steady concentration of PKA over Ca^{2+} concentrations. The results are compared with circuits with PDE1 removal (PDE1$^-$), PDE4 removal (PDE4$^-$), both PDEs removal (PDE1$^-$ & PDE4$^-$) and no removal.

PDE1 and PDE4 display different roles in the regulation, where PDE1 decreases the slope of PKA activation in relation to Ca^{2+} increase since PDE1 is activated by Ca^{2+}/CaM complex synchronised with the activation of AC1/AC8, while PDE4 facilitates a basal inhibition of PKA when Ca^{2+} concentration is low, at which PDE1 is ineffective. The prevention of PKA activation at low Ca^{2+} levels ensure the ineffectiveness of PKA related target sites for LTP to promote the induction of LTD induction.

5.7.2. *Sub-model C: Two levels of PP2B activation*

As shown in Fig. 5.14, the two levels of PP2B activation occurs at different Ca^{2+} concentrations. The basal activation of PP2B occurs at a very low Ca^{2+} concentration and peaks at around $2\,\mu M$ of Ca^{2+}, where the full activation of PP2B starts to accumulate. In comparison, basal activation of PP2B peaks before PKA initialises

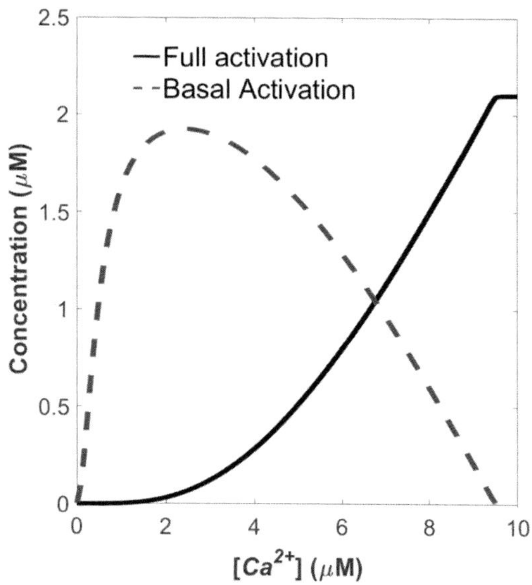

Fig. 5.14.   Steady concentration of basal/full activation of PP2B over Ca^{2+} concentrations.

its activation and PP2B is completely activated before $10\,\mu$M of Ca^{2+}, where PKA is just moderately activated. In both cases, the dephosphorylation becomes dominant on the critical sites, again to promote the induction of LTD over these Ca^{2+} concentrations.

5.7.3. *Sub-model D: Competition on PP1 activation*

The level of active PP1 reflects the relative activation strength between PKA and PP2B. The concentration of the phosphorylated I1, $[I1^{P}PP1]$, which also denotes the concentration of the inhibited PP1 due to the construction of MoNP, provides a quantitative measure of the relative activation strength and is shown in Fig 5.15. At rest, $[I1^{P}PP1]$ is at the peak, approximately $0.4\,\mu$M, as a result of the high basal level of PKA due to Ca^{2+}-independent ACs while PP2B is purely dependent on Ca^{2+} activation. With the increase in Ca^{2+} concentration, $[I1^{P}PP1]$ first decreases to a platform in line with the basal activation of PP2B and then decreases further to a trough that is accounted by exchange of the basal activation to the

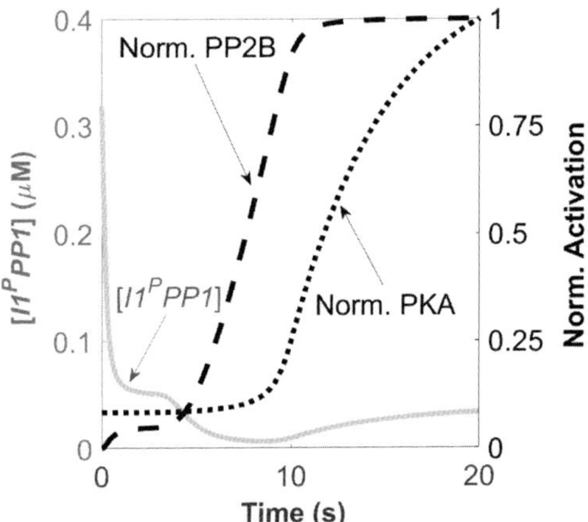

Fig. 5.15. Steady concentration of phosphorylated I1 over Ca^{2+} concentration. The steady concentration of phosphorylated I1, $I1^P PP1$, is aligned with the steady activation of PKA and PP2B, normalised by their maximum activity levels.

full activation of PP2B before PKA is activated. Near to the trough, PKA is activated to lift $[I1^P PP1]$. But, the lift is so weak which is far below the peak at rest. Given the lower catalytic activity and the strong inhibition of PKA compared to those of PP2B as suggest in the literature [41, 58, 92], it is clear that phosphorylation by PKA is hardly competing against the dephosphorylation by PP2B for steady Ca^{2+} stimulations. Hence, a special pattern of stimulation is essential to maximise the effects of PKA on phosphorylating its target sites.

5.8. Understanding Synaptic Plasticity Based on ΔKBI and $\Delta I1^P$

In this section, we first analyse the behaviours of ΔKBI and $\Delta I1^P$ to understand their behaviours in response to different signals. Then, we conduct computational experiments to understand the behaviours of the indices against effective time scales of modulators to draw conclusions of the roles of these modulators in the emergence of synaptic plasticity. At the end, we perform a computational study to understand the implication of the removal of PKA by PP2B in the emergence of LTD.

For the steady state (Fig. 5.16A), the curve of ΔKBI changes its direction with the increase of Ca^{2+} concentration. ΔKBI decreases until $1\,\mu M$ of Ca^{2+}, then rises rapidly to the basal level at approximately $1.5\,\mu M$ of Ca^{2+} and increases smoothly to a steady peak beyond $3\,\mu M$ of Ca^{2+}. These changes reflect different phases of the relative activation strength between CaMKII and phosphatases. The initial drop is caused by the slowing down of Ca^{2+}/CaM complex dissociation from CaNA of PP2B when Ca^{2+} concentration increases [77] as well as the stronger binding affinity of LTD related proteins to Ca^{2+}/CaM complex. These characteristics of LTD related proteins ensure an advantageous activation of LTD related modulators against LTP-related modulators over moderate Ca^{2+} elevations, approximately less than $1\,\mu M$. After the trough reached by the initial drop, the rise is caused by turning on of LTP-related proteins, which dominants the binding to Ca^{2+}/CaM complex. For example, given the abundance in dendritic spine and the enhancement of the binding affinity by the autophosphorylation, CaMKII dominants the binding to Ca^{2+}/CaM complex at high Ca^{2+} levels. The two-level behaviour of $\Delta I1^P$ is exactly the same as discussed in Sec. 5.7.3 following the competition between PP2B and PKA.

For transient over HFS (Fig. 5.16B), ΔKBI dips at very beginning and elevates rapidly afterward to a peak. After the termination of the HFS, ΔKBI remains positive up to nearly $50\,s$ and then stays at the basal level. $\Delta I1^P$ has a great initial drop followed by smooth recovery. At the end of the simulation duration, $\Delta I1^P$ is above the basal level at a peak of 0.2.

For transient over TBS (Fig. 5.16C), the three trains of TBS generates three repeats of a similar pattern to the response of HFS for both ΔKBI and $\Delta I1^P$. One significant difference is that $\Delta I1^P$ peaks at a much higher level, 0.4, at the end of the simulation.

For transient over LFS (Fig. 5.16D), ΔKBI is slightly below the basal level and remains at -0.05 during the simulation. $\Delta I1^P$ changes significantly with an initial drop and remains at a bottom of less than -0.4 during the stimulation. After the termination of the LFS, it recovers and peaks slightly above the basal level indicating a rephosphorylation of its target sites.

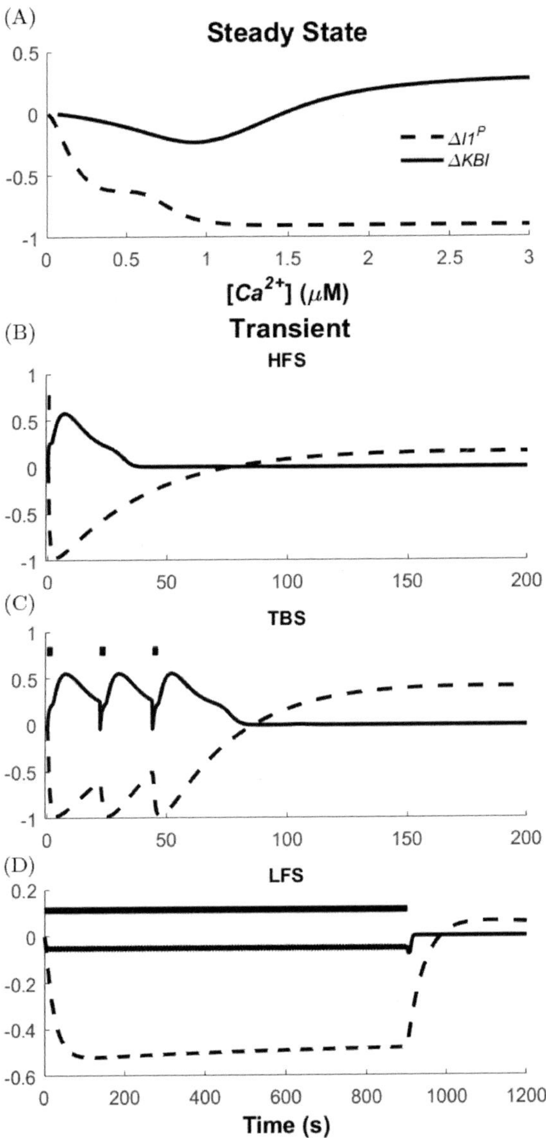

Fig. 5.16. ΔKBI and $\Delta I1^P$ in steady state and transient state. The transient stimulation patterns are shown with black bars indicating their durations in (B), (C) and (D). (A) Behaviours of ΔKBI and $\Delta I1^P$ in the steady state against different Ca^{2+} concentrations. (B) Behaviours of ΔKBI and $\Delta I1^P$ in transient in response to HFS. (C) Behaviours of ΔKBI and $\Delta I1^P$ in transient in response to TBS. (D) Behaviours of ΔKBI and $\Delta I1^P$ in transient in response to LFS.

The transient behaviours of the indices in response to TBS and LFS are in line with the theoretical conditions for the bidirectionality while HFS produces a slightly smaller $\Delta I1^P$ than the requirement. This result is interesting because HFS, with a single train, is unreliable for the induction of the late phase of synaptic plasticity [108], which may require gene expressions which have PKA as the major player [109], and has a greater chance for LTP induction when it lasts for more than 1 s [29]. Hence, we decide to base our later dissuasions on TBS and LFS.

5.8.1. *Effective time scales for the transient activation of modulators*

The rapid changes of the indices within the short simulation duration are underlined by the changes in relative activation strength of the key modulators. To understand the correlation between the activations of the key modulators and the dynamical changes of the indices, the normalised transient activations of the modulators are aligned with the dynamical changes of the indices as shown in Figs. 5.17 and 5.18. The normalisation sets activation level of the modulators into a scale of 0 to 1 based on their corresponding ranges to show the effective time scales of these modulators.

5.8.1.1. *Fast activation and deactivation of PP2B*

Among the modulators, PP2B is activated the fastest in response to the Ca^{2+} elevation following the stimulations, because of its high sensitivity to Ca^{2+}. ΔKBI has a dip (Fig. 5.17) and $\Delta I1^P$ has a great drop (Figs. 5.17 and 5.18), both are driven by the greater activation of PP2B than those of kinases within a transient period following the stimulations. For instance, PP2B is mainly basal activated, indicated by the 5% decrease in KBI, in response to LFS, but a large negative value of $\Delta I1^P$ is shown for the same period because kinases are incapable to be activated under this signal, which is known for LTD induction (Fig. 5.18). On the other hand, PP2B is deactivated the fastest when Ca^{2+} goes back to the basal level. Hence, CaMKII and PKA, which have slower deactivation rates, can maintain a longer

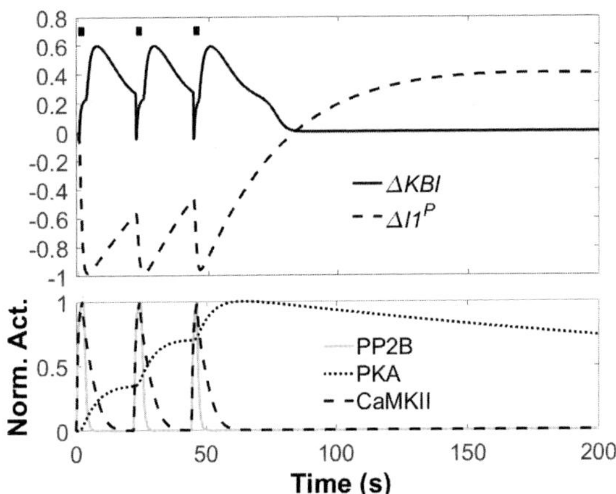

Fig. 5.17.   The effective time scales of the modulators in response to TBS. The black bars show the duration of TBS. Upper panel shows the behaviours of ΔKBI and $\Delta I1^P$ in transient in response to the TBS. The lower panel shows the normalised transient activation levels of PP2B, PKA and CaMKII aligned with the changes of ΔKBI and $\Delta I1^P$ in the upper panel.

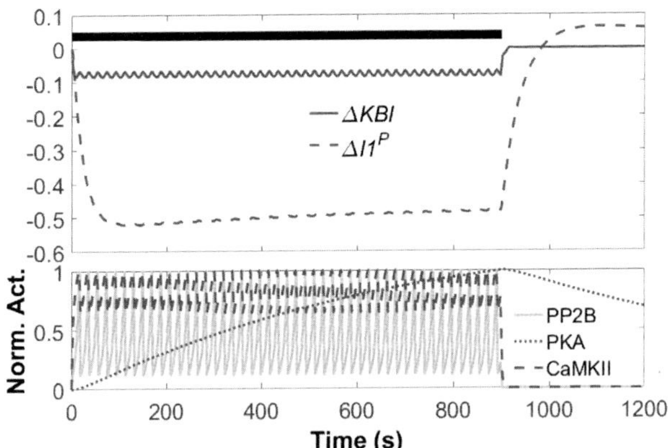

Fig. 5.18.   The effective time scales of the modulators in response to LFS. The black bar shows the duration of LFS. Upper panel shows the behaviours of ΔKBI and $\Delta I1^P$ in transient in response to LFS. Lower panel shows the normalised transient activation levels of PP2B, PKA and CaMKII aligned with the changes of ΔKBI and $\Delta I1^P$ in the upper panel.

phosphorylation level on their target sites for functions related to LTP (Fig. 5.17).

5.8.1.2. *Slower activation and deactivation of CaMKII*

Why has ΔKBI a small dip while $\Delta I1^P$ has a great drop (Fig. 5.17) both in response to TBS? The main reason is the autophosphorylation of CaMKII, which strengthens the binding between CaMKII and Ca^{2+}/CaM complex [105]. Turn-on of the autophosphorylation requires the Ca^{2+} concentration to exceed a threshold [78], resulting in the initial dip, and the autophosphorylation, while it is on, allows CaMKII to remain active and sustain the phosphorylation on its target sites for a much longer period [57, 110–112], resulting in the slow decrease of ΔKBI after the termination of the signal (Figs. 5.17 and 5.19).

5.8.1.3. *Slowest activation and slowest deactivation of PKA*

It is clear that PKA has the slowest activation among the modulators. The initial drops of $\Delta I1^P$ in response to TBS (Fig. 5.17) and LFS (Fig. 5.18) are caused by the relative weak catalytic activity of PKA

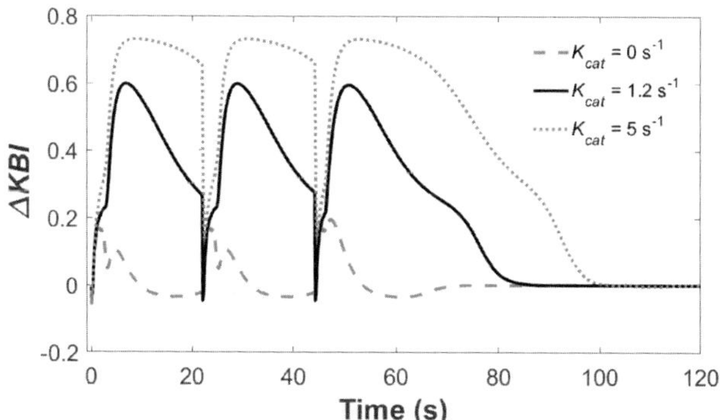

Fig. 5.19. ΔKBI in response to TBS for different autophosphorylation rates of CaMKII in transient. Three turnover rates for the autophosphorylation (K_{cat}) are tested: $0\,s^{-1}$, $1.2\,s^{-1}$ and $5\,s^{-1}$.

against phosphatases. On the other hand, the effective time scale of PKA is much longer than other modulators due to its slowest deactivation. As a result, $\Delta\Pi^P$ exhibits a moderate but lasting increase after the termination of the signals (Figs. 5.17 and 5.18), even though PKA has an insubstantial increase following LFS. Also due to the long effective time scale, PKA can maintain its activation over the long interval between trains to get a higher activation (Fig. 5.17).

5.8.2. *Computational study on removal of PKA by PP2B*

The rephosphorylation of target sites by PKA under LFS, as revealed by $\Delta\Pi^P$ (Fig. 5.18), seems to be unrealistic. The reason is that MoNP does not consider the pathways which prevent the rephosphorylation by PKA. It is clear that PP2B removes AKAP, as well as the PKA attached, from the dendritic spine and the removal is required for LTD [113], but the exact mechanism is unclear. In this section, we perform a computational study to reveal the dynamical behaviours of $\Delta\Pi^P$ when the removal is included in MoNP and to investigate the implication of the removal for LTD.

Since AKPA binds to the regulatory dimer of PKA [114], an assumption is made that the removal of AKAP by PP2B takes away the regulatory dimer as well as the catalytic subunits of PKA, except those are released. For the removed PKA, it is far from cAMP sources hence the binding of the removed PKA to cAMP for the activation is ignored. Hence, three new states are developed, including removed R_2C_2, removed R_2cAMP_2, and removed R_2cAMP_4, with their concentration denoted by $[R_2C_2]_r$, $[R_2C_2cAMP_2]_r$, and $[R_2C_2cAMP_4]_r$, respectively. There is no consideration for the pathway to remove the regulatory dimer which has the catalytic subunits released. The pathway is cut because the concentration of such regulatory dimer is very small during LTD as measured to be less than $0.02\,\mu$M. Now, the reaction rates of the removing $R_2C_2(R_{r1}), R_2cAMP_2(R_{r2})$, and $R_2cAMP_4(R_{r3})$, as well as the transition among (R_{r4} and R_{r5}) are given, respectively, by

$$R_{r1} = k_{rf}[R_2C_2]\frac{[ActivePP2B]}{K_{m17} + [ActivePP2b]} - k_{rb}[R_2C_2]_r, \quad (5.140)$$

$$R_{r2} = k_{rf}[R_2C_2cAMP_2]\frac{[ActivePP2B]}{K_{m17} + [ActivePP2B]}$$
$$-k_{rb}[R_2C_2cAMP_2]_r, \quad (5.141)$$

$$R_{r3} = k_{rf}[R_2C_2cAMP_4]\frac{[ActivePP2B]}{K_{m17} + [ActivePP2B]}$$
$$-k_{rb}[R_2C_2cAMP_4]_r, \quad (5.142)$$

$$R_{r4} = -k_{17b}[R_2C_2cAMP_2]_r, \quad \text{and} \quad (5.143)$$

$$R_{r5} = -k_{18b}[R_2C_2cAMP_4]_r, \quad (5.144)$$

where $[ActivePP2B] = [Ca_4CaNB]/20 + [CaMCaNA]$, k_{rf} and k_{rb} are the rate constants for the removal and recovery of PKA, respectively. K_{m17} takes the $[ActivePP2B]$ at which the removal is half maximum. k_{17b} and k_{18b} are rate constants of the dissociation of cAMP from the removed PKA states. The new ODEs are given by

$$\frac{d[R_2C_2]}{dt} = -R_{16} - R_{r1}, \quad (5.145)$$

$$\frac{d[R_2C_2cAMP_2]}{dt} = R_{16} - R_{17} - R_{r2}, \quad (5.146)$$

$$\frac{d[R_2C_2cAMP_4]}{dt} = R_{17} - R_{18} - R_{r3}, \quad (5.147)$$

$$\frac{d[R_2C_2]_r}{dt} = R_{r1} + R_{r4}, \quad (5.148)$$

$$\frac{d[R_2C_2cAMP_2]_r}{dt} = R_{r2} - R_{r4} + R_{r5}, \quad \text{and} \quad (5.149)$$

$$\frac{d[R_2C_2cAMP_4]_r}{dt} = R_{r3} - R_{r5}, \quad (5.150)$$

where reaction rates, $R_{16}-R_{18}$, are from MoNP.

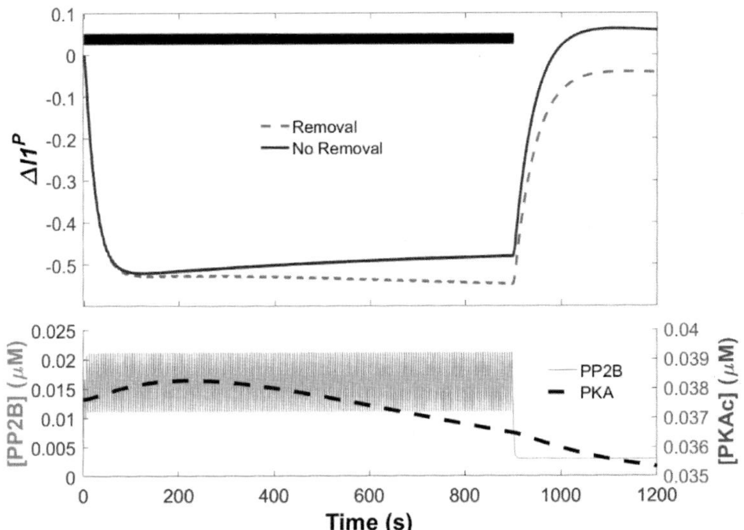

Fig. 5.20. $\Delta I1^P$ in response to LFS with the PKA removal in transient. The black bar shows the duration of LFS. The upper panel shows $\Delta I1^P$ in transient between the removal and no removal of PKA. The lower panel shows the temporal activation levels, for the removal model, of PP2B ($[CaMCaNA] + [Ca_4 CaNB]/20$) and PKA (indicated by the concentration of PKAc) aligned with the changes of $\Delta I1^P$ in the upper panel. Parameter used are: $k_{rf} = 0.25\,\mathrm{s}^{-1}$, $k_{rb} = 1/600\,\mathrm{s}^{-1}$, $K_{m17} = 3\,\mu\mathrm{M}$.

The removal lags behind the drop of $\Delta I1^P$ yielding consistency to experimental observations [113, 115]. As expected, the removal significantly reduces the rephosphorylation by PKA after the termination of LFS without altering the time scale of the rephosphorylation (Fig. 5.20). There is no data available on I1 rephosphorylation, but according to observations on the phosphorylation of AMPAR S845 by PKA, there is a similar drop in the rephosphorylation and unaltered time scale between wild-type and a mutant with reduced PKA removal [113]. Moreover, the mutant has EPSP to recover above the basal level in response to LFS [113].

5.8.3. *Summary*

LTP and LTD are induced as the result of an imbalance of the relative activation strength between kinases and phosphatases. LTD is expressed as the fast activation of phosphatases while keeping kinases at an insignificant level of activation. LTP is expressed as the ability of kinases to maintain a longer activation over the fast deactivation of phosphatases. Hence, LFS provides the moderate activation of PP2B without exceeding the threshold of kinase activation while TBS facilitates the strong activation of kinases along with the minimal activity of phosphatases.

A computational study is performed to understand the implication of the removal PKA by PP2B in LTD. The removal prevents the rephosphorylation by PKA without altering the time scale of the rephosphorylation. Since the modified model may not be adequate for a signal like TBS, which results in large amount of release of catalytic subunits, the removal during TBS is not tested. Further studies are required to look at the removal during TBS; the removal may be minimised by TBS or the removal may take away the regulatory dimer to allow the released catalytic subunits a longer activity duration.

5.9. Significant Factors Revealed by GSA

GSA shows 10 sensitive parameters for both LTP and LTD signals (Table 5.10), while variations in six of them give significant changes on MoNP behaviours. $[PDE4D_T]$ sets the basal regulation strength of PKA; $[CaM_T]$ determines the available CaM; $K_{\text{cat}12}$ and K_{m12} determine the basal concentration of $I1^P PP1$; $n1$ is related to the basal PP2B activation, and K_{m11} is related to the phosphorylation of I1 by PKA.

Three parameters are related to the basal concentration/activity ($[PDE4D_T]$, $K_{\text{cat}12}$ and K_{m12}) because the change of the basal concentration/activity can reflect the remodelling of synaptic property by previous stimulations and hence cause an altered threshold of synaptic plasticity induced by a following stimulation; this phenomenon is called meta-plasticity [116]. However, the coordination of the LTP and LTD expressions may not necessary be in strong correlation to basal concentration/activity. Therefore, we do not discuss the parameters purely related to the basal concentrations.

Table 5.10. Sensitive parameters and their p-value.

Parameters name	p-Value	
	LTD	LTP
[PDE4D$_T$]	0.009	0.009
[CaM$_T$]	<0.001	<0.001
n1	<0.001	<0.001
K_{cat12}	<0.001	0.002
K_{m11}	<0.001	<0.001
K_{m12}	<0.001	<0.001
k_{1f}	<0.001	0.004
k_{3b}	<0.001	0.005
k_{7b}	0.003	0.006
k_{8b}	0.002	<0.001

Parameter in grey are significant parameters.

We summaries the key findings for a further analysis of other three parameters and discuss $[CaM_T]$, which is sensitive to both ΔKBI and $\Delta \Pi^P$, in details at the end.

5.9.1. *Summary of key findings from GSA*

Since behaviours of both ΔKBI and $\Delta \Pi^P$ are considered for the theoretical conditions for the bidirectionality, parameters, which are sensitive to either of them, are marked as sensitive to the bidirectionality. Through a further analysis, we conclude that $n1$ and K_{m11} are both sensitive to $\Delta \Pi^P$, only, and $[CaM_T]$ is sensitive to both ΔKBI and $\Delta \Pi^P$.

$n1$ describes the cooperativity between the successive bindings of four Ca^{2+} ions to CaNB of PP2B. In comparison to the reference value ($n1 = 1.8$), a negative cooperativity ($0 < n1 < 1$) lets $\Delta \Pi^P$ reach a higher peak during LTP and it drops to a higher bottom during LTD, while a positive cooperativity ($n1 > 1$) shows no significant changes of $\Delta \Pi^P$ [89]. Since the basal activation is the premise of the full activation of PP2B, regulation of the cooperativity of the bindings underlying the basal activation may be a potential mechanism to increase susceptibility to LTP induction.

K_{m11} is the Michaelis constant for phosphorylating I1 by PKA. The phosphorylation on I1 is faster or slower than the reference rate when K_{m11} is smaller or greater than its reference value, respectively. K_{m11} has both basal and dynamical effects since PKA concentration is dynamically changing. The basal and dynamical effects cause opposite changes in $\Delta I1^P$ and the combined effect would make LTP or LTD induction easier depending on which one of the two effects is stronger [89].

5.9.2. Ca^{2+}/CaM *complex formation*

A number of parameters related to Ca^{2+}/CaM complex formation are sensitive to the bi-directional behaviour of synaptic plasticity. After inspecting the MoNP behaviours under the variation of these parameters, we conclude that CaM pool size, $[CaM_T]$, is the only extremely significant parameter for the bidirectional behaviour. As shown in Fig. 5.21, $[CaM_T]$ is very sensitive to both ΔKBI (Figs. 5.21A and 5.21C) and $\Delta I1^P$ (Figs. 5.21B and 5.21D). In the following sections, we discuss the unique properties of CaM contributing to the bidirectional behaviour.

5.9.2.1. *Coordination between LTP and LTD expressions by CaM*

For the purpose to understand the influence of the activation of modulators by the competitive binding to Ca^{2+}/CaM complex for the activation, the related competitive pathways are retrieved for the investigation. There are five proteins involved in the competitive pathways: CaNA subunit of PP2B, AC1, AC8, PDE1 and CaMKII, as given by the following reaction schema [32, 33]:

$$Ca_4CaM + iAC1 \underset{k_{kc1b}}{\overset{k_{c1f}}{\rightleftharpoons}} CaMAC1, \tag{5.151}$$

$$Ca_4CaM + iAC8 \underset{k_{c2b}}{\overset{k_{c2f}}{\rightleftharpoons}} CaMAC8, \tag{5.152}$$

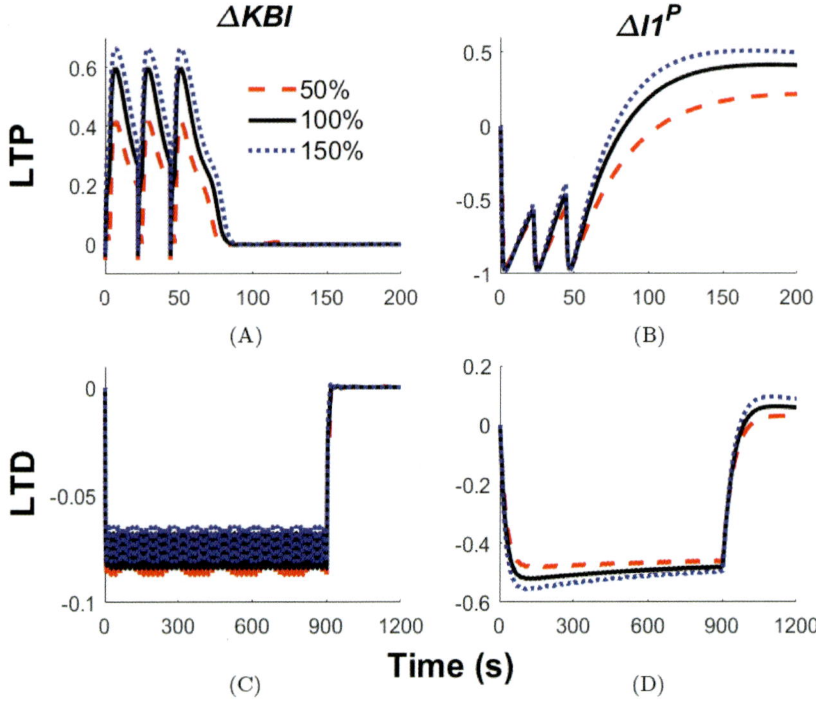

Fig. 5.21. ΔKBI and $\Delta I1^P$ in response to TBS and LFS over the variation of $[CaM_T]$ in transient. Three different values of $[CaM_T]$, 50% (8.85 μM), 100% (17.7 μM) and 150% (26.55 μM) of the reference value are input to MoNP. (A) Behaviours of ΔKBI in transient in response to TBS; (B) behaviours of $\Delta I1^P$ in transient in response to TBS; (C) behaviours of ΔKBI in transient in response to LFS; and (D) behaviours of $\Delta I1^P$ in transient in response to LFS.

$$Ca_4CaM + iPDE1 \underset{k_{c3b}}{\overset{k_{c3f}}{\rightleftharpoons}} CaMPDE1, \qquad (5.153)$$

$$Ca_4CaM + iCaNA \underset{k_{c4b}}{\overset{k_{c4f}}{\rightleftharpoons}} CaMCaNA, \quad \text{and} \qquad (5.154)$$

$$Ca_4CaM + iCaMKII \underset{k_{c5b}}{\overset{k_{c5f}}{\rightleftharpoons}} CaMCaMKII, \qquad (5.155)$$

where the bound complex between a CaM-regulated protein, X, and Ca^{2+}/CaM complex is denoted by the notation $CaMX$ and iX is the unbound form of the protein. The ODEs are given as follows, according to mass action rate law:

$$\frac{d[CaMAC1]}{dt} = k_{c1f}[Ca_4CaM][iAC1] - k_{c1b}[CaMAC1],$$

$$(5.156)$$

$$\frac{d[CaMAC8]}{dt} = k_{c2f}[Ca_4CaM][iAC8] - k_{c2b}[CaMAC8],$$

$$(5.157)$$

$$\frac{d[CaMPDE1]}{dt} = k_{c3f}[Ca_4CaM][iPDE1] - k_{c3b}[CaMPDE1],$$

$$(5.158)$$

$$\frac{d[CaMCaNA]}{dt} = k_{c4f}[Ca_4CaM][iCaNA] - k_{c4b}[CaMCaNA],$$

$$(5.159)$$

$$\frac{d[CaMCaMKII]}{dt} = k_{c5f}[Ca_4CaM][iCaMKII]$$

$$-k_{c5b}[CaMCaMKII], \quad \text{and} \qquad (5.160)$$

$$\frac{d[Ca_4CaM]}{dt} = -\frac{d[CaMAC1]}{dt} - \frac{d[CaMAC8]}{dt}$$

$$-\frac{d[CaMPDE1]}{dt} - \frac{d[CaMCaNA]}{dt}$$

$$-\frac{d[CaMCaMKII]}{dt} \qquad (5.161)$$

where $[iAC1] = [AC1_T] - [CaMAC1], [iAC8] = [AC8_T] - [CaMAC8]$, $[iPDE1] = [PDE1_T] - [CaMPDE1], [iCaNA] = [PP2B_T] - [CaMCaNA]$ and $[iCaMKII] = [CaMKII_T] - [CaMCaMKII]$. The simulation takes input of initial concentrations of Ca^{2+}/CaM complex, $[Ca_4CaM_0]$, from 0.0001 to 30 μM, while keep other proteins fully unbound at initial. A sufficient simulation time is taken to have the proteins reaching their steady states.

To have a contrast for the competitive binding, we calculate the equilibrium concenration of the bound complex, $[CaMX_e]$, when the CaM-regulated protein and Ca^{2+}/CaM complex are mixed in an isolated compartment alone. This equilibrium concentration can be expressed as a function in terms of $[Ca_4 CaM_0]$ [80]:

$$[CaMX_e] = \frac{\Delta - \sqrt{\Delta^2 - 4[X_T][Ca_4CaM_0]}}{2}, \qquad (5.162)$$

where $K_d = k_{cb}/k_{cf}$, $[X_T]$, the total concentration of X, is conserved: $[X_T] = [X_e] + [CaMX_e] = [X_0]$ assuming X is fully unbound initially, and $\Delta = [X_T] + [Ca_4 CaM_0] + K_d$.

We plot the bound portion of each protein to Ca^{2+}/CaM complex against $[Ca_4CaM_0]$ in Fig. 5.22A under the non-competitive and competitive binding. We define $K_{0.5}$ as a measure of competitiveness of a protein in binding to Ca^{2+}/CaM complex if we denote $K_{0.5}$ to be $[Ca_4CaM_0]$ at which the bound portion is $1/2$. (See the top graph of Fig. 5.22A.) Higher $K_{0.5}$ value alludes a high level of competition.

The moderate $[Ca_4CaM_0]$ activate LTD related proteins, including CaNA ($K_{0.5} = 1.11\,\mu M$) and PDE1 ($K_{0.5} = 8.25\,\mu M$); and high $[Ca_4CaM_0]$ activate LTP-related proteins, including AC1 ($K_{0.5} = 11.74\,\mu M$), AC8 ($K_{0.5} = 17.34\,\mu M$) and CaMKII ($K_{0.5} = 17.64\,\mu M$) (Fig. 5.22A). This arrangement follows the order of their binding affinities (K_a) for Ca^{2+}/CaM complex from the strongest to the weakest (Table 5.6). Moreover, the two groups of CaM-regulated proteins, related to LTP or LTD, are separated by a relative large $K_{0.5}$ of $3.5\,\mu M$. The arrangement by the non-competitive binding is mixed: AC8 ($K_{0.5} = 0.365\,\mu M$), CaNA ($K_{0.5} = 1.05\,\mu M$), AC1 ($K_{0.5} = 1.27\,\mu M$), PDE1 ($K_{0.5} = 2.01\,\mu M$), and CaMKII ($K_{0.5} = 10.06\,\mu M$) (Fig. 5.22A).

Analysis of the portions of LTP/LTD related Ca^{2+}/CaM complex bindings indicates that LTD related bindings (PP2B and PDE1) are dominant over low $[Ca_4 CaM_0]$, while LTP-related bindings (AC1, AC8, and CaMKII) dominates over high $[Ca_4CaM_0]$ (Fig. 5.22B); for example, the LTD related bindings trap more than 90% of the bound Ca^{2+}/CaM complex over low $[Ca_4 CaM_0]$ and this number drops to around 20% over high $[Ca_4CaM_0]$. LTD related proteins have stronger affinities for Ca^{2+}/CaM complex so that they occupy more Ca^{2+}/CaM complex when the concentration of Ca^{2+}/CaM complex is insufficient to saturate the binding sites of the proteins. However, the

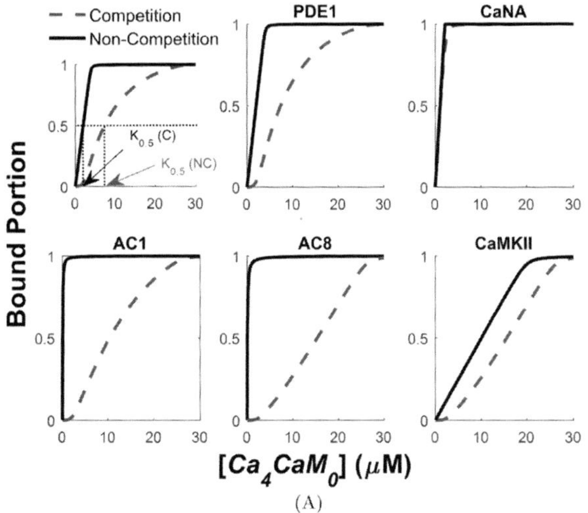

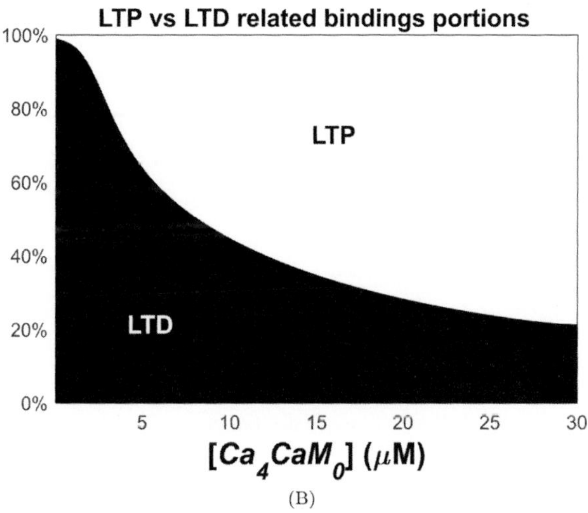

Fig. 5.22. Competitive binding to Ca^{2+}/CaM complex. (A) The steady bound portion of CaM-regulated proteins under non-competitive and competitive bindings to Ca^{2+}/CaM complex with respect to the change in $[Ca_4CaM_0]$. Inset: calculation of $K_{0.5}$, which measures the $[Ca_4CaM_0]$ required for the half-bound of the corresponding CaM-regulated protein. (B) The ratio of the concentrations of Ca^{2+}/CaM complex bound by LTP/LTD related proteins divided by the total concentration of bound Ca^{2+}/CaM complex under the competitive binding. The black and white areas show the change in the percentage of LTD related bindings (PP2B and PDE1) and the change in the percentage of LTP-related bindings (AC1, AC8, and CaMKII) to Ca^{2+}/CaM complex, respectively, with respect to the change in $[Ca_4CaM_0]$.

LTP-related proteins occupy much more Ca^{2+}/CaM complex when the concentration of Ca^{2+}/CaM complex is sufficient, since the LTP-related proteins are much more abundant in the spine, i.e. CaMKII is the most abundant protein in the spine at a concentration up to $100\,\mu M$ [117].

5.9.2.2. *CaM pool size affects the coordination between LTP and LTD expressions*

The CaM concentration has a large variety among different tissues and a much larger concentration of CaM is detected in different areas of the mammalian brain than most other organs [99]. This large amount of CaM in brain facilitates the activation of various CaM-regulated proteins, which are involved in many synaptic functions. In the hippocampus, neurogranin (Ng), a protein kinase C (PKC) substrate, binds to CaM in the synaptic compartment in the absence of Ca^{2+} and releases CaM in a PKC phosphorylated manner in response to Ca^{2+} influx [118]. It is proposed that Ng targets CaM to ideal locations to form CaM pools in order to response fast to transient Ca^{2+} signals [39, 119, 120], given the highly mobile nature of CaM molecules in neuron [120].

Steady state shows that smaller $[CaM_T]$ significantly influences ΔKBI in response to Ca^{2+} (Fig. 5.23). For smaller $[CaM_T]$, the recovery of ΔKBI to the base level after the enhanced LTD related Ca^{2+}/CaM complex binding requires a much higher Ca^{2+} concentration. Moreover, ΔKBI is unable to recover to the base level for very small $[CaM_T]$ (i.e. $[CaM_T] < 5\,\mu M$ in Fig. 5.23). The reason is that it gets harder for LTP-related proteins to trap Ca^{2+}/CaM complex against LTD related proteins when $[CaM_T]$ is insufficient to saturate the binding sites of the proteins. (This is because LTP-related proteins have a much weaker affinity for Ca^{2+}/CaM complex in general.) As a result, smaller $[CaM_T]$ may induce the enhanced LTD expression since a smaller amount of $[CaM_T]$ is insufficient to bind and activate the LTP-related proteins even for high Ca^{2+} concentrations.

Transient shows that smaller $[CaM_T]$ decreases the peaks of ΔKBI and $\Delta I1^P$ in response to TBS (Figs. 5.24A and 5.24B).

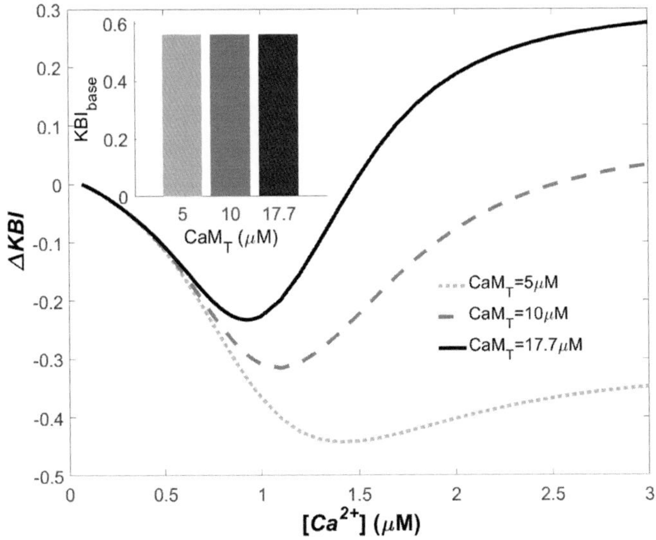

Fig. 5.23. ΔKBI for different $[CaM_T]$ in steady state. The steady patterns of ΔKBI over $[Ca^{2+}]$ are shown for three different $[CaM_T]$: $5\,\mu$M, $10\,\mu$M, and $17.7\,\mu$M. Inset: the basal values of KBI for different $[CaM_T]$ tested.

Moreover, ΔKBI returns faster to the base level for smaller $[CaM_T]$ indicating a deficiency in turning on the autophosphorylation (Fig. 5.24A). The rephosphorylation of $\Delta I1^P$ by PKA becomes weaker for smaller $[CaM_T]$ (Fig. 5.24B). $\Delta I1^P$ stays below the base level all the time for very small $[CaM_T]$. The behaviours of ΔKBI and $\Delta I1^P$ under very small $[CaM_T]$ indicate the domination of LTD expression due to the incapability for the activation of the LTP-related proteins. It is important to note that, although a great reduction of LTD behaviours is revealed for low values of $[CaM_T]$, LTD behaviours are still valid for our LTD conditions (Figs. 5.24C and 5.24D). These results indicate an increased susceptibility to LTD induction for lower $[CaM_T]$.

The reduction in CaM pool may cause a reduction in the activity levels of CaM-regulated downstream targets, for example, reduction in autophosphorylation activity, and hence cause deficits in LTP induction. As shown previously, the LTP expression becomes much harder without the autophosphorylation (Fig. 5.19). Moreover, the

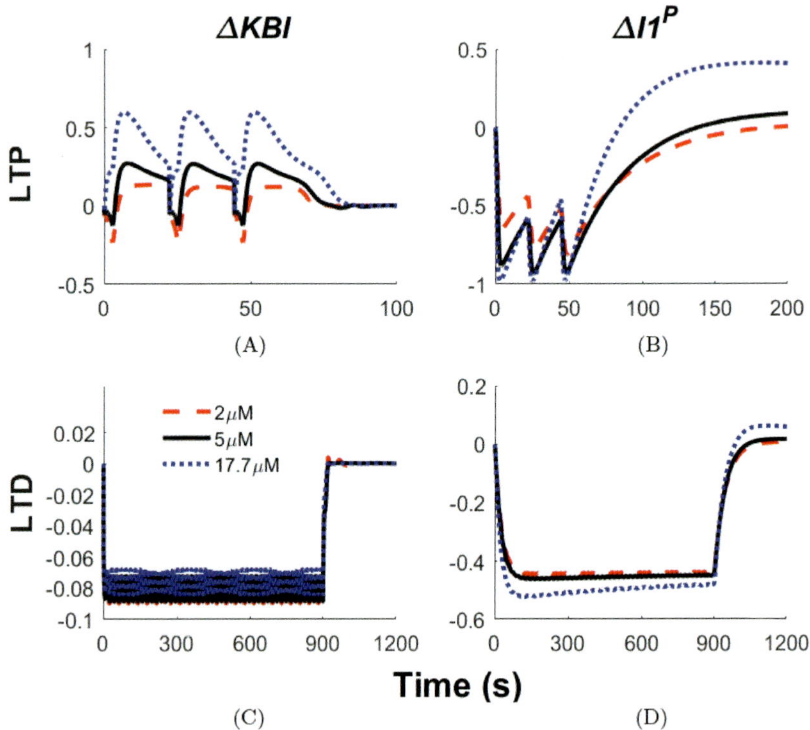

Fig. 5.24. ΔKBI and $\Delta I1^P$ in response to TBS and LFS for different $[CaM_T]$ in transient. Three different values of $[CaM_T]$, $2\,\mu$M, $5\,\mu$M and $17.7\,\mu$M, are input to MoNP. (A) Behaviours of ΔKBI in transient in response to TBS; (B) behaviours of $\Delta I1^P$ in transient in response to TBS; (C) behaviours of ΔKBI in transient in response to LFS; and (D) behaviours of $\Delta I1^P$ in transient in response to LFS.

autophosphorylation couples multiple trains of TBS and maintains the CaMKII activity level for a longer period, while the CaMKII activity level decreases much faster without the autophosphorylation (Fig. 5.19).

5.9.2.3. *CaM pool size and aging?*

Evidence shows the decline of membrane CaM pool size in aged rat [38]; the decline may be caused by the reduced expression of Ng in dendritic segments [39] and the reduced phosphorylation activity of Ng by PKC [40]. Furthermore, CaM undergoes oxidative modification in aging which may cause a reduction in the activity levels of CaM-regulated downstream targets [38, 121]. Recent evidence shows a significant disruption of the oxidative CaM in the dynamics of CaMKII: reductions in the activation, the autophosphorylation and the binding to NMDAR [122].

MoNP predicts the importance of CaM pool size in the induction of LTP, and the CaM pool size is the most significant parameter in GSA studies for bidirectional behaviour. We propose that the experimental scientists may want to investigate the effects of CaM pool size on synaptic plasticity further than what have been done so far, and our computational studies show that CaM pool size is the single most important factor for LTP and therefore memory formation. With the minor impact of CaM pool size on LTD (Fig. 5.24), we hypothesise a potential path, by reducing the CaM pool size, to impair LTP and increase susceptibility to LTD induction, both are impairments due to aging [15–20]. If CaM pool size diminishes with age [38] then abnormal drop in CaM pool size might cause various forms of dementia. This needs to be tested experimentally.

References

[1] Glisky E.L. (2007). *Brain Aging: Models, Methods, and Mechanisms.* ed. Riddle D. Chapter 1 "Changes in cognitive function in human aging" (CRC Press), pp. 3–20.

[2] Craik F.I.M. and Salthouse T.A. (2011). *The Handbook of Aging and Cognition*, 3rd edn. (Psychology Press, New York).

[3] Anderson J.R. (2000). *Learning and Memory*, 2nd edn. (Wiley, New York).

[4] Isaac J.T.R., Nicoll R.A. and Malenka R.C. (1995). Evidence for silent synapses: implications for the expression of LTP. *Neuron*, 15, pp. 427–434.

[5] Bourtchuladze R. *et al.* (1994). Deficient long-term memory in mice with a targeted mutation of the cAMP-responsive element-binding protein. *Cell*, 79, pp. 59–68.

[6] Chang H.P. *et al.* (1999). Impaired memory retention and decreased long-term potentiation in integrin-associated protein-deficient mice. *Learn Mem*, 6, pp. 448–457.

[7] Davis S., Butcher S.P. and Morris R.G. (1992). The NMDA receptor antagonist D-2-amino-5-phosphonopentanoate (D-AP5) impairs spatial learning and LTP *in vivo* at intracerebral concentrations comparable to those that block LTP *in vitro*. *J Neurosci*, 12, pp. 21–34.

[8] Giese K.P., Fedorov N.B., Filipkowski R.K. and Silva A.J. (1998). Autophosphorylation at Thr286 of the α calcium-calmodulin kinase II in LTP and learning. *Science*, 279, pp. 870–873.

[9] Morris R.G. (1989). Synaptic plasticity and learning: selective impairment of learning rats and blockade of long-term potentiation *in vivo* by the *N*-methyl-*D*-aspartate receptor antagonist AP5. *J Neurosci*, 9, pp. 3040–3057.

[10] Silva A.J., Stevens C.F., Tonegawa S. and Wang Y. (1992). Deficient hippocampal long-term potentiation in alpha-calcium-calmodulin kinase II mutant mice. *Science*, 257, pp. 201–206.

[11] Silva A.J., Paylor R., Wehner J.M. and Tonegawa S. (1992). Impaired spatial learning in alpha-calcium-calmodulin kinase II mutant mice. *Science*, 257, pp. 206–211.

[12] Bliss T.V.P. and Collingridge G.L. (1993). A synaptic model of memory: long-term potentiation in the hippocampus. *Nature*, 361, pp. 31–39.

[13] Nabavi S. *et al.* (2014). Engineering a memory with LTD and LTP. *Nature*, 511, pp. 348–352.

[14] Li S. *et al.* (2009). Soluble oligomers of amyloid Beta protein facilitate hippocampal long-term depression by disrupting neuronal glutamate uptake. *Neuron*, 62, pp. 788–801.

[15] Shankar G.M. *et al.* (2008). Amyloid-beta protein dimers isolated directly from Alzheimer's brains impair synaptic plasticity and memory. *Nat Med*, 14, pp. 837–842.

[16] Li S. *et al.* (2011). Soluble Aβ oligomers inhibit long-term potentiation through a mechanism involving excessive activation of extrasynaptic NR2B-containing NMDA receptors. *J Neurosci*, 31, pp. 6627–6638.

[17] Koch G. *et al.* (2012). Impaired LTP- but not LTD-like cortical plasticity in Alzheimer's disease patients. *J Alzheimer's Dis*, 31, pp. 593–599.

[18] Norris C.M., Korol D.L. and Foster T.C. (1996). Increased susceptibility to induction of long-term depression and long-term potentiation reversal during aging. *J Neurosci*, 16, pp. 5382–5392.

[19] Foster T.C. (1999). Involvement of hippocampal synaptic plasticity in age-related memory decline. *Brain Res Rev*, 30, pp. 236–249.

[20] Kumar A. and Foster T.C. (2005). Intracellular calcium stores contribute to increased susceptibility to LTD induction during aging. *Brain Res*, 1031, pp. 125–128.

[21] Mayer M.L., Westbrook G.L. and Guthrie P.B. (1984). Voltage-dependent block by Mg^{2+} of NMDA responses in spinal cord neurones. *Nature*, 309, pp. 261–263.

[22] Nowak L.L. *et al.* (1984). Magnesium gates glutamate-activated channels in mouse central neurones. *Nature*, 307, pp. 462–465.

[23] He Y., Kulasiri D. and Samarasinghe S. (2014). Systems biology of synaptic plasticity: a review on *N*-methyl-*D*-aspartate receptor mediated biochemical pathways and related mathematical models. *Biosystems*, 122, pp. 7–18.

[24] Lee H.-K. *et al.* (2000). Regulation of distinct AMPA receptor phosphorylation sites during bidirectional synaptic plasticity. *Nature*, 405, pp. 955–959.

[25] Lüscher C. and Malenka R.C. (2012). NMDA receptor-dependent long-term potentiation and long-term depression (LTP/LTD). *Cold Spring Harb Perspect Biol*, 4, p. a005710.

[26] Ahmad M. *et al.* (2012). Postsynaptic complexin controls AMPA receptor exocytosis during LTP. *Neuron*, 73, pp. 260–267.

[27] Bear M.F., Cooper L.N. and Ebner F.F. (1987). A physiological basis for a theory of synapse modification. *Science*, 237, pp. 42–48.

[28] Lisman J.E. (1989). A mechanism for the Hebb and the anti-Hebb processes underlying learning and memory. *Proc Natl Acad Sci USA*, 86, pp. 9574–9578.

[29] Yang S., Tang Y. and Zucker R.S. (1999). Selective induction of LTP and LTD by postsynaptic $[Ca^{2+}]i$ Elevation. *J Neurophysiol*, 81, pp. 781–787.

[30] Weng G., Bhalla U.S. and Iyengar R. (1999). Complexity in biological signaling systems. *Science*, 284, pp. 92–96.

[31] Bhalla U.S. and Iyengar R. (1999). Emergent properties of networks of biological signaling pathways. *Science*, 283, pp. 381–387.

[32] Hayer A. and Bhalla U.S. (2005). Molecular switches at the synapse emerge from receptor and kinase traffic. *PLoS Comput Biol*, 1, p. e20.

[33] Kim M., Huang T., Abel T. and Blackwell K.T. (2010). Temporal sensitivity of protein kinase a activation in late-phase long term potentiation. *PLoS Comput Biol*, 6, p. e1000691.

[34] Kim M. *et al.* (2011). Colocalization of protein kinase A with adenylyl cyclase enhances protein kinase A activity during induction of long-lasting long-term-potentiation. *PLoS Comput Biol*, 7, p. e1002084.

[35] Manninen T. *et al.* (2010). Postsynaptic signal transduction models for long-term potentiation and depression. *Front Comput Neurosci*, 4, p. 152.

[36] Shouval H.Z., Wang S.S.-H. and Wittenberg G.M. (2010). Spike timing dependent plasticity: a consequence of more fundamental learning rules. *Front Comput Neurosci*, 4, p. 19.

[37] Graupner M. and Brunel N. (2010). Mechanisms of induction and maintenance of spike-timing dependent plasticity in biophysical synapse models. *Front Comput Neurosci*, 4, p. 136.

[38] Zaidi A., Gao J., Squier T. and Michaelis M. (1998). Age-related decrease in brain synaptic membrane Ca^{2+}-ATPase in F344/BNF1 rats. *Neurobiol Aging*, 19, pp. 487–495.

[39] Mons N., Enderlin V., Jaffard R. and Higueret P. (2001). Selective age-related changes in the PKC-sensitive, calmodulin-binding protein, neurogranin, in the mouse brain. *J Neurochem*, 79, pp. 859–867.

[40] Angenstein F., Buchner K. and Staak S. (1999). Age-dependent differences in glutamate-induced phosphorylation systems in rat hippocampal slices. *Hippocampus*, 9, pp. 173–185.

[41] Zawadzki K.M. and Taylor S.S. (2004). cAMP-dependent protein kinase regulatory subunit type IIbeta: active site mutations define an isoform-specific network for allosteric signaling by cAMP. *J Biol Chem*, 279, pp. 7029–7036.

[42] Esteban J.A. *et al.* (2003). PKA phosphorylation of AMPA receptor subunits controls synaptic trafficking underlying plasticity. *Nat Neurosci*, 6, pp. 136–143.

[43] Lee H.-K. *et al.* (2010). Specific roles of AMPA receptor subunit GluR1 (GluA1) phosphorylation sites in regulating synaptic plasticity in the CA1 region of hippocampus. *J Neurophysiol*, 103, pp. 479–489.

[44] Zheng Z. and Keifer J. (2009). PKA has a critical role in synaptic delivery of GluR1-and GluR4-containing AMPARs during initial stages of acquisition of in vitro classical conditioning. *J Neurophysiol*, 101, pp. 2539–2549.

[45] Blitzer R.D. *et al.* (1998). Gating of CaMKII by cAMP-regulated protein phosphatase activity during LTP. *Science*, 280, pp. 1940–1943.

[46] Huang K.X. and Paudel H.K. (2000). Ser67-phosphorylated inhibitor 1 is a potent protein phosphatase 1 inhibitor. *Proc Natl Acad Sci USA*, 97, pp. 5824–5829.

[47] Cali J.J. *et al.* (1994). Type VIII adenylyl cyclase. A Ca^{2+}/calmodulin-stimulated enzyme expressed in discrete regions of rat brain. *J Biol Chem*, 269, pp. 12190–12195.

[48] Wong S.T. *et al.* (1999). Calcium-stimulated adenylyl cyclase activity is critical for hippocampus-dependent long-term memory and late phase LTP. *Neuron*, 23, pp. 787–798.

[49] Bender A.T. and Beavo J.A. (2006). Cyclic nucleotide phosphodiesterases: molecular regulation to clinical use. *Pharmacol Rev*, 58, pp. 488–520.

[50] Lugnier C. (2006). Cyclic nucleotide phosphodiesterase (PDE) superfamily: a new target for the development of specific therapeutic agents. *Pharmacol Ther*, 109, pp. 366–398.

[51] Stanton P.K., Bramham C. and Scharfman H.E. (2005). *Synaptic Plasticity and Transsynaptic Signaling* (Springer Science & Business Media, New York).

[52] Chao L.H. *et al.* (2011). A mechanism for tunable autoinhibition in the structure of a human Ca^{2+}/calmodulin-dependent kinase II holoenzyme. *Cell*, 146, pp. 732–745.

[53] Barria A. *et al.* (1997). Regulatory phosphorylation of AMPA-type glutamate receptors by CaMKII during long-term potentiation. *Science*, 276, pp. 2042–2045.

[54] Derkach V., Barria A. and Soderling T.R. (1999). Ca^{2+}/calmodulin-kinase II enhances channel conductance of α-amino-3-hydroxy-5-methyl-4-isoxazolepropionate type glutamate receptors. *Proc Natl Acad Sci USA*, 96, pp. 3269–3274.

[55] Opazo P. *et al.* (2010). CaMKII triggers the diffusional trapping of surface AMPARs through phosphorylation of stargazin. *Neuron*, 67, pp. 239–252.

[56] Tomita S. *et al.* (2005). Bidirectional synaptic plasticity regulated by phosphorylation of stargazin-like TARPs. *Neuron*, 45, pp. 269–277.

[57] Lee S.-J.J.R., Escobedo-Lozoya Y., Szatmari E.M. and Yasuda R. (2009). Activation of CaMKII in single dendritic spines during long-term potentiation. *Nature*, 458, pp. 299–304.

[58] Klee C.B., Ren H. and Wang X. (1998). Regulation of the calmodulin-stimulated protein phosphatase, calcineurin. *J Biol Chem*, 273, pp. 13367–13370.

[59] Stemmer P.M. and Klee C.B. (1994). Dual calcium ion regulation of calcineurin by calmodulin and calcineurin B. *Biochemistry*, 33, pp. 6859–6866.

[60] Snyder G.L. *et al.* (2003). Regulation of AMPA receptor dephosphorylation by glutamate receptor agonists. *Neuropharmacology*, 45, pp. 703–713.

[61] Mulkey R.M., Endo S., Shenolikar S. and Malenka R.C. (1994). Involvement of a calcineurin/inhibitor-1 phosphatase cascade in hippocampal long-term depression. *Nature*, 369, pp. 486–488.

[62] Gomez L.L. *et al.* (2002). Regulation of A-kinase anchoring protein 79/150–cAMP-dependent protein kinase postsynaptic targeting by NMDA receptor activation of calcineurin and remodeling of dendritic actin. *J Neurosci*, 22, pp. 7027–7044.

[63] Gorski J.A., Gomez L.L., Scott J.D. and Dell'Acqua M.L. (2005). Association of an A-kinase-anchoring protein signaling scaffold with cadherin adhesion molecules in neurons and epithelial cells. *Mol Biol Cell*, 16, pp. 3574–3590.

[64] Beattie E.C. *et al.* (2000). Regulation of AMPA receptor endocytosis by a signaling mechanism shared with LTD. *Nat Neurosci*, 3, pp. 1291–1300.

[65] Carroll R.C., Beattie E.C., von Zastrow M. and Malenka R.C. (2001). Role of AMPA receptor endocytosis in synaptic plasticity. *Nat Rev Neurosci*, 2, pp. 315–324.

[66] Mulkey R.M., Herron C.E. and Malenka R.C. (1993). An essential role for protein phosphatases in hippocampal long-term depression. *Science*, 261, pp. 1051–1055.

[67] Stemmer P. and Klee C.B. (1991). Serine/threonine phosphatases in the nervous system. *Curr Opin Neurobiol*, 1, pp. 53–64.

[68] Ehlers M.D. (2003). Activity level controls postsynaptic composition and signaling via the ubiquitin-proteasome system. *Nat Neurosci*, 6, pp. 231–242.

[69] Zhabotinsky A.M. (2000). Bistability in the Ca^{2+}/calmodulin-dependent protein kinase-phosphatase system. *Biophys J*, 79, pp. 2211–2221.

[70] Huynh Q.K. and Pagratis N. (2011). Kinetic mechanisms of Ca^{2+}/calmodulin dependent protein kinases. *Arch Biochem Biophys*, 506, pp. 130–136.

[71] Holmes W.R. (2000). Models of calmodulin trapping and CaM kinase II activation in a dendritic spine. *J Comput Neurosci*, 8, pp. 65–86.

[72] Wang H. and Strom D.R. (2003). Calmodulin-regulated adenylyl cyclases: cross-talk and plasticity in the central nervous system. *Mol Pharmacol*, 63, pp. 463–468.

[73] Ferguson G.D. and Storm D.R. (2004). Why calcium-stimulated adenylyl cyclases? *Physiology (Bethesda)*, 19, pp. 271–6.

[74] Polli J.W. and Kincaid R.L. (1994). Expression of a calmodulin-dependent phosphodiesterase isoform (PDE1B1) correlates with brain regions having extensive dopaminergic innervation. *J Neurosci*, 14, pp. 1251–61.

[75] MacKenzie S.J. *et al.* (2002). Long PDE4 cAMP specific phosphodiesterases are activated by protein kinase A-mediated phosphorylation of a single serine residue in Upstream Conserved Region 1 (UCR1). *Br J Pharmacol*, 136, pp. 421–433.

[76] Zhang P. *et al.* (2012). Structure and allostery of the PKA RIIβ tetrameric holoenzyme. *Science*, 335, pp. 712–716.

[77] Quintana A.R., Wang D., Forbes J.E. and Waxham M.N. (2005). Kinetics of calmodulin binding to calcineurin. *Biochem Biophys Res Commun*, 334, pp. 674–680.

[78] Chiba H., Schneider N.S., Matsuoka S. and Noma A. (2008). A simulation study on the activation of cardiac CaMKIIδ-isoform and its regulation by phosphatases. *Biophys J*, 95, pp. 2139–2149.

[79] Keener J. and Sneyd J. (2009). *Mathematical Physiology: II: Systems Physiology*, 2nd edn. (Springer Science & Business Media, New York).

[80] Hulme E.C. and Trevethick M.A. (2010). Ligand binding assays at equilibrium: validation and interpretation. *Br J Pharmacol*, 161, pp. 1219–1237.

[81] Kumar A. (2011). Long-term potentiation at CA3-CA1 hippocampal synapses with special emphasis on aging, disease, and stress. *Front Aging Neurosci*, 3, p. 7.

[82] Dudek S. and Bear M. (1993). Bidirectional long-term modification of synaptic effectiveness in the adult and immature hippocampus. *J Neurosci*, 13, pp. 2910–2918.

[83] Larson J. and Lynch G. (1986). Induction of synaptic potentiation in hippocampus by patterned stimulation involves two events. *Science*, 232, pp. 985–988.

[84] Dudek S.M. and Bear M.F. (1992). Homosynaptic long-term depression in area CA1 of hippocampus and effects of N-methyl-D-aspartate receptor blockade. *Proc Natl Acad Sci USA*, 89, pp. 4363–4367.

[85] Xie Z., Kulasiri D., Samarasinghe S. and Qian J. (2010). An unbiased sensitivity analysis reveals important parameters controlling periodicity of circadian clock. *Biotechnol Bioeng*, 105, pp. 250–259.

[86] Hornberger M.G. and Spear C.R. (1981). *Uncertainty and Forecasting of Water Quality.* eds. Beck M.B. and van Straten G. Chapter "An approach to the analysis of behavior and sensitivity in environmental systems" (Springer, Berlin), pp. 101–116.

[87] Tang Y., Reed P., Wagener T. and van Werkhoven K. (2007). Comparing sensitivity analysis methods to advance lumped watershed model identification and evaluation. *Hydrol Earth Syst Sci*, 11, pp. 793–817.

[88] McKay M.D., Beckman R.J. and Conover W.J. (1979). A Comparison of three methods for selecting values of input variables in the analysis of output from a computer code. *Technometrics*, 21, pp. 239–245.

[89] He Y., Kulasiri D. and Samarasinghe S. (2016). Modelling bidirectional modulations in synaptic plasticity: A biochemical pathway model to understand the emergence of long term potentiation (LTP) and long term depression (LTD). *J Theor Biol*, 403, pp. 159–177.

[90] Mironov S.L. *et al.* (2009). Imaging cytoplasmic cAMP in mouse brainstem neurons. *BMC Neurosci*, 10, p. 29.

[91] Oh M.C., Derkach V.A., Guire E.S. and Soderling T.R. (2006). Extrasynaptic membrane trafficking regulated by GluR1 serine 845 phosphorylation primes AMPA receptors for long-term potentiation. *J Biol Chem*, 281, pp. 752–758.

[92] Hemmings H.C., Nairn A.C. and Greengard P. (1984). DARPP-32, a dopamine- and adenosine 3':5'-monophosphate-regulated neuronal phosphoprotein. II. Comparison of the kinetics of phosphorylation of DARPP-32 and phosphatase inhibitor 1. *J Biol Chem*, 259, pp. 14491–14497.

[93] Sharma R. and Wang J. (1986). Calmodulin and Ca2+-dependent phosphorylation and dephosphorylation of 63-kDa subunit-containing bovine brain calmodulin-stimulated cyclic nucleotide phosphodiesterase isozyme. *J Biol Chem*, 261, pp. 1322–1328.

[94] Wang H. *et al.* (2007). Structures of the four subfamilies of phosphodiesterase-4 provide insight into the selectivity of their inhibitors. *Biochem J*, 408, pp. 193–201.

[95] Hoffmann R. *et al.* (1998). cAMP-specific phosphodiesterase HSPDE4D3 mutants which mimic activation and changes in rolipram inhibition triggered by protein kinase A phosphorylation of Ser-54: generation of a molecular model. *Biochem J*, 333(Pt 1), pp. 139–149.

[96] Bastidas A.C. *et al.* (2012). Role of N-terminal myristylation in the structure and regulation of cAMP-dependent protein kinase. *J Mol Biol*, 422, pp. 215–229.

[97] Su Q. *et al.* (1995). Distribution and activity of calcineurin in rat tissues. Evidence for post-transcriptional regulation of testis-specific calcineurin B. *Eur J Biochem*, 230, pp. 469–474.

[98] Gomperts B.D., Kramer I.M. and Tatham P.E.R. (2009). *Signal Transduction*, 2nd edn. (Elsevier, Oxford).

[99] Kakiuchi S. *et al.* (1982). Quantitative determinations of calmodulin in the supernatant and particulate fractions of mammalian tissues. *J Biochem*, 92, pp. 1041–1048.

[100] Wang H. *et al.* (2003). Type 8 adenylyl cyclase is targeted to excitatory synapses and required for mossy fiber long-term potentiation. *J Neurosci*, 23, pp. 9710–9718.

[101] Hofmann F., Bechtel P.J. and Krebs E.G. (1977). Concentrations of cyclic AMP-dependent protein kinase subunits in various tissues. *J Biol Chem*, 252, pp. 1441–1447.

[102] Andrieu C., Freitas N., de Doucet A. and Jordan M.I. (2003). An introduction to MCMC for machine learning. *Mach Learn*, 50, pp. 5–43.

[103] Haario H., Laine M., Mira A. and Saksman E. (2006). DRAM: efficient adaptive MCMC. *Stat Comput*, 16, pp. 339–354.

[104] Haario H., Saksman E. and Tamminen J. (2001). An adaptive metropolis algorithm. *Bernoulli*, pp. 223–242.

[105] Hudmon A. and Schulman H. (2002). Structure–function of the multifunctional Ca2+/calmodulin-dependent protein kinase II. *Biochem J*, 364, pp. 593–611.

[106] Sheng M. and Hoogenraad C.C. (2007). The postsynaptic architecture of excitatory synapses: a more quantitative view. *Annu Rev Biochem*, 76, pp. 823–847.

[107] Bayer K.U. *et al.* (2006). Transition from reversible to persistent binding of CaMKII to postsynaptic sites and NR2B. *J Neurosci*, 26, pp. 1164–1174.

[108] Bhalla U.S. (2013). *20 Years of Computational Neuroscience.* ed. Bower J.M. Chapter 9 "Still Looking for the memories: molecules and synaptic plasticity" (Springer Science+Business Media, New York), pp. 187–205.

[109] Mayford M., Siegelbaum S.A. and Kandel E.R. (2012). Synapses and memory storage. *Cold Spring Harb Perspect Biol*, 4, p. a005751.

[110] Fong Y.L., Taylor W.L., Means A.R. and Soderling T.R. (1989). Studies of the regulatory mechanism of Ca^{2+}/calmodulin-dependent protein kinase II. Mutation of threonine 286 to alanine and aspartate. *J Biol Chem*, 264, pp. 16759–16763.

[111] Miller S.G., Patton B.L. and Kennedy M.B. (1988). Sequences of autophosphorylation sites in neuronal type II CaM kinase that control Ca^{2+}-independent activity. *Neuron*, 1, pp. 593–604.

[112] Waxham M.N., Aronowski J., Westgate S.A. and Kelly P.T. (1990). Mutagenesis of Thr-286 in monomeric Ca^{2+}/calmodulin-dependent protein kinase II eliminates Ca^{2+}/calmodulin-independent activity. *Proc Natl Acad Sci USA*, 87, pp. 1273–1277.

[113] Sanderson J.L. *et al.* (2012). AKAP150-anchored calcineurin regulates synaptic plasticity by limiting synaptic incorporation of Ca^{2+}-permeable AMPA receptors. *J Neurosci*, 32, pp. 15036–15052.

[114] Gold M.G. *et al.* (2006). Molecular basis of AKAP specificity for PKA regulatory subunits. *Mol Cell*, 24, pp. 383–395.

[115] Smith K.E., Gibson E.S. and Dell'Acqua M.L. (2006). cAMP-dependent protein kinase postsynaptic localization regulated by NMDA receptor activation through translocation of an A-kinase anchoring protein scaffold protein. *J Neurosci*, 26, pp. 2391–2402.

[116] Abraham W.C. and Bear M.F. (1996). Metaplasticity: the plasticity of synaptic plasticity. *Trends Neurosci*, 19, pp. 126–130.

[117] Lisman J.E., Yasuda R. and Raghavachari S. (2012). Mechanisms of CaMKII action in long-term potentiation. *Nat Rev Neurosci*, 13, pp. 169–182.

[118] Prichard L., Deloulme J.C. and Storm D.R. (1999). Interactions between Neurogranin and Calmodulin *in vivo*. *J Biol Chem*, 274, pp. 7689–7694.

[119] Baudier J. *et al.* (1991). Purification and characterization of a brain-specific protein kinase C substrate, neurogranin (p17). Identification of a consensus amino acid sequence between neurogranin and neuromodulin (GAP43) that corresponds to the protein kinase C phosphorylation site and the Calmodulin-binding domain. *J Biol Chem*, 266, pp. 229–237.

[120] Petersen A. and Gerges N.Z. (2015). Neurogranin regulates CaM dynamics at dendritic spines. *Sci Rep*, 5, p. 11135.

[121] Michaelis M.L. *et al.* (1996). Decreased plasma membrane calcium transport activity in aging brain. *Life Sci*, 59, pp. 405–412.

[122] Robison A.J., Winder D.G., Colbran R.J. and Bartlett R.K. (2007). Oxidation of calmodulin alters activation and regulation of CaMKII. *Biochem Biophys Res Commun*, 356, pp. 97–101.

Chapter 6

Uncertainty Quantification of Models Related to Synaptic Plasticity

Environmental stimuli or action potential in a typical presynaptic cell are transmitted to the corresponding postsynaptic cell through the exocytosis of synaptic vesicles, which releases neurotransmitters to the synaptic cleft [1, 2]. The released neurotransmitters diffuse through the protein matrix in the synaptic cleft to reach the postsynaptic terminal, often the dendrite or soma of the postsynaptic neuron [3]. The ligand-gated receptors, such as N-methyl-D-aspartate receptor (NMDAR) and α-amino-3-hydroxy-5-methyl-4-isoxazole-propionic acid receptor (AMPAR), distributed across the postsynaptic density (PSD) of the distal tip of the dendritic spine, are activated by the neurotransmitters to open the ion channels in them. The opening of the ion channels allows an influx of Ca^{2+} triggering the activation of postsynaptic protein cascade [4–7] which eventually results in excitatory postsynaptic potential (EPSP) or excitatory postsynaptic current (EPSC). (If the ion channel permeable to anion, inhibitory postsynaptic potential (IPSP) would result; however, our focus in this Chapter is related to EPSP or EPSC.) The magnitude of the postsynaptic response to an environmental stimulus strongly correlates to the synaptic strength of a synapse or a set of synapses. According to the experimental evidence in the last two decades, memory is stored in synapses [1, 8, 9] as synaptic strength. Environmental stimuli register response patterns as memory which may be retrieved by recognising the stored patterns that match or are sensitive to environmental stimuli [1, 10].

Inside a presynaptic cell, neurotransmitter release by the exocytosis of synaptic vesicles is probabilistic [11–13]; therefore, the postsynaptic Ca^{2+} patterns thus evoked are highly nonlinear. These Ca^{2+} patterns can induce long-term potentiation (LTP) and long-term depression (LTD) depending on the dynamic levels of Ca^{2+} [5, 14–18]. LTP increases EPSP (or EPSC) in response to high frequency stimulations (i.e. high frequency Ca^{2+} patterns) [14, 15, 19] and is an important phenomenon in memory formation; it is strongly correlated to formation of the memory and new synapses [20–22]; spatial memory formation is impaired by blocking LTP [23–29]; and LTP inactivated synapses do not form memory properly [30–32]. The early phase LTP (E-LTP) rapidly potentiates synaptic strength immediately after (a few seconds) the transient stimulation, whereas the late phase LTP (L-LTP) sustains the potentiation during which relevant genes get expressed and the structural changes occur [15, 23, 33]. E-LTP is induced experimentally in CA1 region of hippocampus upon the application of high frequency stimulation [19, 34], which opens NMDARs to allow Ca^{2+} influx into postsynaptic cell. The protein cascade [35–37] then changes the property of AMPAR, a mobile receptor, which increases the number of AMPAR embedded in the membrane and the single channel conductance of AMPAR thereby increasing the total channel conductance. This leads to depolarisation of the membrane leading to the elevated EPSC and EPSP [14, 15, 38–43]. Ca^{2+}/calmodulin (CaM)-dependent protein kinase II (CaMKII) — the "memory molecule" — plays a pivotal role in this cascade by phosphorylating AMPAR [44–46] to increase its ion channel conductance [38] and by anchoring AMPAR into the PSD [47, 48]. Further, CaMKII binds to NMDAR in the cascade forming CaMKII-NMDAR complex which is an essential component of LTP and memory formation [46, 49–57].

As shown by the experiments [58–62], the structural changes in CaMKII are crucial for binding to NMDAR in LTP; therefore a phenomenological model should take them into account. We have already developed a mechanistic model (see [63]) to understand the dynamics of biochemical pathways related to E-LTP, centred on the state transitions (STs) of CaMKII, a holoenzyme with 12 subunits [59]. The model is based on the STs related to conformational changes of the subunits as well as on the holoenzyme states transitions (HSTs) of CaMKII (see Chapter 4) including the translocation to PSD and binding to NMDAR [58, 61, 62]. We denote this model as State Transition Model (STM). STM predicts well the changes in the translocation of T286-mutant CaMKII as observed in the experiments; it also predicts the diminishing ability of T286 mutant to distinguish the frequencies of tetanus. STM also highlights the important roles the autophosphorylation of subunits play in LTP: the autophosphorylation is crucial for the amplification of postsynaptic responses related to external stimuli; the autophosphorylation decodes the frequency of stimulation; and it couples the multiple trains of tetanus with long inter-train intervals [63].

STM, however, has 28 parameters and 5 constants, some of which were estimated by using Markov chain Monte Carlo (MCMC) method [64, 65] and the others were obtained from the literature. The reason to have this many number of parameters is the philosophy behind the model development; the mathematical relationships in this model should be true to the experimental observation and/or to the plausible interpretation of them as much as possible, and they should be based on chemical kinetics. With respect to this approach, Hill [66] states the following: "... of course, experimentally founded 'particular models' are the ultimate goal." However, the caveat of this approach is the uncertainties associated with the parameters as the individual parameters are impossible to measure independently and the reported experimental data also contain uncertainties associated with instrumentation and protocols. (As STM is a deterministic model, we are only concerned with epistemic uncertainty.) What are the effects of parameter uncertainties on the key outputs of STM? Which parameters are the most sensitive ones and why? In STM, we have Ca^{2+} patterns with different level of frequencies as inputs. If a set of parameters are insensitive to the outputs relative to the other parameters of the set, is it possible to reduce the processes in

STM which capture essential biological knowledge with a minimal set of parameters, for E-LTP under high frequency stimulations (HFS)? In this chapter, we discuss such uncertainty quantification with respect to STM using only HFS as they produce E-LTP. (See also the excellent review of uncertainty analysis by Marino *et al.* [67]).

However, STM responds to a variety of high frequency stimulations (HFS) well and helps us to understand the significance of the autophosphorylation induced dynamics in regulating the CaMKII translocation and the CaMKII-NMDAR binding [63] (Multiple trains of tetanus are often used experimentally to induce L-LTP.) In this study, we use a single tetanus of 100 pulses at $100\,\mathrm{Hz}$ as input.

We also extend STM to have the presynaptic action potentials as the input by adding two models upstream to STM in tandem: a vesicle release model (VRM) with four parameters [68] giving the presynaptic release probability of neurotransmitter vesicles due to exocytosis as the output, which acts as the input to the postsynaptic membrane model [69] of NMDAR-mediated Ca^{2+} influx (Ca2M) into the postsynaptic cell. This extended model (ESTM) can be used to investigate the highly nonlinear relationship between presynaptic action potentials and the postsynaptic outputs from STM. ESTM is a much more useful model to understand the behaviours of the outputs as it includes the probabilistic nature of neurotransmitter (glutamate in this study) release [70]. Therefore, we attempt to understand the effects of parameter uncertainties of ESTM on the outputs, and we see that ESTM parameters behave quite differently to STM parameters when globally perturbed.

6.1. A Summary of STM and ESTM Models

We choose the following variables of STM and ESTM as the outputs $(y_i,\ i = 1, 2, 3, 4)$ to explain E-LTP under HFS as in the original model [63]: $y_1 \equiv$ CaMKII-NMDAR complex number in PSD at time $= 300\,\mathrm{s}$ (Output1); $y_2 \equiv$ Peak number of CaMKII holoenzyme in PSD (Output2); $y_3 \equiv$ Peak number of autophosphorylated CaMKII

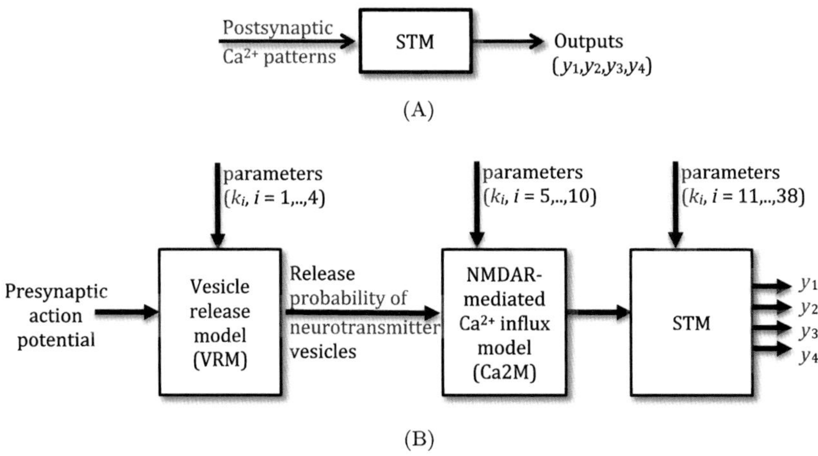

Fig. 6.1. Schematics of inputs and outputs of the models. (A) The state transition model (STM) [63] and the postsynaptic calcium patterns at different frequencies as inputs. (B) The extended model of STM (ESTM) with the vesicle release model (VRM) [68] and NMDAR-mediated Ca^{2+} influx model (Ca2M) [69] added in tandem. The inputs to ESTM are presynaptic action potentials and VRM calculates the release probabilities of vesicles containing neurotransmitters, which are inputs to Ca2M. Ca^{2+} patterns are the outputs of Ca2M, which act as inputs to STM.

subunits (Output3); and $y_4 \equiv$ Peak number of Ca_4CaM (Output4). In Fig. 6.1A, we show these outputs of STM as a schematic diagram for illustrative purposes.

Active CaMKII subunit (Ca_4CaM bound CaMKII subunit) is central to the LTP formation. A CaMKII subunit is activated when a Ca^{2+}/CaM complex binds to the CaM footprint leading to a conformational change which exposes the S site for the catalytic processes [59, 71].

CaMKII is held away from PSD by F-actin [72] and Ca^{2+}/CaM-dependent activation releases the kinase that allows it to translocate into PSD [61]. CaMKII holoenzyme binds to NR2B subunit of NMDAR and the binding anchors CaMKII in PSD [49–51]. The binding locks CaMKII in an active conformation independent of the autophosphorylation [73]. The CaMKII-NMDAR complex, formed by associating the T site of CaMKII to NMDAR, contributes: (1) to the staying of CaMKII in PSD for more than 30 mins [58] that may be a potential mechanism underlying the connection between E-LTP and late phase LTP (L-LTP), and (2) to high associativity to the induction of LTP [52–54]. Further the experimental evidence shows that inhibiting the formation of CaMKII-NMDAR complex impairs LTP [55–57]. Therefore, y_1 is one of the most important outputs for understanding of the LTP formation.

CaMKII in PSD is closer to its target, AMPAR and TARP, and binding partner, NMDAR. Furthermore, a study reports that the autophosphorylation is irreversible in PSD and the irreversibility may have a critical role in LTP [74]. Therefore, the peak number of CaMKII holoenzyme in PSD, y_2, shows the altitude of the irreversible autophosphorylation as well as the potential amount of CaMKII holoenzyme for binding to NMDAR.

The autophosphorylated CaMKII subunits remain at least partially active even when the CaM dissociates; the autophosphorylation maintains the activity of CaMKII for up to one minute after the Ca^{2+} level decreases [75]. Moreover, the autophosphorylation increases the affinity of a subunit for Ca^{2+}/CaM complex binding [76]. Hence, the autophosphorylation prevents the reverting of the active CaMKII subunits into the inhibited state when the Ca^{2+} level decreases [77–79]. Therefore, y_3 is a good indicator for the level of postsynaptic autophosphorylation of CaMKII subunits.

Ca^{2+}/CaM complex activates CaMKII [59, 71]. Therefore, the peak level of Ca^{2+}/CaM complex, y_4, determines the amount of CaMKII to be activated. On the other hand, CaMKII traps Ca^{2+}/CaM complex and the conformational changes of CaMKII decrease the CaM dissociation rate. Hence, the peak level of Ca^{2+}/CaM complex also shows the strength of CaMKII trapping of Ca^{2+}/CaM complex.

VRM and Ca2M are also already published models [68, 69] and the equations are elicited from the publications. The notation of the parameters follows Fig. 6.1B, and their biological meanings, baseline values and the ranges that are used in global sensitivity analysis (GSA) along with the units are listed in Tables 6.1–6.4. All the equations are highly nonlinear and the behaviours of the outputs depend on the nonlinear inputs and the parameter values.

Table 6.1.   Parameters of VRM.

VRM parameters	Biological meaning	Baseline [source]	Range
k_1	Initial depletion (RRP size)	7 [11]	1–12
k_2	Depletion decay time constant	100 ms [80]	50–150
k_3	Magnitude of facilitation	0.2 [11]	0.1–0.3
k_4	Facilitation decay time constant	100 ms [80]	50–150

Table 6.2.   Parameters of Ca2M.

Ca2M parameters	Biological meaning	Baseline [source]	Range
k_5	Postsynaptic membrane potential	-60 mV [69]	from -30 to -90
k_6	Ca^{2+} decay time constant	0.065 s [81]	0.0325–0.0975
k_7	Fraction of NMDARs in the closed state that shift to the open state	0.5 [69]	0.25–0.75
k_8	Relative magnitude of the fast component of the NMDAR current	0.7 (fit to [81])	0.35–1
k_9	Fast exponential decay time constant	0.05 s [69]	0.025–0.075
k_{10}	Slow exponential decay time constant	0.2 s [69]	0.1–0.3

6.2. Nature of Inputs

The inputs for STM and ESTM have the same patterns of stimulation. Each pillar represents one pulse of the stimulation. Hz only specifies the separation between two pulses (0.01 s for 100 Hz, 1 s for 1 Hz, 0.1 s for 10 Hz, and 0.025 s for 40 Hz), but does not tell

Table 6.3.   Constants of STM.

STM constants	Biological meaning	Baseline [source]	Range
k_{11}	Total number of CaMKII subunits	1200 # [70]	600–1800
k_{12}	Total number of NMDAR (or NR2B)	20 # [70, 82]	10–20
k_{13}	Total number of CaM molecule	500 # [63]	250–750
k_{14}	Total number of ATP	240,000 # [83]	120,440–361,330
k_{15}	Total number of PP1	145 # [63]	72–218

how many pulses that is applied. We always apply 100 pulses for the computational experiments in this study. (Output1, Output2 and Output3 reach the maximum against the frequency at 40 Hz in STM, hence the inclusion of this frequency in the computational experiments.) For example, a tetanus at 100 Hz contains 100 pulses delivered with 10 ms separation. This pattern of stimulation is also referred to as HFS for LTP induction. A tetanus at 1 Hz contains 100 pulses delivered with 1 s separation and this pattern of stimulation is also referred to as LFS for LTD induction. A tetanus of multiple trains consists of a number of epochs of pulses, each separated by an inter-train interval.

6.3. Global Sensitivity Analysis (GSA) of STM and ESTM

Most of the parameter values in STM and ESTM are obtained from the literature and some STM parameters are estimated using MCMC method (see [63]). The parameter values based on the experimental data with uncertainties introduce uncertainties in the designated outputs, and these epistemic uncertainties need to be characterized globally using global sensitivity analysis techniques. (See [67] for a summary of these techniques.) In this study, we use Latin hypercube sampling (LHS) and partial rank correlation coefficient (PRCC) for

Table 6.4.   Parameters of STM.

STM parameters	Biological meaning	Baseline [source]	Range
k_{16}	Autophosphorylation rate of CaMKIICaM into CaMKII*	$0.9\,\mathrm{s}^{-1}$ [84]	0.45–1.35
k_{17}	Dephosphorylation rate of CaMKII*	$1.72\,\mathrm{s}^{-1}$ [84]	0.86–2.58
k_{18}	Michaelis constant of autophosphorylation	$1150\,\#^{-1}$ [84]	575–1725
k_{19}	Michaelis constant of dephosphorylation by PP1	$660\,\#^{-1}$ [84]	331–993
k_{20}	Fraction of active CaMKII subunits at half maximal translocation rate	0.39 [63]	0.195–0.585
k_{21}	Fraction of autophosphorylated CaMKII subunits at half maximal dissociation rate	0.019 [63]	0.0095–0.0285
k_{22}	Binding rate of Ca_4CaM and iCaMKII	$0.035\,\#^{-1}\mathrm{s}^{-1}$ [84]	0.0174–0.0523
k_{23}	Dissociation rate of CaMKIICaM into Ca_4CaM and iCaMKII	$0.14\,\mathrm{s}^{-1}$ [84]	0.07–0.21
k_{24}	Translocation rate of EoPSD into PSD	$0.0088\,\mathrm{s}^{-1}$ [63]	0.0044–0.0132
k_{25}	Dissociation rate of EiPSD out of PSD	$0.247\,\mathrm{s}^{-1}$ [63]	0.1235–0.3705
k_{26}	Binding rate of elementary CaMKII to NMDAR at the S site	$0.008\,\#^{-1}\mathrm{s}^{-1}$ [63]	0.004–0.012
k_{27}	Dissociation rate of CaMKIIS into EiPSD and NMDAR	$0.38\,\mathrm{s}^{-1}$ [63]	0.19–0.57
k_{28}	Dissociation rate of CaMKII*S into EiPSD and NMDAR	0.024 [63]	0.012–0.036
k_{29}	Transfer rate of NMDAR binding from S site to T site	$0.01\,\mathrm{s}^{-1}$ [63]	0.005–0.015
k_{30}	Turnover rate of CaMKII	$1/108000\,\mathrm{s}^{-1}$ [85]	1/216000–1/72000

(Continued)

Table 6.4.   (*Continued*)

STM parameters	Biological meaning	Baseline [source]	Range
k_{31}	Binding rate of Ca^{2+} and CaM to form CaCaM	$0.0415\ \#^{-1}\,s^{-1}$ [84]	0.02075–0.06227
k_{32}	Dissociation rate of CaCaM into Ca^{2+} and CaM	$50\,s^{-1}$ [84]	25–75
k_{33}	Binding rate of Ca^{2+} and CaCaM to form Ca_2CaM	$1.45\ \#^{-1}\,s^{-1}$ [84]	0.732–2.198
k_{34}	Dissociation rate of Ca_2CaM into Ca^{2+} and CaCaM	$50\,s^{-1}$ [84]	25–75
k_{35}	Binding rate of Ca^{2+} and CaCaM to form Ca_3CaM	$0.2\ \#^{-1}\,s^{-1}$ [84]	0.1–0.3
k_{36}	Dissociation rate of Ca_3CaM into Ca^{2+} and Ca_2CaM	$1250\,s^{-1}$ [84]	625–1875
k_{37}	Binding rate of Ca^{2+} and CaCaM to form Ca_4CaM	$4.15\ \#^{-1}\,s^{-1}$[84]	2.08–6.23
k_{38}	Dissociation rate of Ca_4CaM into Ca^{2+} and Ca_3CaM	$1250\,s^{-1}$[84]	625–1875

GSA [67, 86] as they are the most popular, reliable, and efficient technique and index, respectively. PRCC is a robust sensitivity measure only when the relationships between the parameters and the outputs are nonlinear and strongly monotonic. If the relationships are non-monotonic, extended Fourier amplitude test (eFAST) provides a more robust sensitivity index [67]. Therefore, it is important to check for monotonicity of the model parameters against the outputs using monotonicity plots [67] before using PRCC.

6.3.1. *Monotonicity plots*

To verify the existence of a monotonic relationship between each parameter and each output, first each parameter is varied from 10%

to 190% of its baseline value and generated the LHS matrix [86] for all the parameters using Matlab LHS function *lhsdesign*. (We assume that the parameters are distributed according to uniform distributions.) To test the level of monotonicity for a specific parameter, the values of this parameter are taken from the LHS matrix while keeping other parameters at their baseline values. We run STM and ESTM using the parameter set for each model thus generated, and plot each output against each parameter. As there are no qualitative differences in the monotonicity plots based on the simulation results of STM and ESTM, we only discuss the results for STM for brevity. The monotonicity plots for Output1, Output2, Output3, and Output4 vs. the parameters $(k_{11}, \ldots, k_{38})$ show for example, at $1\,$Hz stimulation, the observed ranges of all outputs are very small but they monotonically increase against the parameter values. Similarly for other frequencies, the monotonicity plots show neither non-monotonic relationships nor excessively large ranges of outputs, except for k_{11} (total number of CaMKII subunits) which has non-monotonic relationship to Output1, Output2, and Output3 at high frequency stimulations ($F = 40\,$Hz and $100\,$Hz). The turning points appear when k_{11} at 50% of the baseline value (1200). Therefore, we truncate the range for all parameters to 50% to 150% of the baseline value to have monotonicity for calculating PRCC as suggested by Marino *et al.* [67]. The monotonicity plots of all 38 ESTM parameters against four outputs, varying parameter values from 10% to 190% of their baseline values, show similar behaviours (data not shown); and only k_{11} shows non-monotonic relationships against Output1, Output2, and Output3 at high frequency stimulation. Therefore, truncating the range of values for PRCC is needed as in STM.

6.4. Interpretation of PRCC Results

6.4.1. *STM model*

The PRCC for the outputs and the corresponding parameters of STM (k_{11}–k_{38}) for the postsynaptic Ca^{2+} inputs at the frequency (F) of 1, 10, 40 and 100$\,$Hz are calculated and the subsets of parameters for which PRCC > 0.5 and $p < 0.05$ against each of the output for the different F levels are categorized. (Following Marino *et al.* [67], we set the threshold values for PRCC and p-value as 0.5 and 0.05, respectively.) When $F = 1\,\text{Hz}$, for example, CaMKII-NMDAR complex number (at 300 s) is significantly sensitive to 18 parameters; peak number of CaMKII in PSD is sensitive to 15 parameters; peak number of autophosphorylated CaMKII subunits is sensitive to 16 parameters; and peak number of Ca4CaM is sensitive to 12 parameters.

We can summarise our findings as follows:

(1) The set of significantly sensitive parameters for Output4 is a subset of all the other sets of significantly sensitive parameters.

(2) The number of sensitive parameters for each output decreases as the frequency of stimulation increases from 1 to 40 Hz. The number of most sensitive parameters for Output1, for example, are 18 when the frequency of stimulation ($F = 1$) is 1 Hz, 12 when the frequency of stimulation is 10 Hz ($F = 10$), and 9 when $F = 40\,\text{Hz}$. Then it increases to 10 when $F = 100\,\text{Hz}$.

(3) For $F = 1\,\text{Hz}$, the absolute values of PRCC of the first 11 parameters for Output1, all the parameters for Output2, the first 11 parameters for Output3 and the first 12 parameters for Output4 are above 0.65 whereas the rest of the parameters have PRCC values (for Output1 and Output3) slightly above 0.5. k_{22} (binding rate of Ca_4CaM and iCaMKII) is the most sensitive parameter for all four outputs, because at the low frequencies, Ca_4CaM binding to the inhibited CaMKII is a dominant process at low Ca^{2+} levels. The driving mechanism of the binding, Ca^{2+} binding to and dissociating from CaM through the four distinct reactions [63, 87], has a major impact on all four outputs. k_{38} (dissociation rate of Ca_4CaM into Ca^{2+} and Ca_3CaM), which indicates the first step in dissociation of Ca_4CaM, has a very similar absolute PRCC values for all four outputs: for Output1, $k_{38} = -0.7674$; for Output2, -0.7399; for Output3, -0.773; and for Output4, -0.7802. In terms of STM constants, k_{11} (total number of CaMKII subunits) and k_{13} (total number of CaM molecules) have significant PRCC values for all four

outputs, whereas k_{15} (total number of PP1) is only significant for Output3. This is understandable because at low level of Ca^{2+}, there is negligible autophosphorylated Ca_4CaM-bound CaMKII subunits (Output3) which do not influence the other outputs.

(4) It is interesting to note that the STM constants, k_{11} and k_{13}, are significant for Output4 at $F = 40\,Hz$ and $F = 100\,Hz$. As a higher level of Ca^{2+} stimulation results in LTP, k_{13} (total number of CaM molecules) is strongly correlating to all four outputs; the importance of CaM number in LTP has been experimentally observed as well [88].

(5) At $F = 100$ and $40\,Hz$, the parameters related to the kinase pathway (see [63]) are significant in general. k_{24} (translocation rate) is only significant for Output1 and Output2 as it should be at the higher frequencies.

(6) At $F = 40\,Hz$ and $100\,Hz$, almost the same parameters have significant PRCC values for Output1. It is interesting to note however that k_{11} is significant for Output1, only for $F = 100\,Hz$ and $1\,Hz$, not for the in-between frequencies, $10\,Hz$ and $40\,Hz$. This simply means that CaMKII subunit number, even though essential in the reactions, has a lesser global significance in influencing Output1 at those frequencies.

6.4.2. *ESTM*

We evaluate the PRCC and p-values for ESTM with the same frequencies as before, and we have additional four parameters (k_1-k_4) from VRM and six parameters (k_5-k_{10}) from Ca2M increasing the total number of parameters of ESTM to 38. Similarly, we have categorized the subsets of parameters for which $PRCC > 0.5$ and $p < 0.05$ against each of the outputs at $F = 1, 10, 40$ and $100\,Hz$, respectively. k_5 (postsynaptic membrane potential) is the most sensitive parameter across all the frequencies for all four outputs, and almost all the parameters that are significantly sensitive to the outputs of ESTM that belong to VRM and Ca2M except for k_{13} (total number of CaM molecules in STM). Once we have VRM and Ca2M added to STM, the sensitivities of STM parameters are masked by those of VRM and Ca2M showing the dominance of neurotransmitter vesicle release and subsequent Ca^{2+} influx over the processes in the postsynaptic biochemical pathways. This does not mean however that the postsynaptic processes are phenomenologically unimportant, rather they are the most important in producing E-LTP, and any interpretation of GSA should be carefully construed. STM can be reduced to a simpler model retaining the phenomenological integrity in modelling based on the interpretations of GSA on STM parameter space alone. could be used to reduce complex models to simpler ones.

6.5. A Summary of the Possible Simplifications of STM based on GSA and Biological Reasoning

If we are to reduce STM to a simpler one, we could use the information provided by the GSA and PRCC index in a biologically meaningful manner. Many important processes are involved in the mechanisms that connect the Ca^{2+} influx to the formation of CaMKII-NMDAR complex: Ca^{2+}/CaM complex formation; activation of CaMKII subunit; autophosphorylation of CaMKII subunit; translocation of CaMKII holoenzyme into PSD; CaMKII holoenzyme binding to NMDAR at (S site of) a CaM bound subunit; CaMKII holoenzyme binding to NMDAR at (S site of) an autophosphorylated subunit; and S site bound CaMKII transfering to T site bound (see Chapter 4). These processes form building blocks of STM model. We can simplify the processes by evaluating the significance of their associated parameters through PRCC ($|PRCC| > 0.5$). Our judgement on the significance of the parameters, therefore the reactions associated with them, is mainly based on the model response during E-LTP, which displays high associativity to CaMKII-NMDAR complex [52–54]. High frequency stimulation (HFS), the standard protocol for LTP induction [15, 34], is used to simulate the model and record the four outputs chosen. We define that a reaction is significant if it contains at least one significant parameter; and based on the significant reactions and phenomenological importance we propose a new conceptual model.

We can discuss the possible simplifications using the information provided by PRCC indices:

(1) Ca^{2+}/CaM *complex formation.* Ca^{2+}/CaM complex formation involves four sequential steps to attach four Ca^{2+} ions to a CaM molecule. The reaction rates are formulated based on Holmes's model [87]. There are eight parameters associated with this process, $k_{31}-k_{38}$, which are reaction rate constants for the binding and dissociation between Ca^{2+} and CaM. Since none of them are significant in HFS which gives rise to LTP, this process could be simplified. However, Ca^{2+}/CaM complex has to appear in the new model since it is an essential element for the activation of CaMKII.

(2) *Activation of CaMKII subunit.* Activation of a CaMKII subunit frees itself from F-actin binding that allows the translocation of CaMKII into PSD later [61, 62]. The activation requires binding of a Ca^{2+}/CaM complex to an inhibited CaMKII subunit to from an active subunit: Ca^{2+}/CaM bound CaMKII subunit. The activation is formulated by mass action rate laws with two parameters: a binding rate constant k_{22} and a dissociation rate constant k_{23}. Hence k_{23} is significant for Output1 and Output2; we could not simplify this process.

(3) *Autophosphorylation of CaMKII subunit.* An active CaMKII subunit can be autophosphorylated by its active neighbor subunits. An autophosphorylated CaMKII subunit remains at least partially active even when the CaM dissociates [59, 89–92]. The exact relationship between the autophosphorylation and CaMKII-NMDAR complex is unknown. Our computational analysis shows that the delayed dissociation of CaMKII away from PSD by autophosphorylation [61, 62] helps the formation of CaMKII-NMDAR complex in response to transient signals, such as HFS [63]. The autophosphorylation can be reversed by PP1 through dephosphorylation. The autophosphorylation is modelled based on Michaelis Menten kinetics for enzyme reactions [84] with six parameters: the level of ATP (k_{14}), the level of PP1 (k_{15}), the autophosphorylation rate (k_{16}), the dephosphorylation rate (k_{17}), Michaelis constant of

the autophosphorylation (k_{18}) and Michaelis constant of the dephosphorylation (k_{19}). Four of them (k_{15}, k_{16}, k_{17}, and k_{19}) are significant at least for one output. The two insignificant parameters, k_{14} and k_{18}, are related to ATP, which is abundant in all cells (its level is much larger than those of other species of the model). As a result, ATP can be considered as a large constant in the Michaelis Menten kinetics. Hence, we can ignore k_{14} and k_{18}.

(4) *Translocation of CaMKII holoenzyme into PSD.* Translocation is a critical step in moving CaMKII into PSD, where CaMKII's target and binding partner, NMDAR, reside. It is known that the translocation into PSD requires activation of CaMKII and the dissociation of CaMKII out of PSD is delayed by the autophosphorylation [61]. Translocation is modelled by two Hill equations, one describes the translocation of CaMKII into PSD having the fraction of the active CaMKII subunits as the control parameter and the other describes the dissociation of CaMKII out of PSD having the fraction of the autophosphorylated CaMKII subunit as the control parameter [63]. The two equations contain four parameters: translocation rate into PSD (k_{24}), dissociation rate out of PSD (k_{25}), fraction of active CaMKII subunits at half maximal translocation rate into PSD (k_{20}) and fraction of autophosphorylated CaMKII subunits at half maximal translocation rate out of PSD (k_{21}). Since k_{20} and k_{24} are significant, the translocation into PSD is an important process that cannot be simplified.

(5) *CaMKII holoenzyme binds to* NMDAR *at (S site of) a CaM bound subunit or an autophosphorylated subunit.* These processes are intermediate binding processes that lead to the T site binding between a CaMKII subunit and NMDAR later. The CaM bound CaMKII subunit, as the binding point to NMDAR, can be transferred to an autophosphorylated subunit by the autophosphorylation. Since autophosphorylation is irreversible in PSD [74], the autophosphorylated subunit, as the binding point to NMDAR, cannot be reversed into a CaM bound subunit. The binding processes are modelled by mass action kinetics

with three parameters: binding rate of elementary CaMKII to NMDAR at the S site (k_{26}) which is modified by Ω to represent an average binding rate among different conformations of CaMKII holoenzyme; dissociation rate of the binding at a CaM bound subunit (k_{27}); and dissociation rate of the binding at an autophosphorylated subunit (k_{28}). The autophosphorylation is modelled by the same equation (as well the same parameter) as the autophosphorylation of CaMKII subunit, which we do not analyse independently. Only k_{28} is significant. This may mean that the modifier Ω is unnecessary and the binding at a CaM bound subunit becomes the binding at an autophosphorylated subunit since the autophosphorylation is irreversible in PSD. Therefore, we can ignore Ω and combine the two conformations of CaMKII-NMDAR binding at S site into one variable.

(6) *S site bound CaMKII transfers to T site bound.* This step forms CaMKII-NMDAR complex. It is modelled as an irreversible reaction based on mass action law with a significant parameter: transfer rate of NMDAR binding from S site to T site (k_{29}). Therefore, this process is important.

(7) *CaMKII degradation.* Although the degradation rate (k_{30}) is insignificant, the degradation process has to be included, at least for CaMKII-NMDAR complex, for the following reasons: (a) the degradation rate itself is very small, therefore, it does not affect too much on the model dynamics over a short-time scale used in the analysis, and (b) without degradation, the irreversible CaMKII-NMDAR complex can be accumulated to an unrealistically large amount under rest Ca^{2+} level.

6.6. Summary

In this chapter, we show that it is possible to propose simplifications to a highly nonlinear complex model of postsynaptic biochemical pathways related to LTP using global sensitivity analysis and PRCC combined with a good understanding of the biology involved. However, in the reduction process, we need to configure the outputs for the reduced model and the character of the reduced model may differ with the outputs chosen. This is true for the inputs as well. Once the mathematical structure of a model is established, the parameter space plays a pivotal role in giving the mathematical structure an interpretative paradigm.

References

[1] Bear M.F., Connor B.W. and Paradiso M.A. (2007). *Neuroscience: Exploring the Brain*, 3rd edn. (Lippincott Williams & Wilkins, Baltimore).

[2] Purves D. *et al.* (2008). *Neuroscience*, 4th edn. (Sinauer Associates, Inc., Sunderland, MA).

[3] Danbolt N.C. (2001). Glutamate uptake. *Prog Neurobiol*, 65, pp. 1–105.

[4] Citri A. and Malenka R.C. (2007). Synaptic plasticity: multiple forms, functions, and mechanisms. *Neuropsychopharmacology*, 33, pp. 18–41.

[5] Yang S., Tang Y. and Zucker R.S. (1999). Selective induction of LTP and LTD by postsynaptic $[Ca^{2+}]_i$ elevation. *J Neurophysiol*, 81, pp. 781–787.

[6] Soderling T.R. and Derkach V.A. (2000). Postsynaptic protein phosphorylation and LTP. *Trends Neurosci*, 23, pp. 75–80.

[7] Sharma K., Fong D.K. and Craig A.M. (2006). Postsynaptic protein mobility in dendritic spines: long-term regulation by synaptic NMDA receptor activation. *Mol Cell Neurosci*, 31, pp. 702–712.

[8] Isaac J.T.R., Nicoll R.A. and Malenka R.C. (1995). Evidence for silent synapses: implications for the expression of LTP. *Neuron*, 15, pp. 427–434.

[9] Anderson J.R. (2000). *Learning and Memory*, 2nd edn. (Wiley, New York).

[10] Martin S.J., Grimwood P.D. and Morris R.G.M. (2000). Synaptic plasticity and memory: an evaluation of the hypothesis. *Annu Rev Neurosci*, 23, pp. 649–711.

[11] Dobrunz L.E. and Stevens C.F. (1997). Heterogeneity of release probability, facilitation, and depletion at central synapses. *Neuron*, 18, pp. 995–1008.

[12] Branco T. and Staras K. (2009). The probability of neurotransmitter release: variability and feedback control at single synapses. *Nat Rev Neurosci*, 10, pp. 373–383.

[13] Stevens C.F. (2003). Neurotransmitter release at central synapses. *Neuron*, 40, pp. 381–388.

[14] Bliss T.V.P. and Lømo T. (1973). Long-lasting potentiation of synaptic transmission in the dentate area of the anaesthetized rabbit following stimulation of the perforant path. *J Physiol*, 232, pp. 331–356.

[15] Bliss T.V.P. and Collingridge G.L. (1993). A synaptic model of memory: long-term potentiation in the hippocampus. *Nature*, 361, pp. 31–39.

[16] Dudek S.M. and Bear M.F. (1992). Homosynaptic long-term depression in area CA1 of hippocampus and effects of *N*-methyl-*D*-aspartate receptor blockade. *Proc Natl Acad Sci USA*, 89, pp. 4363–4367.

[17] Bear M.F., Cooper L.N. and Ebner F.F. (1987). A physiological basis for a theory of synapse modification. *Science*, 237, pp. 42–48.

[18] Lisman J.E. (1989). A mechanism for the Hebb and the anti-Hebb processes underlying learning and memory. *Proc Natl Acad Sci USA*, 86, pp. 9574–9578.

[19] Bhalla U.S. (2013). *20 Years of Computational Neuroscience.* ed. Bower J.M. Chapter 9 "Still looking for the memories: molecules and synaptic plasticity" (Springer Science+Business Media, New York), pp. 187–205.

[20] Engert F. and Bonhoeffer T. (1999). Dendritic spine changes associated with hippocampal long-term synaptic plasticity. *Nature*, 399, pp. 66–70.

[21] Morgado-Bernal I. (2011). Learning and memory consolidation: linking molecular and behavioral data. *Neuroscience*, 176, pp. 12–19.

[22] Nabavi S. *et al.* (2014). Engineering a memory with LTD and LTP. *Nature*, 511, pp. 348–352.

[23] Abel T. *et al.* (1997). Genetic demonstration of a role for PKA in the late phase of LTP and in hippocampus-based long-term memory. *Cell*, 88, pp. 615–626.

[24] Bourtchuladze R. *et al.* (1994). Deficient long-term memory in mice with a targeted mutation of the cAMP-responsive element-binding protein. *Cell*, 79, pp. 59–68.

[25] Chang H.P. *et al.* (1999). Impaired memory retention and decreased long-term potentiation in integrin-associated protein-deficient mice. *Learn Mem*, 6, pp. 448–457.

[26] Grant S.G.N. and Silva A.J. (1994). Targeting learning. *Trends Neurosci*, 17, pp. 71–75.

[27] Huang A.-M. and Lee E.H.Y. (1995). Role of hippocampal nitric oxide in memory retention in rats. *Pharmacol Biochem Behav*, 50, pp. 327–332.

[28] Silva A.J., Stevens C.F., Tonegawa S. and Wang Y. (1992). Deficient hippocampal long-term potentiation in alpha-calcium-calmodulin kinase II mutant mice. *Science*, 257, pp. 201–206.

[29] Silva A.J., Paylor R., Wehner J.M. and Tonegawa S. (1992). Impaired spatial learning in alpha-calcium-calmodulin kinase II mutant mice. *Science*, 257, pp. 206–211.

[30] Davis S., Butcher S.P. and Morris R.G. (1992). The NMDA receptor antagonist D-2-amino-5-phosphonopentanoate (D-AP5) impairs spatial learning and LTP *in vivo* at intracerebral concentrations comparable to those that block LTP *in vitro*. *J Neurosci*, 12, pp. 21–34.

[31] Morris R.G. (1989). Synaptic plasticity and learning: selective impairment of learning rats and blockade of long-term potentiation *in vivo* by the *N*-methyl-*D*-aspartate receptor antagonist AP5. *J Neurosci*, 9, pp. 3040–3057.

[32] Tsien J.Z. *et al.* (1996). Subregion- and Cell Type-Restricted Gene Knockout in Mouse Brain. *Cell*, 87, pp. 1317–1326.

[33] Frey U., Huang Y.Y. and Kandel E.R. (1993). Effects of cAMP simulate a late stage of LTP in hippocampal CA1 neurons. *Science*, 260, pp. 1661–1664.

[34] Kumar A. (2011). Long-term potentiation at CA3-CA1 hippocampal synapses with special emphasis on aging, disease, and stress. *Front Aging Neurosci*, 3, p. 7.

[35] Lee H.-K. *et al.* (2000). Regulation of distinct AMPA receptor phosphorylation sites during bidirectional synaptic plasticity. *Nature*, 405, pp. 955–959.

[36] Lüscher C. and Malenka R.C. (2012). NMDA receptor-dependent long-term potentiation and long-term depression (LTP/LTD). *Cold Spring Harb Perspect Biol*, 4, p. a005710.

[37] Ahmad M. *et al.* (2012). Postsynaptic complexin controls AMPA receptor exocytosis during LTP. *Neuron*, 73, pp. 260–267.

[38] Derkach V., Barria A. and Soderling T.R. (1999). Ca^{2+}/calmodulin-kinase II enhances channel conductance of α-amino-3-hydroxy-5-methyl-4-isoxazolepropionate type glutamate receptors. *Proc Natl Acad Sci USA*, 96, pp. 3269–3274.

[39] Ehlers M.D. (2000). Reinsertion or degradation of AMPA receptors determined by activity-dependent endocytic sorting. *Neuron*, 28, pp. 511–525.

[40] He K., *et al.* (2009). Stabilization of Ca^{2+}-permeable AMPA receptors at perisynaptic sites by GluR1-S845 phosphorylation. *Proc Natl Acad Sci USA*, 106, pp. 20033–20038.

[41] Oh M.C., Derkach V.A., Guire E.S. and Soderling T.R. (2006). Extrasynaptic membrane trafficking regulated by GluR1 serine 845 phosphorylation primes AMPA receptors for long-term potentiation. *J Biol Chem*, 281, pp. 752–758.

[42] Yang Y., Wang X., Frerking M. and Zhou Q. (2008). Delivery of AMPA receptors to perisynaptic sites precedes the full expression of long-term potentiation. *Proc Natl Acad Sci USA*, 105, pp. 11388–11393.

[43] Banke T.G. *et al.* (2000). Control of GluR1 AMPA receptor function by cAMP-dependent protein kinase. *J Neurosci*, 20, pp. 89–102.

[44] Mammen A.L., Huganir R.L. and O'Brien R.J. (1997). Redistribution and stabilization of cell surface glutamate receptors during synapse formation. *J Neurosci*, 17, pp. 7351–7358.

[45] Barria A., Derkach V. and Soderling T. Identification of the Ca^{2+}/calmodulin-dependent protein kinase II regulatory phosphorylation site in the α-amino-3-hydroxyl-5-methyl4-isoxazole-propionate-type glutamate receptor. *J Biol Chem*, 272, pp. 32727–32730.

[46] Lisman J.E., Yasuda R. and Raghavachari S. (2012). Mechanisms of CaMKII action in long-term potentiation. *Nat Rev Neurosci*, 13, pp. 169–182.

[47] Opazo P. *et al.* (2010). CaMKII triggers the diffusional trapping of surface AMPARs through phosphorylation of stargazin. *Neuron*, 67, pp. 239–252.

[48] Tomita S. *et al.* (2005). Bidirectional synaptic plasticity regulated by phosphorylation of stargazin-like TARPs. *Neuron*, 45, pp. 269–277.

[49] Gardoni F. *et al.* (1998). Calcium/calmodulin-dependent protein kinase II is associated with NR2A/B subunits of NMDA receptor in postsynaptic densities. *J Neurochem*, 71, pp. 1733–1741.

[50] Leonard A.S. *et al.* (1999). Calcium/calmodulin-dependent protein kinase II is associated with the *N*-methyl-*D*-aspartate receptor. *Proc Natl Acad Sci USA*, 96, pp. 3239–3244.

[51] Strack S. and Colbran R.J. (1998). Autophosphorylation-dependent targeting of calcium/calmodulin-dependent protein kinase II by the NR2B subunit of the *N*-methyl-*D*-aspartate receptor. *J Biol Chem*, 273, pp. 20689–20692.

[52] Appleby V.J. *et al.* (2011). LTP in hippocampal neurons is associated with a CaMKII-mediated increase in GluA1 surface expression. *J Neurochem*, 116, pp. 530–543.

[53] Otmakhov N. *et al.* (2004). Persistent accumulation of calcium/calmodulin-dependent protein kinase II in dendritic spines after induction of NMDA receptor-dependent chemical long-term potentiation. *J Neurosci*, 24, pp. 9324–9331.

[54] Zhang Y.-P., Holbro N. and Oertner T.G. (2008). Optical induction of plasticity at single synapses reveals input-specific accumulation of αCaMKII. *Proc Natl Acad Sci USA*, 105, pp. 12039–12044.

[55] Barria A. *et al.* (1997). Regulatory phosphorylation of AMPA-type glutamate receptors by CaM-KII during long-term potentiation. *Science*, 276, pp. 2042–2045.

[56] Halt A.R. *et al.* (2012). CaMKII binding to GluN2B is critical during memory consolidation. *Embo J*, 31, pp. 1203–1216.

[57] Zhou Y. *et al.* (2007). Interactions between the NR2B receptor and CaMKII modulate synaptic plasticity and spatial learning. *J Neurosci*, 27, pp. 13843–13853.

[58] Bayer K.U. *et al.* (2006). Transition from reversible to persistent binding of CaMKII to postsynaptic sites and NR2B. *J Neurosci*, 26, pp. 1164–1174.

[59] Chao L.H. *et al.* (2011). A mechanism for tunable autoinhibition in the structure of a human Ca^{2+}/calmodulin-dependent kinase II holoenzyme. *Cell*, 146, pp. 732–745.

[60] Giese K.P., Fedorov N.B., Filipkowski R.K. and Silva A.J. (1998). Autophosphorylation at Thr286 of the α calcium-calmodulin kinase II in LTP and learning. *Science*, 279, pp. 870–873.

[61] Shen K. and Meyer T. (1999). Dynamic control of CaMKII translocation and localization in hippocampal neurons by NMDA receptor stimulation. *Science*, 284, pp. 162–167.

[62] Shen K. *et al.* (2000). Molecular memory by reversible translocation of calcium/calmodulin-dependent protein kinase II. *Nat Neurosci*, 3, pp. 881–886.

[63] He Y., Kulasiri D. and Samarasinghe S. (2015). Modelling the dynamics of CaMKII-NMDAR complex related to memory formation in synapses: The possible roles of threonine 286 autophosphorylation of CaMKII in long term potentiation. *J Theor Biol*, 365, pp. 403–419.

[64] Haario H., Laine M., Mira A. and Saksman E. (2006). DRAM: efficient adaptive MCMC. *Stat Comput*, 16, pp. 339–354.

[65] Haario H., Saksman E. and Tamminen J. (2001). An adaptive metropolis algorithm. *Bernoulli*, pp. 223–242.

[66] Hill T. (2012). *Free Energy Transduction in Biology: The Steady-State Kinetic and Thermodynamic Formalism* (Elsevier Science).

[67] Marino S., Hogue I.B., Ray C.J. and Kirschner D.E. (2008). A methodology for performing global uncertainty and sensitivity analysis in systems biology. *J Theor Biol*, 254, pp. 178–196.

[68] Maass W. and Zador A.M. (1999). Dynamic stochastic synapses as computational units. *Neural Comput*, 11, pp. 903–917.

[69] Shouval H.Z., Bear M.F. and Cooper L.N. (2002). A unified model of NMDA receptor-dependent bidirectional synaptic plasticity. *Proc Natl Acad Sci USA*, 99, pp. 10831–10836.

[70] Ribrault C., Sekimoto K. and Triller A. (2011). From the stochasticity of molecular processes to the variability of synaptic transmission. *Nat Rev Neurosci*, 12, pp. 375–387.

[71] Lisman J.E., Schulman H. and Cline H. (2002). The molecular basis of CaMKII function in synaptic and behavioural memory. *Nat Rev Neurosci*, 3, pp. 175–190.

[72] Shen K., Teruel M.N., Subramanian K. and Meyer T. (1998). CaMKIIβ functions as an F-actin targeting module that localizes CaMKIIα/β heterooligomers to dendritic spines. *Neuron*, 21, pp. 593–606.

[73] Bayer K.U., Paul De Koninck A., Hell J.W. and Schulman H. (2001). Interaction with the NMDA receptor locks CaMKII in an active conformation. *Nature*, 411, pp. 801–805.

[74] Mullasseril P., Dosemeci A., Lisman J.E. and Griffith L.C. (2007). A structural mechanism for maintaining the "on-state" of the CaMKII memory switch in the post-synaptic density. *J Neurochem*, 103, pp. 357–364.

[75] Lee S.J., Escobedo-Lozoya Y., Szatmari E.M. and Yasuda R. (2009). Activation of CaMKII in single dendritic spines during long-term potentiation. *Nature*, 458, pp. 299–304.

[76] Meyer T., Hanson P.I., Stryer L. and Schulman H. (1992). Calmodulin trapping by calcium-calmodulin-dependent protein kinase. *Science*, 256, pp. 1199–1202.

[77] Fong Y.L., Taylor W.L., Means A.R. and Soderling T.R. (1989). Studies of the regulatory mechanism of Ca^{2+}/calmodulin-dependent protein kinase II. Mutation of threonine 286 to alanine and aspartate. *J Biol Chem*, 264, pp. 16759–16763.

[78] Miller S.G., Patton B.L. and Kennedy M.B. (1988). Sequences of autophosphorylation sites in neuronal type II CaM kinase that control Ca^{2+}-independent activity. *Neuron*, 1, pp. 593–604.

[79] Waxham M.N., Aronowski J., Westgate S.A. and Kelly P.T. (1990). Mutagenesis of Thr-286 in monomeric Ca^{2+}/calmodulin-dependent protein kinase II eliminates Ca^{2+}/calmodulin-independent activity. *Proc Natl Acad Sci USA*, 87, pp. 1273–1277.

[80] Maass W. and Zador A.M. (1999). Computing and learning with dynamic synapses. *Pulsed Neural Networks*, 6, pp. 321–336.

[81] Zhabotinsky A.M. (2000). Bistability in the Ca^{2+}/calmodulin-dependent protein kinase-phosphatase system. *Biophys J*, 79, pp. 2211–2221.

[82] Sheng M. and Hoogenraad C.C. (2007). The postsynaptic architecture of excitatory synapses: a more quantitative view. *Annu Rev Biochem*, 76, pp. 823–847.

[83] Coultrap S.J., Barcomb K. and Bayer K.U. (2012). A significant but rather mild contribution of T286 autophosphorylation to Ca^{2+}/CaM-stimulated CaMKII activity. *PLoS One*, 7, p. e37176.

[84] Chiba H., Schneider N.S., Matsuoka S. and Noma A. (2008). A simulation study on the activation of cardiac CaMKII δ-isoform and its regulation by phosphatases. *Biophys J*, 95, pp. 2139–2149.

[85] Ehlers M.D. (2003). Activity level controls postsynaptic composition and signaling via the ubiquitin-proteasome system. *Nat Neurosci*, 6, pp. 231–242.

[86] McKay M.D., Beckman R.J. and Conover W.J. (1979). A Comparison of three methods for selecting values of input variables in the analysis of output from a computer code. *Technometrics*, 21, pp. 239–245.

[87] Holmes W.R. (2000). Models of calmodulin trapping and CaM kinase II activation in a dendritic spine. *J Comput Neurosci*, 8, pp. 65–86.

[88] Zaidi A., Gao J., Squier T. and Michaelis M. (1998). Age-related decrease in brain synaptic membrane Ca^{2+}-ATPase in F344/BNF1 rats. *Neurobiol Aging*, 19, pp. 487–495.

[89] Lou L.L., Lloyd S.J. and Schulman H. (1986). Activation of the multifunctional Ca^{2+}/calmodulin-dependent protein kinase by autophosphorylation: ATP modulates production of an autonomous enzyme. *Proc Natl Acad Sci USA*, 83, pp. 9497–9501.

[90] Miller S.G. and Kennedy M.B. (1986). Regulation of brain type II Ca^{2+} calmodulin-dependent protein kinase by autophosphorylation: A Ca^{2+}-triggered molecular switch. *Cell*, 44, pp. 861–870.

[91] Saitoh T. and Schwartz J.H. (1985). Phosphorylation-dependent subcellular translocation of a Ca^{2+}/calmodulin-dependent protein kinase produces an autonomous enzyme in Aplysia neurons. *J Cell Biol*, 100, pp. 835–842.

[92] Yang E. and Schulman H. (1999). Structural examination of autoregulation of multifunctional calcium/calmodulin-dependent protein kinase II. *J Biol Chem*, 274, pp. 26199–26208.

Chapter 7

Synaptic Plasticity in Dementia:
Alzheimer's Disease and Role of Calcium

7.1. Introduction

Alzheimer's disease (AD), as the leading cause of dementia, affects about 36 million people worldwide and, unfortunately, is still incurable [1]. AD is characterized by progressive and irreversible loss of memory and cognitive function, while the exact pathophysiology and pathogenesis of the disease are still unknown. Various hypotheses have been proposed and attempted to explain the causes of AD, such as the amyloid deposition hypothesis [2–4], the tau hypothesis [5–7] and the cholinergic hypothesis [8–10]. Although much progress has been made with these hypotheses, they cannot explain all the physiological changes that occur during the development of this disease.

Calcium (Ca^{2+}) dysregulation has been observed in the brains of AD patients before the presence of overt clinical symptoms or the development of the classic biological hallmarks of amyloid plaques and neurofibrillary tangles [11, 12]. Genetic studies have also revealed altered levels of the genes and proteins that are related to intracellular Ca^{2+} signalling pathways in AD cells [13, 14]. Under healthy conditions in the nervous system, Ca^{2+} signalling accounts for diverse functions in brain physiology; for example, neurotransmitter release, neuronal excitability, neuronal development, synaptic plasticity, gene expression, and neuronal survival and death [15].

279

Because of the ubiquitous role Ca^{2+} plays in the nervous system, its dysregulation could induce numerous alterations and cause dysfunctions. Ca^{2+} hypothesis of AD, which was first proposed by Khachaturian, and many subsequent experimental research suggest that the sustained disturbance of intracellular Ca^{2+} signalling may contribute to the major symptoms of AD and may be the predominant cause of neurodegeneration in AD [16–18]. Intracellular Ca^{2+} dynamics is intimately associated with cognitive and memory-related neuronal functions such as neural rhythm and synaptic plasticity.

In pyramidal neurons of the hippocampus, N-methyl-D-aspartate receptor (NMDAR) and α-amino-3-hydroxy-5-methyl-4-isoxazolepropionic acid receptor (AMPAR) at membrane play crucial roles in mediating the Ca^{2+} response of postsynaptic neuron and synaptic plasticity formation [19]. Dysregulations of these membrane receptors have been exclusively studied and are believed to contribute to pathological changes in AD [20]. Inside the neuron, endoplasmic reticulums (ERs) as the largest intracellular Ca^{2+} stores are widely distributed within neurons and participate in Ca^{2+} signalling. ER Ca^{2+} dysregulation in neurons has been shown to further affect downstream events, such as triggering membrane hyperpolarization, reducing electrical excitability and mitochondrial dysfunction [21, 22]. It is believed that alteration of ER Ca^{2+} signalling is a key upstream event in AD pathophysiology that initiates and accelerates other severe events such as amyloid plaque deposition and neuronal apoptosis [23]. Ca^{2+} dysregulation in ER contributes to the persistent rise in cytosolic basal Ca^{2+} level, which in turns affects sleep/wake cycle [24]. In AD, increase in wakefulness significantly disrupts the processes of memory consolidation and erasure during sleep, and contributes to the loss of memory. Upregulation in basal Ca^{2+} level observed in AD transgenic mice is suggested to interrupt mechanisms related to learning and memory [25]. Increase in intracellular Ca^{2+} level shows to enhance long-term depression (LTD), a form of synaptic plasticity which weakens synapses, and may result in memory deficits [26]. Other key features of AD, such as inflammation and oxidative stress, are also related to intracellular Ca^{2+} dysregulation. Chronic neuroinflammation is one of early features of AD, which is critically linked to the disease pathogenesis [27]. Inflammation can be induced by amyloid-β(Aβ) directly or by increasing oxidative stress during aging in the brain [24, 28]. Inflammatory responses of astrocytes and microglia lead to local aberrant Ca^{2+} signalling as well as promote Ca^{2+} signalling in neurons, contributing to memory loss and apoptosis in AD [24]. Beside inflammation, increased generation of reactive oxygen species (ROS) during aging or stimulated by Aβ can affect Ca^{2+} signalling pathway directly [29]. This leads to elevated intracellular Ca^{2+} level and causes excitotoxic effects and neuronal death [30, 31]. Meanwhile, the increased intracellular Ca^{2+} level stimulates mitochondria and induce excessive mitochondria ROS formation, leading to aggravated neuronal oxidative stress [32].

In addition to important roles in enormous intracellular responses, intracellular Ca^{2+} dynamics regulates intercellular communication between neurons and between neurons and other cells [33, 34]. Abnormal intracellular Ca^{2+} in presynaptic terminal impairs the release of neurotransmitters [35]. This will disturb synaptic transmission and ultimately result in aberrant network activity. Moreover, appropriate spino-dendritic compartmentalization provides a foundation for proper neuronal signalling and intercellular communication. In AD transgenic mice, intracellular Ca^{2+} dysregulation is related to the loss of Ca^{2+} compartmentalization and alteration of neuronal morphology, which affect dendritic signal integration and interneuronal communication [25]. Besides, exosome is one of important factors for interneuronal communication, by which lipids, proteins and RNAs can be transferred to target neurons. Its secretion is strongly promoted by increased intracellular Ca^{2+} [36]. In AD, release of pathogenic proteins, such as Aβ, amyloid precursor protein (APP), Tau and cystatin C in AD, is also associated with exosome secretion. This might explain the spreading of the pathology of AD in the nervous system in the brain [36–38]. Furthermore, reduced local connectivity and disturbed global functional organization have been observed in brains of AD patients [39]. It is suggested that these dysregulations on local interneuronal communication may underlie the mechanisms of disturbances on brain synchrony and functional connectivity among distinct brain regions [40, 41].

7.2. NMDAR-Mediated Ca^{2+} Transients

The NMDAR-mediated Ca^{2+} transient in the dendritic spine is initialled by the presynaptic neurotransmitter release induced by the action potential (AP) at the presynaptic terminal. The neurotransmitter released activates postsynaptic receptors such as NMDAR and AMPAR. The Ca^{2+} influx through NMDAR into the cytosol determines various neuronal activities such as synaptic remodelling, synaptic plasticity and cognitive functions [42].

NMDAR has also been found at extrasynaptic locations such as the dendritic spine neck, the dendritic shaft, and in neuron bodies. It is a heterotetramer, mostly comprising two NR1 and two NR2 subunits [43]. There are eight and four splice variants of NR1 and NR2 subunits, respectively. These subunits share similar membrane topologies in that they contain an extracellular N-terminal domain with three transmembrane regions (M1, M3, and M4), a pore lining region (M2), an extracellular N-terminus and an intracellular C-terminus [44]. The agonists bind to the their extracellular binding domains; in NR1 subunits it is the glycine-binding site, whereas in NR2, it is the glutamate-binding sites. To activate NMDAR, all binding sites at the four subunits need to be occupied. Therefore, it requires the binding of two molecules of glutamate to the NR2 subunits and two molecules of agonist to the NR1 subunits. Recent research has found that NMDARs at different locations are gated by different co-agonists: D-serine for the synaptic NMDARs and glycine for the extrasynaptic NMDARs [45]. The affinity for glutamate by NMDAR depends on their NR2 subunit composition. The NMDAR subunit compositions at different locations change during postnatal development [46]. The ratio of NR2A to NR2B increases at the synaptic site and decreases at the extrasynaptic site during postnatal development. In mature synapses, NR2A-NMDARs are predominant at the synaptic sites, which take about 60% of the total synaptic NMDARs [47]. In contrast, NMDARs located outside the synaptic region are mainly NR1/NR2B-NMDARs. They are proposed to play opposite physiological roles in mediating intracellular signalling and death pathways:

activation of synaptic NMDARs shows neuroprotective effects, whereas stimulation of extrasynaptic NMDARs contributes to cell death [48].

In addition to the binding of glutamates and their co-agonists, Ca^{2+} entry through NMDAR requires the relief of Mg^{2+} blocks [49]. The Ca^{2+} permeation of NMDAR is mainly mediated by the M2 and M4 regions. The receptor is voltage-dependent and blocked by physiological concentrations of Mg^{2+} at resting membrane potential. During stimulation, the Mg^{2+} block is relieved by membrane depolarization, which can be produced by the opening of AMPARs. NMDARs are often found to be co-localized with AMPARs at the central synapses [50].

AMPAR is not permeable or less permeable to Ca^{2+} in comparison to NMDAR [51]. The activation of AMPAR by glutamate is fast and leads to a brief depolarization that only lasts for a few milliseconds. When the so-called excitatory postsynaptic potentials (EPSPs) generated by AMPAR are large enough, they will generate an action potential at the postsynaptic neuron. Therefore, the number of AMPAR in PSD is crucial for synaptic transmission and different types of synaptic plasticity formation. Their numbers in PSD are not fixed but are precisely mediated by the trafficking mechanisms [52].

Intraneuronally, NMDAR-dependent Ca^{2+} influx is augmented via Ca^{2+}-induced Ca^{2+} release (CICR) from intracellular Ca^{2+} stores. In the ER, Ca^{2+} homeostasis and signalling are precisely mediated by release and sequestration mechanisms, via Ca^{2+} channels and pumps, to accomplish the regulation of numerous downstream activities [15, 53]. The mechanisms of intracellular Ca^{2+} signalling is briefly illustrated in Fig. 7.1. ER Ca^{2+} handling involves Ca^{2+} release from the ER through two types of Ca^{2+} channels, inositol 1,4,5-trisphosphate receptors (IP_3Rs) and ryanodine receptors (RyRs), and Ca^{2+} uptake via the Ca^{2+} pumps, the sarcoendoplasmic reticulum Ca^{2+} transport ATPase (SERCA) pumps. Two separate mechanisms involved in the process of Ca^{2+} release from the ER via IP_3Rs and RyRs are inositol 1,4,5-trisphosphate (IP_3)-induced Ca^{2+} release and Ca^{2+}-induced Ca^{2+} release (CICR), respectively.

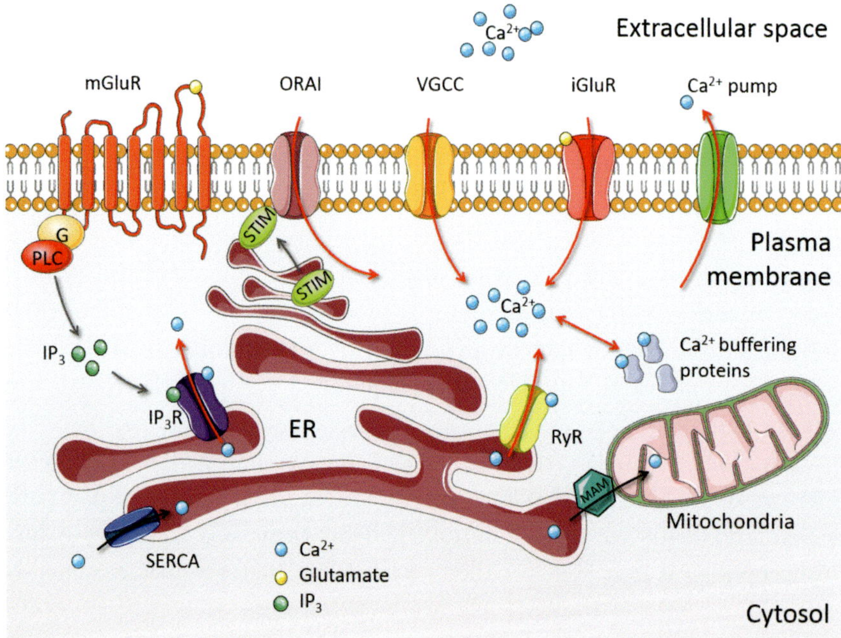

Fig. 7.1. Neuronal ER Ca^{2+} signalling. Intracellular Ca^{2+} concentration is regulated by ion channels (e.g. voltage-gated Ca^{2+} channels (VGCCs) and ionotropic glutamate receptors (iGluRs)) and transporters in the plasma membrane. Intracellularly, the regulation of Ca^{2+} homeostasis involves Ca^{2+} ions bind to Ca^{2+} buffer proteins and release from organelles, especially ER. Ca^{2+} ions release from ER via IP₃Rs and RyRs and uptake by ER via SERCA. Activation of IP₃Rs requires IP₃, which is generated by stimulation of metabotropic glutamate receptors (mGluRs) in the plasma membrane. Depletion in ER Ca^{2+} level promotes Ca^{2+} influx via membrane store-operated Ca^{2+} channels, such as ORAI, by translocation of the stromal interaction molecules (STIMs) into the ER–plasma membrane junctions. Besides, Ca^{2+} ions in ER can be transported into mitochondria via mitochondria-associated ER membrane (MAM) to stimulate mitochondrial metabolism. This figure is produced using Servier Medical Art (http://www.servier.com/Powerpoint-image-bank).

IP₃R is a ligand-gated Ca^{2+} channel embedded in the ER membrane and is activated by the intracellular second messenger, IP₃ [54]. IP₃-mediated Ca^{2+} release is initiated by the agonist stimulation of G protein-coupled plasma membrane receptors, such as mGluRs. Subsequently, mGluRs catalyze the activation of G-proteins, which, in turn, activate phospholipase C (PLC). PLC catalyzes a membrane phospholipid, phosphatidyl inositol-bisphosphate (PIP_2), to release IP_3 and diacyl glycerol (DAG) [55]. In addition, IP_3Rs are also regulated by cytosolic Ca^{2+} after activation, in a rapidly stimulated and slowly inhibited, fashion [56, 57]. They can interact locally with nearby channels to coordinate Ca^{2+} release in clusters through the CICR process [58]. This will generate the hierarchical recruitment of elementary Ca^{2+} release events and lead to the propagation of regenerative Ca^{2+} waves through cells in response to strong extracellular stimulation [56, 59].

The other Ca^{2+} channel, RyR, is the largest ion channel protein to date and exists in three isoforms, RyR1, RyR2, and RyR3, all of which can be found in the brain [60]. These RyR subtypes seem to have different involvement in memory processing, synaptic plasticity and other events in the brain [60, 61]. Ca^{2+} release through RyRs is stimulated by cytosolic Ca^{2+} ions via the process of CICR. Although IP_3Rs can also be activated by Ca^{2+} ions, as mentioned above, CICR is more usually related to RyR-mediated Ca^{2+} release. As an inherent, positive feedback mechanism, CICR can greatly amplify the initial Ca^{2+} signals and contribute to subsequent neuronal events, such as neurotransmitter release and Ca^{2+}-dependent gene expression [62].

The SERCA pump of the ER has the highest affinity for Ca^{2+} and plays an opposite role to the Ca^{2+} channels: it refills the ER by pumping Ca^{2+} ions back from the cytosol [63]. SERCAs, together with Ca^{2+}-buffering proteins, Ca^{2+} pumps and transporters on other organelles or plasma membrane, form a Ca^{2+} buffering system which contributes to buffer cytosolic Ca^{2+} transients and restores the cytosolic Ca^{2+} concentration back to resting levels [64]. SERCA pumps are suggested to play an interdependent role to refilling ER Ca^{2+} stores and maintaining the CICR process [65]. Furthermore, SERCA pumps are one of the key factors in store-operated Ca^{2+} entry (SOCE), a process involving extracellular Ca^{2+} influx via the plasma membrane Ca^{2+} channels in response to the depletion of intracellular Ca^{2+} stores [66]. The tight coupling between the ER refilling and SOCE is managed by SERCA pumps [67].

The ER is physically in contact with other cytoplasmic organelles and the plasma membrane. Through these contact sites, ER Ca^{2+} release is closely coupled with other cellular events via communication with other cytoplasmic organelles, especially mitochondria, and with Ca^{2+} channels in the plasma membrane [68]. Depletion in the ER Ca^{2+} level will cause SOCE and Ca^{2+} refilling via store-operated Ca^{2+} channels (SOCs) located in the plasma membrane. In the ER, a family of Ca^{2+} sensor proteins, the stromal interaction molecules (STIMs), will be activated by the decreased ER Ca^{2+} level and translocate into the ER–plasma membrane junctions. STIMs can then activate SOCs, such as the ORAI channels, and induce the Ca^{2+} influx via them (reviewed in [69]). The ER is physically connected to the mitochondria by a lipid, raft-like structure, called mitochondria-associated ER membrane (MAM) [70]. The MAM plays important roles in multiple cellular functions, including maintaining intracellular Ca^{2+} homeostasis. It allows selective transmission of Ca^{2+} ions from the ER to the mitochondria and, thereby, regulates the functions of the mitochondria and cell survival. This process is regulated by a set of proteins localized in the MAMs (reviewed in [71]).

7.3. Amyloid Hypothesis

Aging is the most important risk factor for developing AD. Based on the age of onset, AD is often divided into two categories: early-onset or familial AD (FAD) and late-onset or sporadic AD (SAD). Fewer than 5% of all AD cases are FADs, which are inherited as autosomal dominant mutations. In contrast, non-inherited SAD with unestablished causes accounts for the vast majority of total AD cases (90–95%) and has an age of onset of over 65 [1]. Noticeably, although FAD and SAD have different onset times and development speeds, they present similar neurodegenerative processes, such as a loss of memory and neuronal cell death [66].

AD is characterised by the abnormal accumulation of harmful proteins, $A\beta$ protein plaques and neurofibrillary tangles in the nervous system [72]. It is widely believed that these hallmarks are in connection with a massive loss of neurons and synapses in the patients' brains. AD has been identified for more than 100 years, however, to date, the cause of AD is still not fully understood. Scientists believe that AD is a result of multiple factors and to explain the underlying mechanisms of AD pathophysiology, various hypotheses have been proposed [2, 9, 73, 74], among which the amyloid hypothesis is the dominant one hypothesis [3].

The amyloid hypothesis was first proposed in the early 1990s [2, 73]. It suggests that AD is initiated by $A\beta$ oligomers and $A\beta$ plaques in brain tissue. $A\beta$ deposition may directly injure brain neurons and contribute to neuronal toxicity, neuronal death and the subsequent degeneration and cognitive deficiencies in AD patients. During the past 20 years, $A\beta$ deposition has been widely accepted to be the leading cause of AD by the majority of researchers and a large number of experimental research projects are being carried out based on the amyloid hypothesis.

$A\beta$ is produced from a sequence beginning with the cleavage of its precursor, called the amyloid precursor protein (APP), by certain secretase enzymes [75, 76]. The process of APP cleavage is illustrated in Fig. 7.2. APP is a transmembrane protein with a long extracellular N-terminal domain, a transmembrane domain, and a short intracellular C-terminal domain [77]. APP can be cleaved by two different kinds of enzymes, α-secretase and β-secretase, at different sites to produce two different peptides. The α-secretase cleaves APP within the critical $A\beta$ region that lies in part of the extracellular N-terminal and transmembrane domains. The cleavage releases a large soluble amino-terminal fragment, sAPPα, and a smaller C-terminal fragment, which will be further cleaved by γ-secretase. Cleavage by α-secretase prevents the generation and release of the $A\beta$ peptide and, thereby, it is called the non-amyloidogenic pathway in APP processing. In the alternative amyloidogenic pathway, APP is cleaved by β-secretase outside the critical $A\beta$ region, releasing sAPPβ and a smaller C-terminal fragment. The γ-secretase further cleaves the C-terminal fragment releasing the $A\beta$. The majority of $A\beta$ peptides are 40 residues in length ($A\beta_{40}$), while a small proportion ($\approx 10\%$) are 42 residues in length ($A\beta_{42}$) [78]. Compared with $A\beta_{40}$, $A\beta_{42}$ is more

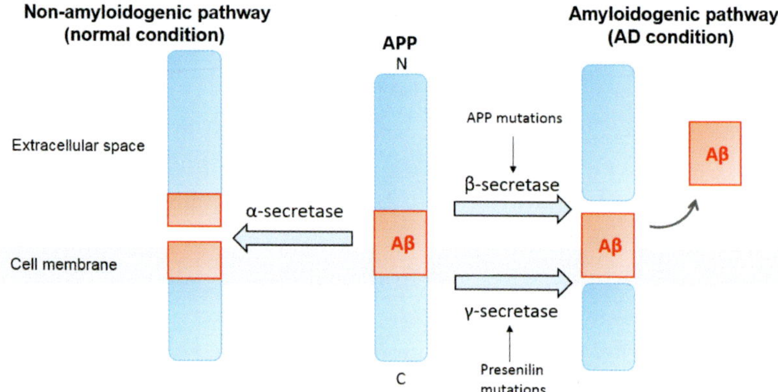

Fig. 7.2. The process of APP cleavage. The transmembrane protein APP can undergo a series of proteolytic cleavage by different secretase enzymes. When it is cleaved by α-secretase in the middle of the Aβ domain, it is not amyloidogenic. However, when APP is cleaved by β- and γ-secretase enzymes, neurotoxic Aβ peptides are released, which can accumulate into oligomer aggregates. Mutations in the APP gene tend to inhibit cleavage by α-secretase and, consequently, enable preferential cleavage by β-secretase. Mutations in the presenilin-1 and presenilin-2 genes, which are components of the γ-secretase complex, increase cleavage by γ-secretase at this site. In both situations, the result is excess Aβ peptide production.

hydrophobic and readily forms amyloid plaques, which are considered to play a causal role in the progression of AD [75]. Therefore, Aβ_{42} is believed to be the most neurotoxic form of the Aβ peptides.

Under normal physiological conditions, Aβ peptides can be removed from brains quickly by its clearance mechanisms. For example, Aβ peptides can be degraded proteolytically by multiple enzymes, especially neprilysin (NEP) and insulin degrading enzyme (IDE) [79]. The balance between production and clearance of Aβ determines steady levels of Aβ. An imbalance will result in the deposition of Aβ and formation of amyloid plaques. In FAD, mutations of APP, presenilins 1 and 2 genes will lead into increasing production of Aβ and the ratio of Aβ_{42}/Aβ_{40} increases the cleavage activity of β-secretase and γ-secretase [76, 78]. In SAD, rather than an overproduction of Aβ, the formation of amyloid plaques is more likely attributed to the failure of Aβ clearance. The initial cause of this failure is still unclear.

$A\beta$ has the ability to self-associate and exists in many different assembly forms, ranging from oligomers to fibrils and amyloid plaques. Initially, fibrillar $A\beta$ that constitutes amyloid plaque was assumed to be toxic. However, recent research suggests that $A\beta$ oligomers are the most toxic form, and may result in nerve damage [80, 81]. $A\beta$ oligomers at pathophysiological levels are considered to contribute to the activation of inflammatory neuronal cascades, oxidative stress, mitochondrial disturbances, Ca^{2+} metabolism deregulation, tau protein phosphorylation and neuronal apoptosis [18, 82–85]. Moreover, $A\beta$ oligomers can inhibit hippocampal long-term potentiation (LTP) and facilitate hippocampal long-term depression (LTD) and, thus, cause impairments in synaptic plasticity, learning, and memory [86].

7.4. Ca^{2+} Hypothesis

$A\beta$ is proposed to cause neuronal toxicity via a variety of pathways, and several hypotheses extending from the amyloid hypothesis attempt to further elucidate the alterations resulting from amyloid pathology. The most important, and feasible, pathway is through the Ca^{2+} signalling systems by triggering deregulation of Ca^{2+} homeostasis. The Ca^{2+} hypothesis that links altered Ca^{2+} signalling to $A\beta$-induced neuronal dysfunction is becoming more popular, and is supported by accumulating experimental evidence [87]. The deregulation of Ca^{2+} homeostasis is suggested to account for the decline in cognitive function in the early stages of AD, and neuronal loss in the late stage of AD. Therefore, the focus of this research is mainly on the amyloid and Ca^{2+} hypotheses.

$A\beta$ oligomers may interact with neuronal membranes and result in a significant elevation of intracellular Ca^{2+} levels in AD. High levels of Ca^{2+} are toxic and will induce neural apoptosis, the most common form of programmed cell death, leading to neuronal dysfunction and neurodegeneration in both SAD and FAD [18]. Meanwhile, alteration of Ca^{2+} signalling may accelerate $A\beta$ formation, thus, a degenerative feed-forward cycle of $A\beta$ formation and dysregulation of Ca^{2+} signalling is formed.

Since 1987, when Khachaturian [88] first proposed the Ca^{2+} hypothesis, a growing body of evidence reveals that dysregulation of Ca^{2+} signalling plays a crucial role in the initiation and development of AD [18, 89, 90]. A rise in the resting level of cytosolic Ca^{2+} in neurons has been reported in both transgenic AD animal models and autopsies of brains from patients who have died from AD, which suggests it is a causal factor in neuronal excitotoxicity, synaptic loss and cell death during the development of AD [91].

The extracellular Aβ oligomers have been shown to impair intracellular Ca^{2+} homeostasis by promoting an influx of extracellular Ca^{2+} through the plasma membrane [92, 93]. This can disrupt the plasma membrane Ca^{2+} permeability, particularly by forming Ca^{2+}-permeable pores that affect certain Ca^{2+}-permeable channels and/or interact with membrane lipids and affect their integrity. In this section, we review dysregulation related to the NMDAR-mediated Ca^{2+} response in the dendritic spine as observed under AD conditions.

An excessive extracellular glutamate concentration will cause a mild, chronic activation of glutamate receptors, especially NMDARs, and lead to neurodegeneration, which is also called excitotoxicity c. Several studies have shown that a dysfunctional glutamatergic system in AD could be a critical upstream event, which leads to the overactivation of NMDARs and pathological NMDAR-mediated Ca^{2+} influx into the neuron that, subsequently, triggers apoptotic pathways [26, 48]. An elevated resting level of extracellular glutamate has been found in AD patients [94]. Researchers have linked Aβ to upregulation in glutamate availability and excitotoxicity in AD, suggesting that Aβ is the culprit behind enhanced excitotoxicity, either by increasing glutamate release or by reducing glutamate uptake by transporters [20, 95, 96].

Aβ is suggested to promote glutamate release from presynaptic terminal [97–99]. Aβ may participate and act as a positive regulator in the glutamate release under healthy conditions [100]. In aged neurons, Aβ_{23-25} has been found to augment glutamate release [101]. Puzzo *et al.* [102] reported that a picomolar level of Aβ_{42} could enhance LTP by increasing neurotransmitter release, whereas low nanomolar levels of Aβ_{42} lead to synaptic depression. Palop and Mucke [35] proposed a bell-shaped relationship between extracellular Aβ and the synaptic transmission that low and high levels of Aβ depress the postsynaptic transmission while intermediate levels facilitate the presynaptic vesicle release.

In addition, Aβ is suggested to potentiate the release of glutamate from glia cells [103, 104]. Aβ induces excitotoxic levels of glutamate release from astrocytes, which can lead to the activation of extrasynaptic NMDARs and synaptic loss [94, 105]. This may be a result of the abnormal Ca^{2+} signalling in the glia cells observed in AD [106].

In contrast, Aβ oligomers may disturb glutamate clearance mechanisms by affecting the glutamate transporters [107–111]. A number of studies have revealed reduced levels of vesicular glutamate transporters (VGLUT) and excitatory amino acid transporters (EAAT) in AD [112–114]. Moroever, the decreased glutamate transporter activity found in AD is suggested to be associated with this excitotoxicity and neurodegeneration [112]. The lack of transporters and/or reduced activity of these transporters contributes to the increase in glutamate availability in the synaptic cleft and extrasynaptic space [20]. Persistent activation of postsynaptic NMDAR may lead to receptor desensitisation and affect synaptic functions, such as synaptic plasticity [26], whereas prolonged extrasynaptic NMDAR promotes neuronal Aβ production [115, 116].

7.5. Dysregulation on NMDAR in AD

Several research studies have shown the colocalisation of Aβ oligomers with NMDARs, suggesting that they may directly interact with NMDARs [117–119]. Aβ oligomers are reported to directly interact with NMDARs [120] and activate NR2B-NMDAR, leading to an increase in cytosolic Ca^{2+} levels [121]. The precise underlying mechanisms are yet to be uncovered. Besides, Aβ may also cause a slowed desensitisation by disturbing the cellular prion protein-mediated NMDAR activity [122, 123]. Furthermore, the increase of NMDAR activity may inhibit the α-cleavage of APP and promotes the amyloidogenic processing of APP [124–126].

Aβ oligomers may also disturb the distribution of NMDARs [127]. Interestingly, the full-length APP is found to increase surface NR2B containing NMDAR (but not NR2A-containing one) and decrease its internalisation in primary hippocampal neurons [126]. However, following exposure of Aβ_{25-35} and full-length Aβ_{1-40} deposits, a decrease in the number of NMDARs positive cells is observed in the immediate surrounding of Aβ_{25-35} and full-length Aβ_{1-40} deposits in mature hippocampal slice cultures [128]. Pretreatment of cultured hippocampal neurons with Aβ_{1-42} significantly reduces surface expression of NR1, without affecting the total NR1 expression [129]. Treatment of mature hippocampal cells with Aβ oligomers rapidly decreases the expression of surface NMDARs [130]. In cultured cortical neurons, Aβ_{1-42} has been shown to reduce the surface expression of NR1 and NR2B subunits of the whole cell as well as at the synaptic site, without any changes in the total level of synaptic NMDARs (including both internal and surface receptors) [127]. Similarly, a decrease in the surface expression of synaptic NR2B containing NMDAR is observed in primary hippocampal neurons pre-incubated with Aβ oligomers and in APP transgenic mice [117]. In contrast, extrasynaptic NMDARs are not affected by the presence of Aβ oligomers [127].

In addition, Aβ-induced surface expression of NMDAR can be restored by the γ-secretases inhibitor [127]. Taken all together, it is suggested that Aβ may play a role in mediating the trafficking of NMDARs, especially by promoting synaptic NR2B-NMDAR endocytosis. Aβ can bind to 7α-nicotinic acetylcholine receptors (nAchRs) with high affinity [131]. Kessels *et al.* [127] hypothesised that the Aβ-dependent endocytosis of NMDARs is initiated by Aβ binding to the nAchRs, which leads to activation of protein phosphatase 2B and dephosphorylation and activation of tyrosine phosphatase STEP. The activation of STEP may promote dephosphorylation of the phosphotyrosine residue Tyr1472 of NMDAR, which locates in a region which regulates NMDAR endocytosis and interaction with synaptic scaffolding proteins. Kessels *et al.* [132] showed that Aβ oligomers promote the switch in subunit composition from NR2B- to NR2A-NMDAR, which normally occurs during development.

The loss of synaptic NMDARs may also contribute to the depression of glutamatergic transmission and reductions in memory formation. Aβ is reported to decrease NMDAR-mediated synaptic responses and inhibit NMDAR-dependent LTP [133, 134]. However, to some extent, the internalisation of synaptic NMDARs may be a neuroprotective mechanism against the glutamate-induced neurotoxicity and excessive Ca^{2+} influx [129, 135].

7.6. Dysregulation of ER Ca^{2+} Handling in AD

The ER, as the major Ca^{2+} storage compartment, has been studied with the focus on its effects in Ca^{2+} signalling in AD [136, 137]. Research carried on stabilising ER Ca^{2+} signalling in AD cells or animal models have offered new strategies for the prevention of disease progression [17, 138–140]. The mechanism causing these alterations in ER Ca^{2+} signalling is not fully understood, while current research is mainly on the disturbing effects from mutations in presenillin or/and deposition of Aβ oligomers.

Aβ can affect the ER Ca^{2+} signalling by promoting an influx of extracellular Ca^{2+} through the plasma membrane [92, 93]. For example, there is growing evidence that links overactivation of NMDARs to ER Ca^{2+} dysregulation [121, 141, 142]. Aβ oligomers can induce ER stress through interaction with NMDARs, resulting in a massive Ca^{2+} influx [121, 142]. This will subsequently activate downstream NADPH oxidase (NOX)-mediated superoxide production and, consequently, impair ER Ca^{2+} homeostasis and even trigger ER stress-mediated apoptotic pathway [143]. Costa *et al.* [142] have demonstrated that Aβ-induced ER stress can be effectively inhibited by ifenprodil, an antagonist of NR2B subunits, which could be a potential therapeutic strategy for AD.

Intracellular Aβ oligomers are also shown to directly disrupt Ca^{2+} handling of intracellular Ca^{2+} stores, especially ERs (Fig. 7.3) [144]. They are believed to be produced from APP residing on the ER and in other intracellular compartments, as well as by uptake from the extracellular space [145]. The accumulation of intracellular Aβ oligomers has been observed prior to the presence of

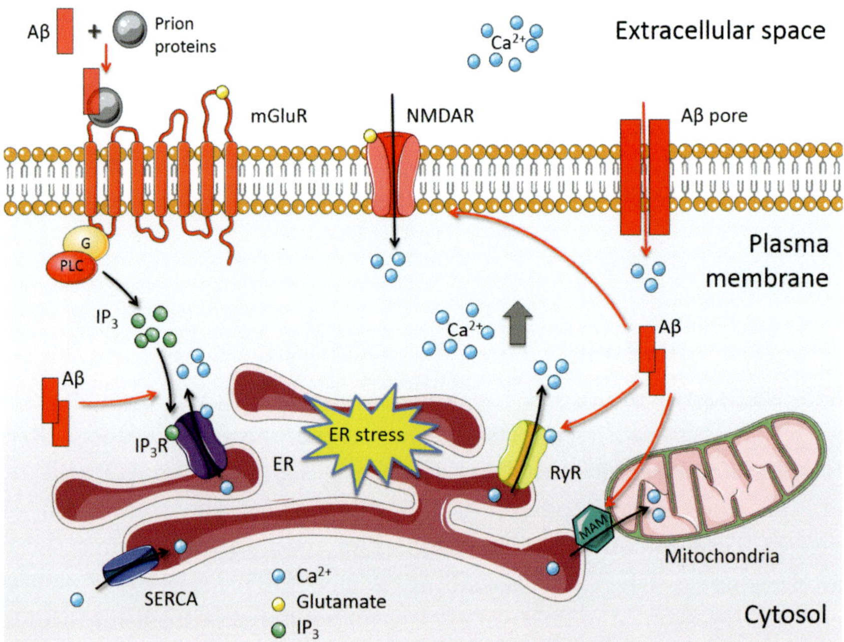

Fig. 7.3. Mechanisms that might contribute intracellular Ca^{2+} dysregulation in AD. Evidences show that $A\beta$ oligomers may disturb intracellular Ca^{2+} homeostasis and result in ER stress, by disturbing membrane conductance, forming Ca^{2+}-permeable pores as well as perturbing ER Ca^{2+} handling. $A\beta$ oligomers may directly modulate activity of IP$_3$Rs and RyRs. $A\beta$ oligomers are also suggested to accelerate the production of IP$_3$, therefore, induce IP$_3$Rs opening and Ca^{2+} release from ER. Besides, $A\beta$ oligomers may promote ER Ca^{2+} transmission to mitochondria via MAM and lead to mitochondrial dysfunction. Furthermore, alteration of intracellular Ca^{2+} homeostasis may in turn promote the production of $A\beta$ peptides. This figure is produced using Servier Medical Art (http://www.servier.com/Powerpoint-image-bank).

extracellular $A\beta$ deposition in the cerebral cortex and hippocampus in AD brains, suggesting that intracellular $A\beta_{42}$-induced Ca^{2+} dysregulation may play a crucial role in early AD pathogenesis [11, 12].

$A\beta_{42}$ oligomer pores can also be formed in the membranes of intracellular organelles and result in Ca^{2+} leakage into the cytosol; the disruption is comparatively mild [92, 144]. Most research has demonstrated that $A\beta$ oligomers enhance Ca^{2+} liberation from the ER, mainly by altering Ca^{2+} release via IP3Rs and RyRs [13, 146–148]. Pereira's group suggested an ER-specific apoptotic pathway trigged by $A\beta$-induced perturbation of ER Ca^{2+} homeostasis [143]. Their research showed that continuous production could induce ER stress, perturb ER Ca^{2+} homeostasis and lead to increased levels of cytosolic Ca^{2+}. Together with other $A\beta$-induced alterations, such as increased activity of caspase-3, it would further affect mitochondrial function and, ultimately, trigger the apoptotic pathway and lead to neuronal death.

$A\beta$ exposure has been shown to disturb IP_3R-regulated Ca^{2+} handling by promoting IP_3-mediated Ca^{2+} release from the ER in neurons [148]. Similarly, Lopez *et al.* observed that the accumulation of intracellular $A\beta$ in a 3xTg-AD mouse model enhanced Ca^{2+} release via IP_3R on the ER as well as increased Ca^{2+} influx on the plasma membrane, which potentially can lead to a large elevation of resting cytosolic free Ca^{2+} level [91]. In addition, the $A\beta$ has also been reported to affect Ca^{2+} flux via IP_3Rs indirectly by modulating upstream events. For example, $A\beta$ directly interacts with IP_3 and promotes binding of IP_3 ligands to their receptors in rats [149]. Increased levels of both astrocytic mGlu5 receptor mRNA and protein have been observed in $A\beta40$ treatment *in vitro* [150]. Likewise, Renner *et al.* reported that $A\beta$ oligomers altered the lateral diffusion of mGluR5 within the plasma membrane and stimulated their signalling activity which, therefore, affected downstream IP_3 production [151]. $A\beta$ oligomers can bind to cellular prion proteins and active mGluR5, which may also lead to increased production of IP_3 [152]. In addition, recent research has reported that intracellular $A\beta_{42}$ oligomers induce Ca^{2+} liberation from the ER via IP_3Rs by stimulating PLC-mediated IP_3 production in *Xenopus* oocytes [144]. Another study demonstrated that $A\beta_{42}$-induced Ca^{2+} release from ER may mostly depend on activating PLC and, subsequent, IP3 generation rather than by direct interaction with IP_3Rs [153].

Similarly, intracellular $A\beta_{42}$ oligomers also disrupt RyR-regulated Ca^{2+} signals. $A\beta_{42}$ oligomers appear to increase the channel open probability of RyRs and change their gating kinetics, leading to enhanced Ca^{2+} liberation from the ER. Shtifman *et al.*

demonstrated that $A\beta$ can directly modify and activate RyR type1 by reconstituting RyR in a planar lipid bilayer to study inclusion body myositis, a skeletal muscle disorder, which may share a similar pathogenesis with AD [154]. This is consistent with their earlier study on the inclusion body myositis showing that $A\beta$ overexpression increases RyRs' sensitivity to the agonist caffeine and susceptibility to CICR [155]. In addition, the elevation of RyR expression has also been reported to result in the accumulation of intracellular $A\beta$ oligomers. For example, Supnet *et al.* reported increased levels of RyR type3 in non-TgCRND8 neurons after treatment with $A\beta_{42}$ oligomers, suggesting that $A\beta_{42}$ may potentiate Ca^{2+} release from the ER and disturb intracellular Ca^{2+} homeostasis by directly increasing RyR expression [156, 157].

Moreover, $A\beta$ oligomers also upregulate the MAM function in post-mortem AD brains and APPSwe/Lon mice [158]. Increased expression of MAM-associated proteins and a number of contact sites have been observed in primary hippocampal neurons. This may, consequently, promote Ca^{2+} transmission from the ER to mitochondria and may lead to mitochondrial dysfunction and AD pathology.

Interestingly, the APP intracellular domain (AICD), which is generated together with $A\beta$ by γ-secretases cleavage, may also be associated with ER signalling. AICD has been suggested to be involved in regulating IP_3-mediated Ca^{2+} signalling, although the precise mechanism is unknown [159]. AICD may promote the expression of IP_3R based on its transcriptional activity [160]. Alternatively, in another study, AICD was shown to activate glycogen synthase kinase-3β (GSK3β), a modulator of the IP3R expression, which may be a potential mechanism by which AICD mediated ER Ca^{2+} signalling [161]. In addition, Oule's *et al.* observed disrupted ER Ca^{2+} homeostasis contributed to the increased expression of RyR levels in APP-overexpressing transgenic mice, suggesting AICD may also mediate RyR expression [162].

Presenilins (PSs), a family of highly conserved membrane proteins with multi-transmembrane domains, are primarily located in the ER membrane and can be found in the cell body and in the dendrites of neurons [163]. Missense mutations in the genes encoding presenilin-1 (*PSEN1*) and presenilin-2 (*PSEN1*) account for the majority of the inherited forms of AD, FAD [164, 165]. *PSEN1* mutations are the most common genetic cause and, to date, 185 *PSEN1* mutations have been identified in patients with FAD, listed on the Alzheimer Disease & Frontotemporal Dementia Mutation Database [166]. In contrast, the *PSEN2* mutation is a relatively rare cause of FAD, with only 13 mutations have been identified [166]. Mutations in PS genes are suggested to play a causal role in the cognitive impairment and pathogenesis of FAD [167]. Much research has reported that both mutations in both *PSEN1* and *PSEN2* are associated with the dysregulation of Ca^{2+} signalling, disturbance of Aβ production and neuronal death, although the underlying mechanisms are still under investigation.

However, unlike FAD, causes of SAD are much more complex, involving both genetic and environmental risk factors. Therefore, it remains a great challenge to study the key factors in SAD [168]. Research on microRNA (miRNA) may be helpful to explain the pathogenesis of SAD as well as FAD. miRNAs are small non-coding RNAs, which negatively regulate gene expression at the posttranscriptional level [169]. Dysregulation of several miRNAs have been observed in AD brains and are suggested to relate to AD pathology [170]. Increase in hallmark genes such as *APP* and *PSEN1* observed in SAD brains may be a result of dysregulation in miRNAs which target these genes (reviewed in [171]). Thus, abnormal level of miRNAs may indirectly affect Ca^{2+} signalling and Aβ production in AD. Therefore, identify specific miRNAs for AD, especially SAD, will be useful for modelling SAD in the laboratory, which remains underdeveloped. Besides miRNAs may serve as potential biomarkers in AD, which may contribute to the early diagnosis and therapeutic research [147].

7.7. ER Alteration may Influence Aβ Production

Alteration of intracellular Ca^{2+} homeostasis shows conflicting effects on the production of Aβ peptides. Sustained elevation in the cytosolic

Ca^{2+} level has been reported to accelerate $A\beta$ formation, which suggests that reduction of Ca^{2+} release from the ER could be a potential neuroprotective strategy against $A\beta$ peptide neurotoxicity [146, 172, 173]. Increased expression and activity of RyRs have been found in the brains of patients in the early stages of AD and also in mice models, and this may contribute to the progression of AD pathogenesis [13, 174, 175]. Early research showed that the elevation of intracellular Ca^{2+} levels after treatment with Ca^{2+} ionophore led to increased $A\beta$ generation in human cell lines [172]. Later, the same group reported that RyR activation by caffeine resulted in an elevation of intracellular Ca^{2+} level, which further enhanced the release of $A\beta$ from APP in HEK293 cells [176]. Likewise, a study of tissues from post mortem AD brains has shown the loss of RyRs is correlated with amyloid deposition [13]. Recent research on Tg2576 mouse models has revealed that reduced RyR-mediated Ca^{2+} release after treatment with dantrolene, a RyR inhibitor, could reduce both intracellular and extracellular $A\beta$ loads and slow down the loss of memory and cognitive function. This possibly resulted from the decreasing Thr-668-dependent APP phosphorylation and β- and γ-secretases activities by interaction with Ca^{2+} [162]. However, there are conflicting results from other groups who reported that knockout of RyR type3 neurons or ones after long-term exposure to RyR inhibitor, dantrolene, presented with an increased amyloid pathologic condition, suggesting a potential neuroprotective role for RyR [157, 177].

Likewise, IP_3R and its regulation of cytosolic Ca^{2+} signalling may also affect the generation of $A\beta$. Cheung *et al.* reported reduced $A\beta$ production in IP_3R-deficient cells, which suggested that amyloidogenic processing of APP might be closely related to IP_3R-mediated Ca^{2+} release [23]. Pierrot *et al.* suggested increased Ca^{2+} liberation from the ER alone may not be enough to trigger intraneuronal $A\beta$ production, which requires influx of extracellular Ca^{2+} and a sustained increase in cytosolic Ca^{2+} concentration [178]. This is consistent with results from a recent study, which investigated the interactions between genes of different Ca^{2+} channels in the AD pathway from a genetic perspective using an online database [179]. In their study, Gene *RYR3* and *CACNA1C* have shown to interact significantly with each other, suggesting this interaction may result in Ca^{2+} dysregulation, leading to enhanced $A\beta$ production and deposition.

However, there are studies which shows increase in the cytosolic Ca^{2+} level reduces $A\beta$ production. The role of IP_3R in $A\beta$ generation is controversial. Severely reduced levels of both IP_3 and IP_3Rs have been observed in AD brains, and the loss of IP_3Rs is suggested to be important for amyloid production and deposition by altering Ca^{2+} homeostasis [180]. Other studies have demonstrated that the elevation of intracellular Ca^{2+} can decrease amyloid production and promote the anti-amyloidogenic processing of APP [181, 182]. For example, an increased level of cytosolic Ca^{2+} followed treatment with high levels of thapsigargin, an inhibitor of SERCA, leading to decreased $A\beta$ formation [181]. This is consistent with results from a recent study, which suggests increase Ca^{2+} influx via SOCE significantly by overexpression of STIM1 can actually inhibit $A\beta$ secretion [183]. These contradicting observations may result from different experimental conditions, such as cell lines chosen, or different detection methods [180].

7.8. Modelling Ca^{2+} Dynamics in Dendritic Spines

A number of models for the Ca^{2+} dynamics in the dendritic spines have been proposed for different types of neurons. Some of them have been modified and extended to study Ca^{2+}-mediated downstream events, such as the induction of different types of synaptic plasticity [184]. These Ca^{2+} models generally contain similar components, such as Ca^{2+} influxes from extracellular specs and/or release from an internal Ca^{2+} source, Ca^{2+} extrusion from the cytosol and interactions between Ca^{2+} and various intracellular proteins. In multi-compartment models, which take the geometry of the spine head and dendrite into consideration and simulate it as a series of compartments, the diffusion of Ca^{2+} between neighbouring compartments. In this section, we review the existing models related to the NMDAR-mediated Ca^{2+} response in dendritic spines regarding these components.

There are a number of models for presynaptic neurons that contain components related to glutamate vesicle release with different levels of detail [185, 186] that are beyond the scope of this thesis. In contrast, most models of the postsynaptic Ca^{2+} response to the presynaptic stimulation assume the release of glutamate from a vesicle is instantaneous and, thus, uses a constant glutamate concentration per release [187]. Other models use simple equations to represent glutamate release from a vesicle as a non-instantaneous process [188].

The spatiotemporal profile of the glutamate concentration in the synaptic cleft and the extrasynaptic space following release are due to their diffusion and uptake. The transient of a single pulse of glutamate at certain locations can be simply modelled as a process with a fast rise time and a slow decay time or a square-wave pulse [189]. Complex models make an approximation of the synaptic cleft and the extracellular space. They then simulate the gluamate concentration using diffusion equations, which are either partial differential equations or analytical solutions for them. The synaptic cleft can be approximated as a two-dimensional disc or a three-dimensional flat cylinder whereas the extrasynaptic space is a spherical, isotropic porous medium containing obstacles [187, 188]. The diffusion space is divided into different small diffusion compartments, such as concentric cylindrical or spherical shells [190] and, in the diffusion compartments of the extrasynaptic space, the gluatmate concentration with time is simulated as a diffusion–reaction process, which accounts for the diffusion of glutamate as well as the uptake of glutamate by gluatmate transporters in the adjacent glia. The kinetics of binding glutamate to different glutamate transporters can be modelled as a chain reaction [188]; or follow the proposed schemes that contain more reaction details [191].

The Ca^{2+} influx depends on the number of opened Ca^{2+} channels as well as the difference between the membrane potential and the reversal potential. Markov's kinetic models are mostly used to describe the state-transition of AMPAR and NMDAR by glutamates.

The receptor between the different states includes a closed state, a ligand-binding state (single-bound and double-bound, respectively), a desensitisation state and an open state, which are determined by various rate constants. The values of these rate constants are estimated by fitting the specific model schemes to the single-channel experimental data. For example, there are six-state [192], seven-state [193] and twelve-state [194] models for AMPAR, and a five-state model [195] for NMDAR. In these models, the fully bound receptors by two glutamate can directly transition to the open state. In contrast, there are a number of NMDAR models that take into consideration the conformational changes of the fully bound receptors by including two desensitised states before moving to the open state [196–200]. Most NMDAR models only consider the glutamate binding-steps and exclude the coagonist binding-steps by fitting the data obtained from experiments under the condition of a saturated concentration of glycine [201].

The relief of the Mg^{2+} blockage of NMDAR is generally treated as a instantaneous event and the model proposed by [202] is widely used for simulating it as a membrane voltage dependent process. Moreover, there are few models that simulate the relief of Mg^{2+} blockage as a slow component and include it into their kinetic models [203, 204].

ER as an important compartment appears in numerous synaptic and neuronal models, because of its crucial role of internal Ca^{2+} source and its ubiquitous distribution across the neuron. In synaptic level, models mostly include Ca^{2+} ions flux via proteins (such as NMDARs and sodium Ca^{2+} exchangers) on the post synaptic plasma membrane and Ca^{2+}-mediated membrane electrical activity as the aspect of contribution from Ca^{2+} dynamics [205–207]. In spite of this, several synaptic models include ER as an internal Ca^{2+} source to study its response to synaptic Ca^{2+} fluctuation during the signal transduction process in synapses [208–210]. In contrast, ER as an internal compartment is generally includes in the most models to study its contribution on cytosolic events such as intracellular propagation of Ca^{2+} waves [211, 212].

In Ca^{2+} signalling models of the ER, the uptake mechanism is basically governed by modelling the gating behavior of its two

Ca^{2+} channels, IP_3R and RyR. At molecular level, different kinetic schemas have been developed for these channels.

Models of the IP_3R-regulated Ca^{2+} release have been the focus of many research on account of their significance in Ca^{2+} oscillations and synaptic plasticity. Modelling of the IP_3R shows more complexity in its dynamics as a result of its activation and inhibition, and the involvement of a G protein-coupled receptor-induced IP_{3-} production process. The first model for IP_3-induced Ca^{2+} release was proposed by De Young and Keizer, which assumed that there were three equivalent and independent subunits of IP_3R involved and all of them have to be in particular conducting states to open the channel [213]. Similar to De Young and Keizer's four-state model, there were other models that used Markov kinetic schemes to describe transitions between states. They are based on the assumption that each subunit of the IP_3R receptor has different multiple transitional states; for example, three [214] and seven transitional states [210]. A well-known simplified two-state model based on De Young and Keizer's model was developed by Li and Rinzel, who used Hodgkin–Huxley style equations based on the assumption of instantaneous activation following slow inactivation of IP_3R by Ca^{2+} [215]. Many follow-up models have been developed based on, or influenced by, the De Young–Keizer model or the Li-Rinzel simplified model. In addition, the models have been also developed based on other different assumptions such as sequential binding behavior [216], receptor conformation changes [217], adaptation of IP_3R and saturable fashion of IP_3 [217], or built to investigate different aspects, such as open probability of IP_3R on steady states [218, 219] and IP_3 metabolism [220]. The development of IP_3R models has been reviewed in detail by Sneyd and Falcke [221].

RyR-regulated Ca^{2+} signalling has been ignored in some models of intracellular Ca^{2+} dynamics, since IP_3R itself is enough to be responsible for CICR as it is sensitive to both IP_3 and Ca^{2+} ions. However, it is important to include RyRs to simulate the Ca^{2+} signalling more realistically. For example, by using Chemesis, a neural simulation software, Blackwell and Kotaleski demonstrated that the involvement of RyR channels affects the model simulation behavior of Ca^{2+} wave propagation [222]. The models, which included RyR-regulated Ca^{2+} signalling, have considered that both RyR and IP$_3$R contribute to Ca^{2+} release from the ER and the CICR is governed by RyR [184, 223]. Similar strategies to IP$_3$R modelling have been applied in RyR models. For example, there are models of RyRs using Markov kinetic schemes which describe the transitions between activated and inactivated states or multi-states [224–226].

Ca^{2+} pumps include plasma membrane Ca^{2+} pumps, which remove Ca^{2+} from the cytosol, and SERCA pumps, which are the only contributors to the elevation of ER Ca^{2+} levels. These pumps have been present in cytosolic Ca^{2+} signalling models and been shown to modify Ca^{2+} transient and oscillation. In most models, they have been modelled as unidirectional pumps that are activated by the increase in cytosolic Ca^{2+} and defined by a simple Hill equation or Michaelis–Menten kinetics with a Hill coefficient [227–230]. A few complex models have taken into consideration the multiple transition states of SERCA activation [231, 232] or the effect of Ca^{2+}-buffering on SERCA [233]. A leak influx of Ca^{2+} through the membrane into the cytosol is always used to balance the Ca^{2+} efflux through the pumps in the resting state. The membrane leakage is modelled as either a constant leak influx [227] or as a process with a constant rate and in proportion to the concentration gradient across the membrane [234].

Ca^{2+} ions also interact with a variety of buffer proteins, such as calmodulin and calbindin, and fluorescent dyes in both the cytosolic space and the ER lumina [53]. In order to study the buffering effects on Ca^{2+} dynamics and make the simulation more reliable, various assumptions relating to buffering behavior have been made and added to the Ca^{2+} signalling model. It would obviously make the model more complex and larger to include details of binding events, such as the effects from different buffers. Therefore, simplified theories for Ca^{2+} buffering have been applied. For example, excess buffer approximation (EBA) and rapid buffer approximation (RBA) are two important simplifications, which assume the mobile buffers are exceeded and cannot be saturated, and binding of Ca^{2+} ions to buffers is much faster compared with the changes in cytosolic

Ca^{2+} concentration, respectively [189]. Besides, diffusion of Ca^{2+} ions has been included in some models which consider the effect of concentration gradients from geometry facts [209]. The intracellular compartment is divided into multi-pool and movement of Ca^{2+} ions between two adjacent subcompartments is driven by Brownian motion [189]. Radial and/or longitudinal diffusion will be considered based on model assumptions. Either deterministic or stochastic strategies can be applied to simulation of diffusion processing by using partial differential equations [235].

7.9. Modelling Intracellular Signalling Related to AD

Modelling intracellular signalling in computational neuroscience has been used to study complex temporal and spatial characteristics of nerve systems for decades. Well-established modelling methods of Ca^{2+} signalling provide good foundations for modelling studies of AD and Ca^{2+}-related disease. However, only a few models have been developed to simulate or partly simulated the dysregulation of Ca^{2+} signalling in neurons related to AD. There has been some progress in the mathematical modelling of $A\beta$-induced alterations as a function of ion channels in neuronal plasma membranes. For example, based on the observation that $A\beta$ induced a voltage-dependent decrease in membrane conductance in rat hippocampal neurons, Good and Murphy developed a mathematical model of $A\beta$-mediated blockages of fast-inactivating K^+ channels [236]. The simulation results show good agreement with experiments in the aspect of stimulation responses and Ca^{2+} buffering capacity. Through simulation they hypothesized that blockages of a fast-inactivating K^+ current by $A\beta$ as an early event plays a crucial role in the neurotoxicity of $A\beta$ in AD and could lead to increased intracellular Ca^{2+} levels and membrane excitability resulting in neuronal neurotoxicity and, eventually, death. A recently published model based on Good and Murphy's model has been developed to further explore $A\beta$-induced blockages of fast-inactivating K+ channels as well as to study the $A\beta$-induced increase of membrane conductance during $A\beta$-neuron interaction in a short time-scale [237]. This model included the voltage-clamp and the membrane conductance mechanisms, which enabled it to simulate Aβ-neuron interactions under various experimental conditions, make comparisons with available data and generate predictions. Good's group also utilized mathematical kinetic analysis methods to study the potential mechanism underlying Aβ-induced G protein activation [238]. By using experimental rate data to examine the proposed underlying mechanisms, four possible different mechanisms have been tested to find the best fit of the rate expression with experimental data. Based on the results, they suggested that Aβ oligomers activate the G protein while excessive Aβ aggregation might inhibit further GTPase activity. In addition, models using simulation software packages such as NEURON have been constructed to simulate the effects of Aβ on the neuronal membrane. For example, Kidd and Sattelle used NEURON to construct a model neuron to study the blockage effects by Aβ on A-type K^+ currents of Drosophila larval cholinergic neurons and made predictions on their firing properties related to the alteration of the steady-state properties of the A-type K+ current [239]. Similarly, Morse *et al.* also used NEURON to analyze the hyper-excitability induced by Aβ via blocking A-type K^+ currents observed in proximal dendrites in AD animal models [240]. Through simulation they hypothesized the disruption Aβ-induced blockage of A-type K^+ currents at oblique branches may be most vulnerable and may play an important role in the decline of cognitive function in the early stage of AD.

Although dysregulation of intracellular Ca^{2+} signalling has been widely studied experimentally, little work has been done on computational model development in this area. Tiveci *et al.* developed a model of brain energy metabolism by including Ca^{2+} dynamics to an existing hemodynamic model of the brain [241]. This model is able to investigate the effects of the existence of Ca^{2+} dynamics on the blood oxygenation level dependent (BOLD) signal based on experimental observations. They also used this model to simulate AD cases and, based on the simulation results, they suggested the cerebral blood flow changes observed in AD cases might be the cause of negative BOLD effects and increased cytosolic Ca^{2+} level. As well as modelling Ca^{2+} dynamics in neurons, there are also several models of Ca^{2+} signalling in astrocytes, the predominant glial cells in the central nervous system. Toivari *et al.* developed a computational stochastic model of Ca^{2+} signalling in rat cortical astrocytes [242]. This model was used to study the effects of $A\beta_{25-35}$ and transmitters on intracellular Ca^{2+} signalling in astrocytes. The simulation results were consistent with their experimental findings on $A\beta_{25-35}$ and transmitters induced Ca^{2+} transient. Another example is a model constructed by Riera *et al.* This model was based on spontaneous Ca^{2+} oscillations in astrocytes and used it to simulate Ca^{2+} dynamics in wild-type (WT) and Tg2576 mice [243]. Through simulations, they suggested an increased Ca^{2+} influx from the extracellular space in APP transgenic mice might be activate astrocytes and play a crucial role in intracellular Ca^{2+} dysregulation. Furthermore, De Caluwé and Dupont recently developed a simple theoretical model to study the positive feedback loop between $A\beta$ and cytosolic Ca^{2+} [244]. Despite this being a simplified model, which has excluded detailed molecular mechanisms, it was still able to reveal a bistable switch between "healthy" and the "pathological" states, depending on the concentration of $A\beta$ and cytosolic Ca^{2+}.

References

[1] Thies W. and Bleiler L. (2013). Alzheimer's disease facts and figures. *Alzheimers Dement*, 9, pp. 208–245.

[2] Hardy J.A. and Higgins G.A. (1992). Alzheimer's disease: the amyloid cascade hypothesis. *Science*, 256, pp. 184–185.

[3] Hardy J. and Selkoe D.J. (2002). The amyloid hypothesis of Alzheimer's disease: progress and problems on the road to therapeutics. *Science*, 297, pp. 353–356.

[4] Karran E., Mercken M. and De Strooper B. (2011). The amyloid cascade hypothesis for Alzheimer's disease: an appraisal for the development of therapeutics. *Nat Rev Drug Discov*, 10, pp. 698–712.

[5] Braak H. and Braak E. (1991). Neuropathological stageing of Alzheimer-related changes. *Acta Neuropathol*, 82, pp. 239–259.

[6] Iqbal K. *et al.* (2005). Tau pathology in Alzheimer disease and other tauopathies. *Biochim Biophys Acta — Mol Basis Dis*, 1739, pp. 198–210.

[7] Wischik C.M., Wischik D.J., Storey J.M.D. and Harrington C.R. (2010). *Emerging Drugs and Targets for Alzheimer's Disease: Volume 1: Beta-Amyloid, Tau Protein and Glucose Metabolism.* ed. Martinez A. Chapter 11 "Rationale for tau-aggregation inhibitor therapy in Alzheimer's disease and

other tauopathies" (The Royal Society of Chemistry, Cambridge, UK), pp. 210–232.

[8] Bartus R.T., Dean 3rd R.L., Beer B. and Lippa A.S. (1982). The cholinergic hypothesis of geriatric memory dysfunction. *Science*, 217, pp. 408–414.

[9] Francis P.T., Palmer A.M., Snape M. and Wilcock G.K. (1999). The cholinergic hypothesis of Alzheimer's disease: a review of progress. *J Neurol Neurosurg Psychiatry*, 66, pp. 137–147.

[10] Craig L.A., Hong N.S. and McDonald R.J. (2011). Revisiting the cholinergic hypothesis in the development of Alzheimer's disease. *Neurosci Biobehav Rev*, 35, pp. 1397–1409.

[11] Gouras G.K. *et al.* (2000). Intraneuronal Abeta42 accumulation in human brain. *Am J Pathol*, 156, pp. 15–20.

[12] Oddo S. *et al.* (2006). A dynamic relationship between intracellular and extracellular pools of Abeta. *Am J Pathol*, 168, pp. 184–194.

[13] Kelliher M. *et al.* (1999). Alterations in the ryanodine receptor calcium release channel correlate with Alzheimer's disease neurofibrillary and beta-amyloid pathologies. *Neuroscience*, 92, pp. 499–513.

[14] Emilsson L., Saetre P. and Jazin E. (2006). Alzheimer's disease: mRNA expression profiles of multiple patients show alterations of genes involved with calcium signalling. *Neurobiol Dis*, 21, pp. 618–625.

[15] Verkhratsky A. (2005). Physiology and pathophysiology of the calcium store in the endoplasmic reticulum of neurons. *Physiol Rev*, 85, pp. 201–279.

[16] Khachaturian Z.S. (1994). Calcium hypothesis of Alzheimer's disease and brain aging. *Ann N Y Acad Sci*, 747, pp. 1–11.

[17] Thibault O., Gant J.C. and Landfield P.W. (2007). Expansion of the calcium hypothesis of brain aging and Alzheimer's disease: minding the store. *Aging Cell*, 6, pp. 307–317.

[18] Berridge M.J. (2010). Calcium hypothesis of Alzheimer's disease. *Pflugers Arch*, 459, pp. 441–449.

[19] Regehr W.G. and Tank D.W. (1990). Postsynaptic NMDA receptor-mediated calcium accumulation in hippocatnpal CAI pyramidal cell dendrites. *Nature*, 345.

[20] Danysz W. and Parsons C.G. (2012). Alzheimer's disease, beta-amyloid, glutamate, NMDA receptors and memantine — searching for the connections. *Br J Pharmacol*, 167, pp. 324–352.

[21] Stutzmann G.E., Caccamo A., LaFerla F.M. and Parker I. (2004). Dysregulated IP3 signalling in cortical neurons of knock-in mice expressing an Alzheimer's-linked mutation in presenilin1 results in exaggerated Ca^{2+} signals and altered membrane excitability. *J Neurosci*, 24, pp. 508–513.

[22] Schon E.A. and Area-Gomez E. (2013). Mitochondria-associated ER membranes in Alzheimer disease. *Mol Cell Neurosci*, 55, pp. 26–36.

[23] Cheung K.H. *et al.* (2008). Mechanism of Ca^{2+} disruption in Alzheimer's disease by presenilin regulation of InsP3 receptor channel gating. *Neuron*, 58, pp. 871–883.

[24] Berridge M.J. (2014). Calcium regulation of neural rhythms, memory and Alzheimer's disease. *J Physiol*, 592, pp. 281–293.

[25] Kuchibhotla K.V. *et al.* (2008). Aβ plaques lead to aberrant regulation of calcium homeostasis *in vivo* resulting in structural and functional disruption of neuronal networks. *Neuron*, 59, pp. 214–225.

[26] Li S. *et al.* (2009). Soluble oligomers of amyloid Beta protein facilitate hippocampal long-term depression by disrupting neuronal glutamate uptake. *Neuron*, 62, pp. 788–801.

[27] Ferreira S.T., Clarke J.R., Bomfim T.R. and De Felice F.G. (2014). Inflammation, defective insulin signalling, and neuronal dysfunction in Alzheimer's disease. *Alzheimer's Dement*, 10, pp. S76–S83.

[28] Floyd R.A. and Hensley K. (2002). Oxidative stress in brain aging: Implications for therapeutics of neurodegenerative diseases. *Neurobiol Aging*, 23, pp. 795–807.

[29] Christen Y. (2000). Oxidative stress and Alzheimer disease. *Am J Clin Nutr*, 71, pp. 621–629.

[30] Ermak G. and Davies K.J.A. (2002). Calcium and oxidative stress: from cell signalling to cell death. *Mol Immunol*, 38, pp. 713–721.

[31] Uttara B., Singh A.V., Zamboni P. and Mahajan R.T. (2009). Oxidative stress and neurodegenerative diseases: a review of upstream and downstream antioxidant therapeutic options. *Curr Neuropharmacol*, 7, pp. 65–74.

[32] Müller M., Cheung K.-H. and Foskett J.K. (2011). Enhanced ROS generation mediated by Alzheimer's disease presenilin regulation of InsP(3)R Ca(2+) signaling. *Antioxid Redox Signal*, 14, pp. 1225–1235.

[33] Pasti L., Volterra A., Pozzan T. and Carmignoto G. (1997). Intracellular calcium oscillations in astrocytes: a highly plastic, bidirectional form of communication between neurons and astrocytes *in situ*. *J Neurosci*, 17, pp. 7817–7830.

[34] Augustine G.J. (2001). How does calcium trigger neurotransmitter release? *Curr Opin Neurobiol*, 11, pp. 320–326.

[35] Palop J.J. and Mucke L. (2010). Amyloid-β induced neuronal dysfunction in Alzheimer's disease: from synapses toward neural networks. *Nat Neurosci*, 13, pp. 812–818.

[36] Chivet M. *et al.* (2012). Emerging role of neuronal exosomes in the central nervous system. *Front Physiol*, 3, p. 145.

[37] Rajendran L. *et al.* (2006). Alzheimer's disease β-amyloid peptides are released in association with exosomes. *Proc Natl Acad Sci USA*, 103, pp. 11172–11177.

[38] Ghidoni R. *et al.* (2011). Cystatin C is released in association with exosomes: a new tool of neuronal communication which is unbalanced in Alzheimer's disease. *Neurobiol Aging*, 32, pp. 1435–1442.

[39] Supekar K. *et al.* (2008). Network analysis of intrinsic functional brain connectivity in Alzheimer's disease. *PLoS Comput Biol*, 4, p. e1000100.

[40] Arendt T. (2009). Synaptic degeneration in Alzheimer's disease. *Acta Neuropathol*, 118, pp. 167–179.

[41] Sanz-Arigita E.J. *et al.* (2010). Loss of "small-world" networks in Alzheimer's disease: graph analysis of FMRI resting-state functional connectivity. *PLoS One*, 5, p. e13788.

[42] Sabatini B.L., Maravall M. and Svoboda K. (2001). Ca^{2+} signalling in dendritic spines. *Curr Opin Neurobiol*, 11, pp. 349–356.

[43] Furukawa H., Singh S.K., Mancusso R. and Gouaux E. (2005). Subunit arrangement and function in NMDA receptors. *Nature*, 438, pp. 185–192.

[44] Cull-Candy S.G. (2007). *NMDA Receptors* (John Wiley & Sons).

[45] Papouin T. *et al.* (2012). Synaptic and extrasynaptic NMDA receptors are gated by different endogenous coagonists. *Cell*, 150, pp. 633–646.

[46] Petralia R.S. (2012). Distribution of extrasynaptic NMDA receptors on neurons. *Sci World J*, 2012, p. 11.

[47] Cheng D. *et al.* (2006). Relative and absolute quantification of postsynaptic density proteome isolated from rat forebrain and cerebellum. *Mol Cell Proteomics*, 5, pp. 1158–1170.

[48] Hardingham G.E. and Bading H. (2010). Synaptic versus extrasynaptic NMDA receptor signalling: implications for neurodegenerative disorders. *Nat Rev Neurosci*, 11, pp. 682–696.

[49] Mayer M.L., Westbrook G.L. and Guthrie P.B. (1984). Voltage-dependent block by Mg^{2+} of NMDA responses in spinal cord neurones. *Nature*, 309, pp. 261–263.

[50] Dingledine R., Borges K., Bowie D. and Traynelis S.F. (1999). The glutamate receptor ion channels. *Pharmacol Rev*, 51, pp. 7–62.

[51] Michaelis E.K. (1998). Molecular biology of glutamate receptors in the central nervous system and their role in excitotoxicity, oxidative stress and aging. *Prog Neurobiol*, 54, pp. 369–415.

[52] Collingridge G.L., Isaac J.T.R. and Wang Y.T. (2004). Receptor trafficking and synaptic plasticity. *Nat Rev Neurosci*, 5, pp. 952–962.

[53] Berridge M.J., Bootman M.D. and Roderick H.L. (2003). Calcium signalling: dynamics, homeostasis and remodelling. *Nat Rev Mol Cell Biol*, 4, pp. 517–529.

[54] Bezprozvanny I. (2005). The inositol 1,4,5-trisphosphate receptors. *Cell Calcium*, 38, pp. 261–272.

[55] Mattson M.P. *et al.* (2000). Calcium signalling in the ER: its role in neuronal plasticity and neurodegenerative disorders. *Trends Neurosci*, 23, pp. 222–229.

[56] Foskett J.K., White C., Cheung K.H. and Mak D.O. (2007). Inositol trisphosphate receptor Ca^{2+} release channels. *Physiol Rev*, 87, pp. 593–658.

[57] Taufiq Ur R., Skupin A., Falcke M. and Taylor C.W. (2009). Clustering of InsP3 receptors by InsP3 retunes their regulation by InsP3 and Ca^{2+}. *Nature*, 458, pp. 655–659.

[58] Berridge M.J. (1997). Elementary and global aspects of calcium signalling. *J Physiol*, 499(Pt 2), pp. 291–306.

[59] Rahman T. and Taylor C.W. (2009). Dynamic regulation of IP3 receptor clustering and activity by IP3. *Channels (Austin)*, 3, pp. 226–232.

[60] Galeotti N., Vivoli E., Bartolini A. and Ghelardini C. (2008). A gene-specific cerebral types 1, 2, and 3 RyR protein knockdown induces an antidepressant-like effect in mice. *J Neurochem*, 106, pp. 2385–2394.

[61] Adasme T. *et al.* (2011). Involvement of ryanodine receptors in neurotrophin-induced hippocampal synaptic plasticity and spatial memory formation. *Proc Natl Acad Sci USA*, 108, pp. 3029–3034.

[62] Hidalgo C., Bull R., Behrens M.I. and Donoso P. (2004). Redox regulation of RyR-mediated Ca2+ release in muscle and neurons. *Biol Res*, 37, pp. 539–552.

[63] Green K.N. *et al.* (2008). SERCA pump activity is physiologically regulated by presenilin and regulates amyloid beta production. *J Cell Biol*, 181, pp. 1107–1116.

[64] Wuytack F., Raeymaekers L. and Missiaen L. (2002). Molecular physiology of the SERCA and SPCA pumps. *Cell Calcium*, 32, pp. 279–305.

[65] Buchholz J.N. *et al.* (2012). *Calcium Regulation in Neuronal Function with Advancing Age: Limits of Homeostasis.* ed. Nagata T. "Senescence" *(InTech, Rijeka)*, pp. 531–558.

[66] Stutzmann G.E. and Mattson M.P. (2011). Endoplasmic reticulum Ca^{2+} handling in excitable cells in health and disease. *Pharmacol Rev*, 63, pp. 700–727.

[67] Manjarres I.M., Rodriguez-Garcia A., Alonso M.T. and Garcia-Sancho J. (2010). The sarco/endoplasmic reticulum Ca^{2+} ATPase (SERCA) is the third element in capacitative calcium entry. *Cell Calcium*, 47, pp. 412–418.

[68] Elbaz Y. and Schuldiner M. (2011). Staying in touch: the molecular era of organelle contact sites. *Trends Biochem Sci*, 36, pp. 616–623.

[69] Soboloff J., Rothberg B.S., Madesh M. and Gill D.L. (2012). STIM proteins: dynamic calcium signal transducers. *Nat Rev Mol Cell Biol*, 13, pp. 549–565.

[70] Rowland A.A. and Voeltz G.K. (2012). Endoplasmic reticulum-mitochondria contacts: function of the junction. *Nat Rev Mol Cell Biol*, 13, pp. 607–625.

[71] Hayashi T., Rizzuto R., Hajnoczky G. and Su T.P. (2009). MAM: more than just a housekeeper. *Trends Cell Biol*, 19, pp. 81–88.

[72] Mattson M.P. (2004). Pathways towards and away from Alzheimer's disease. *Nature*, 430, pp. 631–639.

[73] Selkoe D.J. (1991). The molecular pathology of Alzheimer's disease. *Neuron*, 6, pp. 487–498.

[74] Spires-Jones T.L. *et al.* (2009). Tau pathophysiology in neurodegeneration: a tangled issue. *Trends Neurosci*, 32, pp. 150–159.

[75] Patterson C. *et al.* (2008). Diagnosis and treatment of dementia: 1. Risk assessment and primary prevention of Alzheimer disease. *Can Med Assoc J*, 178, pp. 548–556.

[76] Duyckaerts C., Delatour B. and Potier M.-C. (2009). Classification and basic pathology of Alzheimer disease. *Acta Neuropathol*, 118, pp. 5–36.

[77] Turner P.R., O'Connor K., Tate W.P. and Abraham W.C. (2003). Roles of amyloid precursor protein and its fragments in regulating neural activity, plasticity and memory. *Prog Neurobiol*, 70, pp. 1–32.

[78] LaFerla F.M., Green K.N. and Oddo S. (2007). Intracellular amyloid-beta in Alzheimer's disease. *Nat Rev Neurosci*, 8, pp. 499–509.

[79] Jiang Q. *et al.* (2008). ApoE promotes the proteolytic degradation of Abeta. *Neuron*, 58, pp. 681–693.

[80] Shankar G.M. and Walsh D.M. (2009). Alzheimer's disease: synaptic dysfunction and Abeta. *Mol Neurodegener*, 4, p. 48.

[81] Crews L. and Masliah E. (2010). Molecular mechanisms of neurodegeneration in Alzheimer's disease. *Hum Mol Genet*, 19, pp. 12–20.

[82] Golde T.E. (2003). Alzheimer disease therapy: can the amyloid cascade be halted? *J Clin Invest*, 111, pp. 11–18.

[83] Sanz-Blasco S. *et al.* (2008). Mitochondrial Ca2+ overload underlies Aβ oligomers neurotoxicity providing an unexpected mechanism of neuroprotection by NSAIDs. *PLoS One*, 3, p. e2718.

[84] Jin M. *et al.* (2011). Soluble amyloid β-protein dimers isolated from Alzheimer cortex directly induce Tau hyperphosphorylation and neuritic degeneration. *Proc Natl Acad Sci USA*, 108, pp. 5819–5824.

[85] Mucke L. and Selkoe D.J. (2012). Neurotoxicity of amyloid β-protein: synaptic and network dysfunction. *Cold Spring Harb Perspect Med*, 2, p. a006338.

[86] Malchiodi-Albedi F. *et al.* (2011). Amyloid oligomer neurotoxicity, calcium dysregulation, and lipid rafts. *Int J Alzheimers Dis*, 2011, pp. 906–964.

[87] Van Dam D. and De Deyn P.P. (2006). Drug discovery in dementia: the role of rodent models. *Nat Rev Drug Discov*, 5, pp. 956–970.

[88] Khachaturian Z.S. (1987). Hypothesis on the regulation of cytosol calcium concentration and the aging brain. *Neurobiol Aging*, 8, pp. 345–346.

[89] LaFerla F.M. (2002). Calcium dyshomeostasis and intracellular signalling in alzheimer's disease. *Nat Rev Neurosci*, 3, pp. 862–872.

[90] Berridge M.J. (2013). Dysregulation of neural calcium signalling in Alzheimer disease, bipolar disorder and schizophrenia. *Prion*, 7, pp. 2–13.

[91] Lopez J.R. *et al.* (2008). Increased intraneuronal resting [Ca^{2+}] in adult Alzheimer's disease mice. *J Neurochem*, 105, pp. 262–271.

[92] Demuro A. *et al.* (2005). Calcium dysregulation and membrane disruption as a ubiquitous neurotoxic mechanism of soluble amyloid oligomers. *J Biol Chem*, 280, pp. 17294–17300.

[93] Kawahara M. *et al.* (2011). Membrane incorporation, channel formation, and disruption of calcium homeostasis by Alzheimer's β-amyloid protein. *Int J Alzheimers Dis*, 2011.

[94] Talantova M. *et al.* (2013). Aβ induces astrocytic glutamate release, extrasynaptic NMDA receptor activation, and synaptic loss. *Proc Natl Acad Sci USA*, 110, pp. 2518–2527.

[95] Butterfield D.A. and Pocernich C.B. (2003). The glutamatergic system and Alzheimer's disease. *CNS Drugs*, 17, pp. 641–652.

[96] Ondrejcak T. *et al.* (2010). Alzheimer's disease amyloid beta-protein and synaptic function. *Neuromolecular Med*, 12, pp. 13–26.

[97] Bobich J.A., Zheng Q. and Campbell A. (2004). Incubation of nerve endings with a physiological concentration of Aβ1-42 activates CaV2.2(N-Type)-voltage operated calcium channels and acutely increases glutamate and noradrenaline release. *J Alzheimer's Dis*, 6, pp. 243–255.

[98] Chin J.H., Ma L., MacTavish D. and Jhamandas J.H. (2007). Amyloid β protein modulates glutamate-mediated neurotransmission in the rat basal forebrain: involvement of presynaptic neuronal nicotinic acetylcholine and metabotropic glutamate receptors. *J Neurosci*, 27, pp. 9262–9269.

[99] Kabogo D. *et al.* (2010). β-amyloid-related peptides potentiate K$^+$-evoked glutamate release from adult rat hippocampal slices. *Neurobiol Aging*, 31, pp. 1164–1172.

[100] Abramov E. *et al.* (2009). Amyloid-beta as a positive endogenous regulator of release probability at hippocampal synapses. *Nat Neurosci*, 12, pp. 1567–1576.

[101] Arias C., Arrieta I. and Tapia R. (1995). β-amyloid peptide fragment 25–35 potentiates the calcium-dependent release of excitatory amino acids from depolarized hippocampal slices. *J Neurosci Res*, 41, pp. 561–566.

[102] Puzzo D. *et al.* (2008). Picomolar amyloid-β positively modulates synaptic plasticity and memory in hippocampus. *J Neurosci*, 28, pp. 14537–14545.

[103] Noda M., Nakanishi H. and Akaike N. (1999). Glutamate release from microglia via glutamate transporter is enhanced by amyloid-beta peptide. *Neuroscience*, 92, pp. 1465–1474.

[104] Orellana J.A. *et al.* (2011). Amyloid β-induced death in neurons involves glial and neuronal hemichannels. *J Neurosci*, 31, pp. 4962–4977.

[105] Barger S.W. and Basile A.S. (2001). Activation of microglia by secreted amyloid precursor protein evokes release of glutamate by cystine exchange and attenuates synaptic function. *J Neurochem*, 76, pp. 846–854.

[106] Kuchibhotla K.V., Lattarulo C.R., Hyman B.T. and Bacskai B.J. (2009). Synchronous hyperactivity and intercellular calcium waves in astrocytes in Alzheimer mice. *Science*, 323, pp. 1211–1215.

[107] Harris M.E. *et al.* (1995). Beta-Amyloid peptide-derived, oxygen-dependent free radicals inhibit glutamate uptake in cultured astrocytes: implications for Alzheimer's disease. *Neuroreport*, 6, pp. 1875–1879.

[108] Harris M.E. *et al.* (1996). Amyloid β peptide (25–35) inhibits Na+-dependent glutamate uptake in rat hippocampal astrocyte cultures. *J Neurochem*, 67, pp. 277–286.

[109] Parpura-Gill A., Beitz D. and Uemura E. (1997). The inhibitory effects of β-amyloid on glutamate and glucose uptakes by cultured astrocytes. *Brain Res*, 754, pp. 65–71.

[110] Fernández-Tomé P., Brera B., Arévalo M.-A. and de Ceballos M.L. (2004). β-Amyloid 25-35 inhibits glutamate uptake in cultured neurons and astrocytes: modulation of uptake as a survival mechanism. *Neurobiol Dis*, 15, pp. 580–589.

[111] Matos M., Augusto E., Oliveira C.R. and Agostinho P. (2008). Amyloid-beta peptide decreases glutamate uptake in cultured astrocytes: involvement of oxidative stress and mitogen-activated protein kinase cascades. *Neuroscience*, 156, pp. 898–910.

[112] Masliah E. *et al.* (1996). Deficient glutamate tranport is associated with neurodegeneration in Alzheimer's disease. *Ann Neurol*, 40, pp. 759–766.

[113] Masliah E. *et al.* (2000). Abnormal glutamate transport function in mutant amyloid precursor protein transgenic mice. *Exp Neurol*, 163, pp. 381–387.

[114] Kirvell S.L., Esiri M. and Francis P.T. (2006). Down-regulation of vesicular glutamate transporters precedes cell loss and pathology in Alzheimer's disease. *J Neurochem*, 98, pp. 939–950.

[115] Bordji K., Becerril-Ortega J., Nicole O. and Buisson A. (2010). Activation of extrasynaptic, but not synaptic, NMDA receptors modifies amyloid precursor protein expression pattern and increases amyloid-β production. *J Neurosci*, 30, pp. 15927–15942.

[116] Bordji K., Becerril-Ortega J. and Buisson A. (2011). Synapses, NMDA receptor activity and neuronal Abeta production in Alzheimer's disease. *Rev Neurosci*, 22, pp. 285–294.

[117] Dewachter I. *et al.* (2009). Deregulation of NMDA-receptor function and down-stream signalling in APP[V717I] transgenic mice. *Neurobiol Aging*, 30, pp. 241–256.

[118] Alberdi E. *et al.* (2010). Amyloid beta oligomers induce Ca^{2+} dysregulation and neuronal death through activation of ionotropic glutamate receptors. *Cell Calcium*, 47, pp. 264–272.

[119] Texidó L. *et al.* (2011). Amyloid β peptide oligomers directly activate NMDA receptors. *Cell Calcium*, 49, pp. 184–190.

[120] De Felice F.G. *et al.* (2007). Aβ oligomers induce neuronal oxidative stress through an N-methyl-D-aspartate receptor-dependent mechanism that is blocked by the Alzheimer drug memantine. *J Biol Chem*, 282, pp. 11590–11601.

[121] Ferreira I.L. *et al.* (2012). Amyloid beta peptide 1–42 disturbs intracellular calcium homeostasis through activation of GluN2B-containing N-methyl-D-aspartate receptors in cortical cultures. *Cell Calcium*, 51, pp. 95–106.

[122] Stys P.K., You H. and Zamponi G.W. (2012). Copper-dependent regulation of NMDA receptors by cellular prion protein: implications for neurodegenerative disorders. *J Physiol*, 590, pp. 1357–1368.

[123] You H. *et al.* (2012). Aβ neurotoxicity depends on interactions between copper ions, prion protein, and N-methyl-D-aspartate receptors. *Proc Natl Acad Sci USA*, 109, pp. 1737–1742.

[124] Kamenetz F. *et al.* (2003). APP processing and synaptic function. *Neuron*, 37, pp. 925–937.

[125] Lesné S. *et al.* (2005). NMDA receptor activation inhibits α-secretase and promotes neuronal amyloid-β production. *J Neurosci*, 25, pp. 9367–9377.

[126] Hoe H.-S. *et al.* (2009). The effects of amyloid precursor protein on postsynaptic composition and activity. *J Biol Chem*, 284, pp. 8495–8506.

[127] Snyder E.M. *et al.* (2005). Regulation of NMDA receptor trafficking by amyloid-beta. *Nat Neurosci*, 8, pp. 1051–1058.

[128] Johansson S. *et al.* (2006). Modelling of amyloid β-peptide induced lesions using roller-drum incubation of hippocampal slice cultures from neonatal rats. *Exp brain Res*, 168, pp. 11–24.

[129] Goto Y. *et al.* (2006). Amyloid β-peptide preconditioning reduces glutamate-induced neurotoxicity by promoting endocytosis of NMDA receptor. *Biochem Biophys Res Commun*, 351, pp. 259–265.

[130] Lacor P.N. *et al.* (2007). Aβ oligomer-induced aberrations in synapse composition, shape, and density provide a molecular basis for loss of connectivity in Alzheimer's disease. *J Neurosci*, 27, pp. 796–807.

[131] Wang H.-Y. *et al.* (2000). β-amyloid1–42 binds to α7 nicotinic acetylcholine receptor with high affinity implications for Alzheimer's disease pathology. *J Biol Chem*, 275, pp. 5626–5632.

[132] Kessels H.W., Nabavi S. and Malinow R. (2013). Metabotropic NMDA receptor function is required for beta-amyloid-induced synaptic depression. *Proc Natl Acad Sci USA*, 110, pp. 4033–4038.

[133] Chen Q.-S., Wei W.-Z., Shimahara T. and Xie C.-W. (2002). Alzheimer amyloid β-peptide inhibits the late phase of long-term potentiation through calcineurin-dependent mechanisms in the hippocampal dentate gyrus. *Neurobiol Learn Mem*, 77, pp. 354–371.

[134] Li S. *et al.* (2011). Soluble Aβ oligomers inhibit long-term potentiation through a mechanism involving excessive activation of extrasynaptic NR2B-containing NMDA receptors. *J Neurosci*, 31, pp. 6627–38.

[135] Chuang D.M., Gao X.-M. and Paul S.M. (1992). N-methyl-D-aspartate exposure blocks glutamate toxicity in cultured cerebellar granule cells. *Mol Pharmacol*, 42, pp. 210–216.

[136] Popugaeva E. and Bezprozvanny I. (2013). Role of endoplasmic reticulum Ca2+ signalling in the pathogenesis of Alzheimer disease. *Front Mol Neurosci*, 6, p. 29.

[137] Liang J., Kulasiri D. and Samarasinghe S. (2015). Ca^{2+} dysregulation in the endoplasmic reticulum related to Alzheimer's disease: a review on experimental progress and computational modeling. *Biosystems*, 134, pp. 1–15.

[138] Ito E. *et al.* (1994). Internal Ca^{2+} mobilization is altered in fibroblasts from patients with Alzheimer disease. *Proc Natl Acad Sci USA*, 91, pp. 534–538.

[139] Lee S.Y. *et al.* (2006). PS2 mutation increases neuronal cell vulnerability to neurotoxicants through activation of caspase-3 by enhancing of ryanodine receptor-mediated calcium release. *FASEB J*, 20, pp. 151–153.

[140] Chakroborty S. *et al.* (2012). Stabilizing ER Ca^{2+} channel function as an early preventative strategy for Alzheimer's disease. *PLoS One*, 7, p. e52056.

[141] Kelly B.L. and Ferreira A. (2006). Beta-amyloid-induced dynamin 1 degradation is mediated by N-methyl-D-aspartate receptors in hippocampal neurons. *J Biol Chem*, 281, pp. 28079–28089.

[142] Costa R.O. *et al.* (2012). Endoplasmic reticulum stress occurs downstream of GluN2B subunit of *N*-methyl-*D*-aspartate receptor in mature hippocampal cultures treated with amyloid-β oligomers. *Aging Cell*, 11, pp. 823–833.

[143] Ferreiro E. *et al.* (2006). An endoplasmic-reticulum-specific apoptotic pathway is involved in prion and amyloid-beta peptides neurotoxicity. *Neurobiol Dis*, 23, pp. 669–678.

[144] Demuro A. and Parker I. (2013). Cytotoxicity of intracellular abeta42 amyloid oligomers involves Ca^{2+} release from the endoplasmic reticulum by stimulated production of inositol trisphosphate. *J Neurosci*, 33, pp. 3824–3833.

[145] Mohamed A. and Posse de Chaves E. (2011). Aβ internalization by neurons 13 and glia. *Int J Alzheimers Dis*. 2011, 127984. http://doi.org/10.4061/2011/127984

[146] Ferreiro E., Oliveira C.R. and Pereira C. (2004). Involvement of endoplasmic reticulum Ca^{2+} release through ryanodine and inositol 1,4,5-triphosphate receptors in the neurotoxic effects induced by the amyloid-beta peptide. *J Neurosci Res*, 76, pp. 872–880.

[147] Stutzmann G.E. *et al.* (2006). Enhanced ryanodine receptor recruitment contributes to Ca^{2+} disruptions in young, adult, and aged Alzheimer's disease mice. *J Neurosci*, 26, pp. 5180–5189.

[148] Schapansky J., Olson K., Van Der Ploeg R. and Glazner G. (2007). NF-κB activated by ER calcium release inhibits Aβ-mediated expression of CHOP protein: enhancement by AD-linked mutant presenilin 1. *Exp Neurol*, 208, pp. 169–176.

[149] Cowburn R.F., Wiehager B. and Sundström E. (1995). β-amyloid peptides enhance binding of the calcium mobilising second messengers, inositol(1,4,5)trisphosphate and inositol-(1,3,4,5)tetrakisphosphate to their receptor sites in rat cortical membranes. *Neurosci Lett*, 191, pp. 31–34.

[150] Casley C.S. *et al.* (2009). Up-regulation of astrocyte metabotropic glutamate receptor 5 by amyloid-beta peptide. *Brain Res*, 1260, pp. 65–75.

[151] Renner M. *et al.* (2010). Deleterious effects of amyloid beta oligomers acting as an extracellular scaffold for mGluR5. *Neuron*, 66, pp. 739–754.

[152] Um J.W. *et al.* (2013). Metabotropic glutamate receptor 5 is a coreceptor for Alzheimer abeta oligomer bound to cellular prion protein. *Neuron*, 79, pp. 887–902.

[153] Allan L.E. *et al.* (2013). Alzheimer's Disease-associated peptide Aß42 mobilises ER Ca^{2+} via InsP3R-dependent and -independent mechanisms. *Front Mol Neurosci*, 6.

[154] Shtifman A. *et al.* (2010). Amyloid-beta protein impairs Ca^{2+} release and contractility in skeletal muscle. *Neurobiol Aging*, 31, pp. 2080–2090.

[155] Christensen R.A. *et al.* (2004). Calcium dyshomeostasis in beta-amyloid and tau-bearing skeletal myotubes. *J Biol Chem*, 279, pp. 53524–53532.

[156] Supnet C. *et al.* (2006). Amyloid-beta-(1-42) increases ryanodine receptor-3 expression and function in neurons of TgCRND8 mice. *J Biol Chem*, 281, pp. 38440–38447.

[157] Supnet C. *et al.* (2010). Up-regulation of the type 3 ryanodine receptor is neuroprotective in the TgCRND8 mouse model of Alzheimer's disease. *J Neurochem*, 112, pp. 356–365.

[158] Hedskog L. *et al.* (2013). Modulation of the endoplasmic reticulum-mitochondria interface in Alzheimer's disease and related models. *Proc Natl Acad Sci USA*, 110, pp. 7916–7921.

[159] Smith I.F., Green K.N. and LaFerla F.M. (2005). Calcium dysregulation in Alzheimer's disease: recent advances gained from genetically modified animals. *Cell Calcium*, 38, pp. 427–437.

[160] Kasri N.N. *et al.* (2006). Up-regulation of inositol 1,4,5-trisphosphate receptor type 1 is responsible for a decreased endoplasmic-reticulum Ca^{2+} content in presenilin double knock-out cells. *Cell Calcium*, 40, pp. 41–51.

[161] Ryan K.A. and Pimplikar S.W. (2005). Activation of GSK-3 and phosphorylation of CRMP2 in transgenic mice expressing APP intracellular domain. *J Cell Biol*, 171, pp. 327–335.

[162] Oule's B. *et al.* (2012). Ryanodine receptor blockade reduces amyloid-beta load and memory impairments in Tg2576 mouse model of Alzheimer disease. *J Neurosci*, 32, pp. 11820–11834.

[163] Brunkan A.L. and Goate A.M. (2005). Presenilin function and γ-secretase activity. *J Neurochem*, 93, pp. 769–792.

[164] Shen J. and Kelleher 3rd R.J. (2007). The presenilin hypothesis of Alzheimer's disease: evidence for a loss-of-function pathogenic mechanism. *Proc Natl Acad Sci USA*, 104, pp. 403–409.

[165] Supnet C. and Bezprozvanny I. (2011). Presenilins function in ER calcium leak and Alzheimer's disease pathogenesis. *Cell Calcium*, 50, pp. 303–309.

[166] Cruts M., Theuns J. and Van Broeckhoven C. (2012). Locus-specific mutation databases for neurodegenerative brain diseases. *Hum Mutat*, 33, pp. 1340–1344.

[167] Saura C.A. *et al.* (2004). Loss of presenilin function causes impairments of memory and synaptic plasticity followed by age-dependent neurodegeneration. *Neuron*, 42, pp. 23–36.

[168] Young J.E. and Goldstein L.S.B. (2012). Alzheimer's disease in a dish: promises and challenges of human stem cell models. *Hum Mol Genet*, 21, pp. R82–R89.

[169] He L. and Hannon G.J. (2004). MicroRNAs: small RNAs with a big role in gene regulation. *Nat Rev Genet*, 5, pp. 522–531.

[170] Van den Hove D.L. *et al.* (2014). Epigenetically regulated microRNAs in Alzheimer's disease. *Neurobiol Aging*, 35, pp. 731–745.

[171] Maes O.C., Chertkow H.M., Wang E. and Schipper H.M. (2009). MicroRNA: implications for Alzheimer disease and other human CNS disorders. *Curr Genomics*, 10, pp. 154–168.

[172] Querfurth H.W. and Selkoe D.J. (1994). Calcium ionophore increases amyloid beta peptide production by cultured cells. *Biochemistry*, 33, pp. 4550–4561.

[173] Isaacs A.M. *et al.* (2006). Acceleration of amyloid beta-peptide aggregation by physiological concentrations of calcium. *J Biol Chem*, 281, pp. 27916–27923.

[174] Bruno A.M. *et al.* (2012). Altered ryanodine receptor expression in mild cognitive impairment and Alzheimer's disease. *Neurobiol Aging*, 33, pp. 1001.e1–1001.e6.

[175] Chakroborty S. *et al.* (2012). Early presynaptic and postsynaptic calcium signalling abnormalities mask underlying synaptic depression in presymptomatic Alzheimer's disease mice. *J Neurosci*, 32, pp. 8341–8353.

[176] Querfurth H.W., Jiang J., Geiger J.D. and Selkoe D.J. (1997). Caffeine stimulates amyloid beta-peptide release from beta-amyloid precursor protein-transfected HEK293 cells. *J Neurochem*, 69, pp. 1580–1591.

[177] Zhang H. *et al.* (2010). Role of presenilins in neuronal calcium homeostasis. *J Neurosci*, 30, pp. 8566–8580.

[178] Pierrot N., Ghisdal P., Caumont A.-S. and Octave J.-N. (2004). Intraneuronal amyloid-β1-42 production triggered by sustained increase of cytosolic calcium concentration induces neuronal death. *J Neurochem*, 88, pp. 1140–1150.

[179] Koran M.E., Hohman T.J. and Thornton-Wells T.A. (2014). Genetic interactions found between calcium channel genes modulate amyloid load measured by positron emission tomography. *Hum Genet*, 133, pp. 85–93.

[180] Kurumatani T. *et al.* (1998). Loss of inositol 1,4,5-trisphosphate receptor sites and decreased PKC levels correlate with staging of Alzheimer's disease neurofibrillary pathology. *Brain Res*, 796, pp. 209–221.

[181] Buxbaum J.D. *et al.* (1994). Calcium regulates processing of the Alzheimer amyloid protein precursor in a protein kinase C-independent manner. *Proc Natl Acad Sci USA*, 91, pp. 4489–4493.

[182] Dreses-Werringloer U. *et al.* (2008). A polymorphism in CALHM1 influences Ca^{2+} homeostasis, Abeta levels, and Alzheimer's disease risk. *Cell*, 133, pp. 1149–1161.

[183] Zeiger W. *et al.* (2013). Ca^{2+} influx through store-operated Ca^{2+} channels reduces Alzheimer disease beta-amyloid peptide secretion. *J Biol Chem*, 288, pp. 26955–26966.

[184] De Schutter E. and Smolen P. (1998). *Calcium Dynamics in Large Neuronal Models*. eds. Koch C. and Segev I. "Methods in neuronal modeling: from ions to networks", 2nd edn. (The MIT Press, Cambridge), pp. 211–250.

[185] Senn W., Markram H. and Tsodyks M. (2001). An algorithm for modifying neurotransmitter release probability based on pre-and postsynaptic spike timing. *Neural Comput*, 13, pp. 35–67.

[186] Nadkarni S., Bartol T.M., Sejnowski T.J. and Levine H. (2010). Modelling Vesicular Release at Hippocampal Synapses. *PLoS Comput Biol*, 6, p. e1000983.

[187] Barbour B. and Häusser M. (1997). Intersynaptic diffusion of neurotransmitter. *Trends Neurosci*, 20, pp. 377–384.

[188] Rusakov D.A. and Kullmann D.M. (1998). Extrasynaptic glutamate diffusion in the hippocampus: ultrastructural constraints, uptake, and receptor activation. *J Neurosci*, 18, pp. 3158–3170.

[189] Sterratt D., Graham B., Gillies A. and Willshaw D. (2011). *Principles of Computational Modelling in Neuroscience* (Cambridge University Press, New York).

[190] Rusakov D.A. (2001). The role of perisynaptic glial sheaths in glutamate spillover and extracellular Ca^{2+} depletion. *Biophys J*, 81, pp. 1947–1959.

[191] Grewer C. *et al.* (2008). Glutamate forward and reverse transport: from molecular mechanism to transporter-mediated release after ischemia. *IUBMB Life*, 60, pp. 609–619.

[192] Patneau D.K. and Mayer M.L. (1991). Kinetic analysis of interactions between kainate and AMPA: evidence for activation of a single receptor in mouse hippocampal neurons. *Neuron*, 6, pp. 785–798.

[193] Jonas P., Major G. and Sakmann B. (1993). Quantal components of unitary EPSCs at the mossy fibre synapse on CA3 pyramidal cells of rat hippocampus. *J Physiol*, 472, pp. 615–663.

[194] Nielsen T.A., DiGregorio D.A. and Silver R.A. (2004). Modulation of glutamate mobility reveals the mechanism underlying slow-rising AMPAR EPSCs and the diffusion coefficient in the synaptic cleft. *Neuron*, 42, pp. 757–771.

[195] Lester R.A. and Jahr C.E. (1992). NMDA channel behavior depends on agonist affinity. *J Neurosci*, 12, pp. 635–643.

[196] Banke T.G. and Traynelis S.F. (2003). Activation of NR1/NR2b NMDA receptors. *Nat Neurosci*, 6, pp. 144–152.

[197] Popescu G., Robert A., Howe J.R. and Auerbach A. (2004). Reaction mechanism determines NMDA receptor response to repetitive stimulation. *Nature*, 430, pp. 790–793.

[198] Erreger K. *et al.* (2005). Subunit-specific gating controls rat NR1/NR2A and NR1/NR2B NMDA channel kinetics and synaptic signalling profiles. *J Physiol*, 563, pp. 345–358.

[199] Schorge S., Elenes S. and Colquhoun D. (2005). Maximum likelihood fitting of single channel NMDA activity with a mechanism composed of independent dimers of subunits. *J Physiol*, 569, pp. 395–418.

[200] Kussius C.L. and Popescu G.K. (2009). Kinetic basis of partial agonism at NMDA receptors. *Nat Neurosci*, 12, pp. 1114–1120.

[201] Ambert N. *et al.* (2010). Computational studies of NMDA receptors: differential effects of neuronal activity on efficacy of competitive and non-competitive antagonists. *Open Access Bioinformatics*, 2, pp. 113–125.

[202] Jahr C.E. and Stevens C.F. (1990). Voltage dependence of NMDA-activated macroscopic conductances predicted by single-channel kinetics. *J Neurosci*, 10, pp. 3178–3182.

[203] Kampa B.M., Clements J., Jonas P. and Stuart G.J. (2004). Kinetics of Mg(2+) unblock of NMDA receptors: implications for spike-timing dependent synaptic plasticity. *J Physiol*, 556, pp. 337–345.

[204] Vargas-Caballero M. and Robinson H.P.C. (2004). Fast and slow voltage-dependent dynamics of magnesium block in the NMDA receptor: the asymmetric trapping block model. *J Neurosci*, 24, pp. 6171–6180.

[205] Holmes W.R. and Levy W.B. (1990). Insights into associative long-term potentiation from computational models of NMDA receptor-mediated calcium influx and intracellular calcium concentration changes. *J Neurophysiol*, 63, pp. 1148–1168.

[206] Holcman D., Schuss Z. and Korkotian E. (2004). Calcium Dynamics in Dendritic Spines and Spine Motility. *Biophys J*, 87, pp. 81–91.

[207] Rubin J.E., Gerkin R.C., Bi G.Q. and Chow C.C. (2005). Calcium time course as a signal for spike-timing-dependent plasticity. *J Neurophysiol*, 93, pp. 2600–2613.

[208] Schiegg A., Gerstner W., Ritz R. and van Hemmen J.L. (1995). Intracellular Ca^{2+} stores can account for the time course of LTP induction: a model of Ca^{2+} dynamics in dendritic spines. *J Neurophysiol*, 74, pp. 1046–1055.

[209] Volfovsky N., Parnas H., Segal M. and Korkotian E. (1999). Geometry of dendritic spines affects calcium dynamics in hippocampal neurons: theory and experiments. *J Neurophysiol*, 82, pp. 450–462.

[210] Doi T., Kuroda S., Michikawa T. and Kawato M. (2005). Inositol 1,4,5-trisphosphate-dependent Ca^{2+} threshold dynamics detect spike timing in cerebellar Purkinje cells. *J Neurosci*, 25, pp. 950–961.

[211] Sneyd J., Girard S. and Clapham D. (1993). Calcium wave propagation by calcium-induced calcium release: an unusual excitable system. *Bull Math Biol*, 55, pp. 315–344.

[212] Fink C.C. *et al.* (2000). An image-based model of calcium waves in differentiated neuroblastoma cells. *Biophys J*, 79, pp. 163–183.

[213] De Young G.W. and Keizer J. (1992). A single-pool inositol 1,4,5-trisphosphate-receptor-based model for agonist-stimulated oscillations in Ca^{2+} concentration. *Proc Natl Acad Sci USA*, 89, pp. 9895–9899.

[214] Gin E., Kirk V. and Sneyd J. (2006). A bifurcation analysis of calcium buffering. *J Theor Biol*, 242, pp. 1–15.

[215] Li Y.-X. and Rinzel J. (1994). Equations for InsP3 receptor-mediated $[Ca^{2+}]i$ oscillations derived from a detailed kinetic model: a Hodgkin–Huxley like formalism. *J Theor Biol*, 166, pp. 461–473.

[216] Bezprozvanny I. (1994). Theoretical analysis of calcium wave propagation based on inositol (1,4,5)-trisphosphate (InsP3) receptor functional properties. *Cell Calcium*, 16, pp. 151–166.

[217] Dawson A.P., Lea E.J. and Irvine R.F. (2003). Kinetic model of the inositol trisphosphate receptor that shows both steady-state and quantal patterns of Ca^{2+} release from intracellular stores. *Biochem J*, 370, pp. 621–629.

[218] Kaftan E.J., Ehrlich B.E. and Watras J. (1997). Inositol 1,4,5-trisphosphate (InsP3) and calcium interact to increase the dynamic range of InsP3 receptor-dependent calcium signalling. *J Gen Physiol*, 110, pp. 529–538.

[219] Mak D.O., McBride S. and Foskett J.K. (2001). Regulation by Ca^{2+} and inositol 1,4,5-trisphosphate (InsP3) of single recombinant type 3 InsP3

receptor channels. Ca^{2+} activation uniquely distinguishes types 1 and 3 insp3 receptors. *J Gen Physiol*, 117, pp. 435–446.

[220] LeBeau A.P., Yule D.I., Groblewski G.E. and Sneyd J. (1999). Agonist-dependent phosphorylation of the inositol 1,4,5-trisphosphate receptor: a possible mechanism for agonist-specific calcium oscillations in pancreatic acinar cells. *J Gen Physiol*, 113, pp. 851–872.

[221] Sneyd J. and Falcke M. (2005). Models of the inositol trisphosphate receptor. *Prog Biophys Mol Biol*, 89, pp. 207–245.

[222] Blackwell K.T. and Kotaleski J. (2003). *Neuroscience Databases: A Practical Guide.* ed. Kötter R. Chapter 5 "Modeling the dynamics of second messenger pathways" (Springer, New York), pp. 63–79.

[223] Nakano T., Yoshimoto J. and Doya K. (2013). A model-based prediction of the calcium responses in the striatal synaptic spines depending on the timing of cortical and dopaminergic inputs and post-synaptic spikes. *Front Comput Neurosci*, 7.

[224] Sachs F., Qin F. and Palade P. (1995). Models of Ca^{2+} release channel adaptation. *Science*, 267, pp. 2010–2011.

[225] Keizer J. and Levine L. (1996). Ryanodine receptor adaptation and Ca^{2+}-induced Ca^{2+} release-dependent Ca^{2+} oscillations. *Biophys J*, 71, pp. 3477–3487.

[226] Saftenku E., Williams A.J. and Sitsapesan R. (2001). Markovian models of low and high activity levels of cardiac ryanodine receptors. *Biophys J*, 80, pp. 2727–2741.

[227] Zador A., Koch C. and Brown T.H. (1990). Biophysical model of a Hebbian synapse. *Proc Natl Acad Sci USA*, 87, pp. 6718–6722.

[228] Blaustein M.P. and Lederer W.J. (1999). Sodium/calcium exchange: its physiological implications. *Physiol Rev*, 79, pp. 763–854.

[229] Schuster S., Marhl M. and Hofer T. (2002). Modelling of simple and complex calcium oscillations. From single-cell responses to intercellular signalling. *Eur J Biochem*, 269, pp. 1333–1355.

[230] Bondarenko V.E. *et al.* (2004). Computer model of action potential of mouse ventricular myocytes. *Am J Physiol Hear Circ Physiol*, 287, pp. H1378–403.

[231] Matsuoka S. *et al.* (2003). Role of individual ionic current systems in ventricular cells hypothesized by a model study. *Jpn J Physiol*, 53, pp. 105–123.

[232] Yano K., Petersen O.H. and Tepikin A.V. (2004). Dual sensitivity of sarcoplasmic/endoplasmic Ca^{2+}-ATPase to cytosolic and endoplasmic reticulum Ca^{2+} as a mechanism of modulating cytosolic Ca^{2+} oscillations. *Biochem J*, 383, pp. 353–360.

[233] Higgins E.R., Cannell M.B. and Sneyd J. (2006). A buffering SERCA pump in models of calcium dynamics. *Biophys J*, 91, pp. 151–163.

[234] Shannon T.R., Ginsburg K.S. and Bers D.M. (2000). Reverse mode of the sarcoplasmic reticulum calcium pump and load-dependent cytosolic calcium decline in voltage-clamped cardiac ventricular myocytes. *Biophys J*, 78, pp. 322–333.

[235] Blackwell K.T. (2013). Approaches and tools for modeling signalling pathways and calcium dynamics in neurons. *J Neurosci Methods*, 220, pp. 131–140.

[236] Good T.A. and Murphy R.M. (1996). Effect of beta-amyloid block of the fast-inactivating K^+ channel on intracellular Ca^{2+} and excitability in a modeled neuron. *Proc Natl Acad Sci USA*, 93, pp. 15130–15135.

[237] Wilson N.P., Gates B. and Castellanos M. (2013). Modeling the short time-scale dynamics of β-amyloid–neuron interactions. *J Theor Biol*, 331, pp. 28–37.

[238] Wang S.S.S., Kazantzi V. and Good T.A. (2003). A kinetic analysis of the mechanism of β-amyloid induced G protein activation. *J Theor Biol*, 221, pp. 269–278.

[239] Kidd J. and Sattelle D. (2006). The effects of amyloid peptides on A-type K+ currents of Drosophila larval cholinergic neurons: modeled actions on firing properties. *Invertebr Neurosci*, 6, pp. 207–213.

[240] Morse T.M. *et al.* (2010). Abnormal excitability of oblique dendrites implicated in early Alzheimer's: a computational study. *Front Neural Circuits*, 4.

[241] Tiveci S. *et al.* (2005). Modelling of calcium dynamics in brain energy metabolism and Alzheimer's disease. *Comput Biol Chem*, 29, pp. 151–162.

[242] Toivari E. *et al.* (2011). Effects of transmitters and amyloid-beta peptide on calcium signals in rat cortical astrocytes: Fura-2AM measurements and stochastic model simulations. *PLoS One*, 6, p. e17914.

[243] Riera J. *et al.* (2011). Quantifying the uncertainty of spontaneous Ca^{2+} oscillations in astrocytes: particulars of Alzheimer's disease. *Biophys J*, 101, pp. 554–564.

[244] De Caluwé J. and Dupont G. (2013). The progression towards Alzheimer's disease described as a bistable switch arising from the positive loop between amyloids and Ca^{2+}. *J Theor Biol*, 331, pp. 12–18.